DATE

MOLECULAR VIROLOGY

MOLECULAR VIROLOGY

MOLECULAR AND MEDICAL ASPECTS OF DISEASE-CAUSING VIRUSES OF MAN AND ANIMALS

YECHIEL BECKER
Carolyn Jane Bendheim Professor of Molecular Virology
Department of Molecular Virology
Institute of Microbiology, Faculty of Medicine
Hebrew University of Jerusalem, Israel

JULIA HADAR, Associate Editor

MARTINUS NIJHOFF PUBLISHERS
THE HAGUE/BOSTON/LONDON

Distributors:

for the United States and Canada

Kluwer Boston, Inc.
190 Old Derby Street
Hingham, MA 02043

for all other Countries

Kluwer Academic Publishers Group
Distribution Center
P.O. Box 322
3300 AH Dordrecht
The Netherlands

Library of Congress Cataloging in Publication Data

Becker, Yechiel.
Molecular virology.

Bibliography: p.
1. Virology. 2. Virus diseases. I. Hadar, Julia. II. Title. [DNLM: 1. Molecular biology. 2. Viruses. QW 160 B396m]
QR360.B415 1982 616'.0194 82-14258
ISBN 90-247-2742-1

To my wife Miriam, for her continuous support and interest

CONTENTS

PREFACE

This book was written during a period when the technologies of genetic engineering were being applied to the study of animal viruses and when the organization and function of individual virus genes were being elucidated. This book, which uses human and animal viruses as models, aims to understand the developments in molecular virology during the last 20 years. Although molecular virology could also be taught by means of bacteriophages or plant viruses, the advantage of using animal viruses is in their ability to cause human and animal diseases as well as to transform cells, a primary problem in medicine.

For the sake of clarity and convenience, not all the individual contributors to the various aspects of molecular virology were cited in the text. Instead, the reader is referred to review articles or key papers that list the numerous excellent publications that have contributed to clarification of the various molecular processes. Thus the end-of-chapter bibliographies will guide the reader to the publications in which the original contributing authors are quoted. References given under the heading Recommended Reading are intended to assist those interested in pursuing a given subject further.

I hope that this book will fulfill the purpose for which it is designed, and I urge readers to contact me if errors are found or updating is required.

ACKNOWLEDGMENTS

I wish to thank all those who contributed information and advice during the writing of this book—particularly the following:

Dr. Julia Hadar, for her excellent editing of the book, for her careful reading and checking of the material, and for her most valuable criticisms
Mrs. Esther Herskovics, for excellence and care in typing the manuscript and for her helpful comments
Dr. Daniel Dekegel, Pasteur Institute, Brussels, for contributing his electron micrographs of many viruses
Dr. Walter Doerfler, Institut für Genetik, Köln, for comments on the adenovirus chapter
Dr. John Coffin, Tufts University, Boston, for comments on the retrovirus chapter
Dr. Yosef Aloni, Weizmann Institute of Science, Rehovot, for comments on the papovavirus chapter
Dr. Donald H. Gilden, University of Pennsylvania, Philadelphia, for comments on viruses affecting the nervous system.

Thanks are due all the scientists who gave me permission to use figures from their publications, and to the editors of the scientific journals who granted permission to use published material. The authors and journals are acknowledged in the text.

I. MOLECULAR ASPECTS OF ANIMAL VIRUSES

1. STEPS IN THE DEVELOPMENT OF VIRUS RESEARCH

VIRUSES AS DISEASE-CAUSING AGENTS

Evidence that even many centuries ago viruses were human pathogens causing epidemics and death can be obtained from examining sources that have survived through the ages. These include ancient scripts, art, and archaeology. Epidemics in humans and animals were vividly described in the Bible and other historic writings, but our ability to distinguish between an epidemic caused by a pathogenic bacterium and one originating from a virus did not develop until the end of the nineteenth century and gained momentum during the first half of the twentieth century.

One of the first indications that viruses caused diseases in humans several thousand years ago can be found in Egyptian archaeology. Examination of the 3,000-year-old mummy of Pharaoh Ramses V revealed pockmarks similar to those found on the faces of persons infected with smallpox virus (figure 1). This suggests that the Egyptian king may have contracted smallpox. Another example is recorded on a wall covering dating from the period of 1500 B.C., in which an Egyptian prince is depicted leaning on a cane because his left leg is thinner and shorter than the right one (figure 2). Comparing the picture of the ancient prince to a contemporary victim of paralytic poliomyelitis leads us to deduce that infantile paralysis has been prevalent in North Africa, and possibly the Mediterranean basin, for at least 3,500 years.

The Persian physician, Muhamad ibn Zakariyya (Rhazes), who was born August 27, 865 A.D. and lived in Baghdad, described smallpox as an epidemic

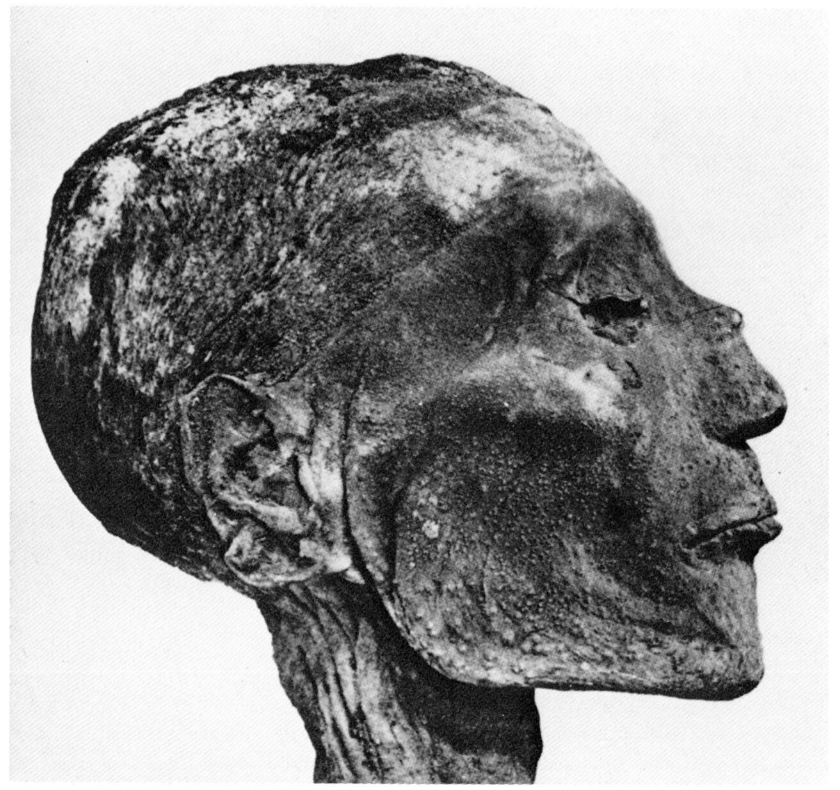

Figure 1. Pockmarks on the 3,000-year-old mummified head of Egyptian Pharaoh Ramses V testify to the length of time that smallpox has scourged humanity.
(WHO/11617, General, Health Historical, SM 2-3, 1977. Reprinted by permission from World Health Organization, Geneva, Switzerland.)

brought into Arabia during the sixth century A.D., most probably by troops arriving from Abyssinia. In his writings, Rhazes also referred to the fact that smallpox had been described by the physician Galen in the second century A.D. and by the Hebrew physician El Yehudi who lived in the seventh century A.D. Smallpox spread through Africa during the sixteenth century and, with the advent of slavery, infected Africans brought the disease to the New World.

During the sixteenth century, smallpox also spread in England, and during the eighteenth century the disease wrought havoc in many parts of the world among children under the age of ten. Rhazes wrote a treatise on smallpox and measles but did not recognize them as separate diseases. The physician Avicenna, in his book, *Canon,* written 115 years later, recognized that smallpox and measles were two different diseases (Elgood 1951).

Figure 2. An Egyptian youth with a withered leg paralyzed by what may have been poliomyelitis is depicted in a sculpture that is 3,500 years old.
(WHO/8063, General, Health Historical, SM 2-3, 1977. Reprinted by permission from World Health Organization, Geneva, Switzerland.)

Viruses before the dawn of human history

Viruses infect all prokaryotic and eukaryotic cells in nature. Although the ancestors of viruses and their evolution are not known (chapter 21), it is possible to assume that viruses paralleled the evolutionary processes in nature. Viruses can be looked upon as primitive forms of evolution that have lost their ability to live an independent life and have attained the form of parasitic nucleic acids. Studies on viruses of vertebrates and mammals revealed relatedness between more ancient viruses (such as rodent viruses) and viruses of monkeys higher on the evolutionary ladder. It is assumed that viruses adapted themselves to infect and colonize more advanced forms of life in order to maintain themselves in nature and avoid extinction together with the disappearing species of animals that could not adapt to the changing environmental conditions.

Attempts to immunize humans against smallpox and measles

In the eighteenth century, smallpox was widespread in Europe, especially in England. A medical procedure called *variolation* was developed in Turkey to immunize humans against the disease. In this procedure, material taken from a vesicle on the skin of a smallpox patient was inserted into a scarification in the skin of the individual being immunized. At the site of inoculation, a vesicle usually formed within a few days, and the vaccinated person invariably fell ill, but with a milder form of the disease than that contracted naturally. However, some of the people immunized did develop a devastating infection and died. This method of variolation was practiced in England starting in 1721.

Francis Home attempted to immunize humans against measles by scarification of the skin with blood of infected individuals. The procedure was partially successful.

The concept of *virus* (meaning poison)

The English physician Edward Jenner was the first to realize that vesicles that appeared on the hands of farm workers in contact with cows protected them from smallpox. Jenner noted that the cows had vesicles on their udders which were similar to those on the hands of the laborers and concluded that the reason for the farm workers' resistance to smallpox was their exposure to infection with cowpox. To test his hypothesis, Jenner used fluid from vesicles on the hands of a milkmaid infected with cowpox to inoculate an eight-year-old boy and subsequently infected the child with material taken from vesicles of a smallpox patient, thus challenging him with the virulent form of the disease. The boy developed a mild illness after the second procedure and recovered. Thus Jenner, using the procedure of variolation, demonstrated that cowpox is capable of protecting humans against the more deadly smallpox.

In his study published in 1798 (Brock 1961) Jenner used the term *virus* to describe the fluid taken from the skin vesicles. He assumed that the fluid contained a poison (virus in Greek), and the transfer of cowpox material to a nonimmunized person was named *vaccination* (*vacca* being the Latin word for cow). Both expressions introduced by Jenner have remained in use, even though the nature of viruses began to be understood only about 40 years ago. Jenner's method of vaccination against smallpox is employed today almost without modification.

The concept of virus as a filterable agent not visible in the light microscope

The groundwork of immunization against many viral diseases was laid by Louis Pasteur in his efforts to produce a vaccine against the rabies virus (chapter 13). Pasteur injected a brain homogenate from a rabid dog infected with street virus into the brains of rabbits and then used the rabbit brains and spinal cords to produce a vaccine by gradual desiccation. This was done without knowledge of the nature of the causative agent of the disease or what effect drying would have on the agent. Pasteur hoped that injection of the desiccated material into humans would lead to the production of neutralizing antibodies that would protect against the disease. Pasteur's technique, in modified form, was used for many years; but today vaccine is prepared from inactivated virus particles grown in cultured cells under in vitro conditions.

During Pasteur's era, researchers became aware that bacteria give rise to sickness, but Pasteur suggested that agents smaller than bacteria might also trigger human diseases. These tiny agents are below the resolution of the light microscope and pass through filters that retain bacteria. This theory subsequently proved to be true when tissue homogenates of infected animals were passed through a filter that retained bacteria and cell debris. Injection of the filtrate into susceptible animals led to development of a disease. The term used for these filterable agents was *virus,* as suggested by Edward Jenner 100 years earlier.

The use of filters that retain bacteria caused two other agents to be classified with the viruses. The microorganisms *Chlamydia* and *Mycoplasma* were subsequently demonstrated to be prokaryotic cellular organisms that are obligate parasites of eukaryotic cells.

Mosaic disease of tobacco plants

In 1887, A.E.M. Mayer described the mosaic disease of tobacco plants and demonstrated that the disease could be transferred from sick to healthy plants. D.I. Ivanovski showed in 1892 (Lechevalier 1972) that, after passing a homogenate of infected leaves through a filter that retains bacteria, the filtrate could still cause disease in healthy plants. Ivanovski suggested that the causative agent was a toxin. Beijerinck repeated Ivanovski's experiment in 1899 and concluded that the cause of tobacco mosaic disease was a virus. Studies on this virus by W.M. Stanley 40 years later made the breakthrough into understanding the role of nucleic acid as the carrier of virus genetic information (Brock 1961).

Foot and mouth disease in cattle

The first written description of foot and mouth disease in cattle was provided in 1514 by Hyronimus Proctorius, who lived in northern Italy. The disease-causing agent was not known then or 240 years later when the disease appeared in Germany.

In 1898, F. Loeffler and P. Frosch (Brock 1961) published their report on the nature of foot and mouth disease. They used fluid from vesicles of infected cattle to demonstrate the absence of bacteria and the presence of an agent capable of infecting calves and heifers via the mucous membrane of the lips. They showed that bacteria-free filtrates of lymph from diseased animals were as infectious as the unfiltered material and that the agent was active at a higher dilution than that found with tetanus toxins. They speculated that other diseases like smallpox, cowpox, measles, and so on might also be caused by these filterable agents (viruses) capable of replication in the infected host. Subsequently, in 1901, J. Caroll discovered the virus of yellow fever.

Viruses cause cancer

Using the filtration technique to separate filterable agents from tissue cells, the investigators V. Ellerman and O. Bang of Copenhagen were able to transmit leukemia of hens in 1908 (Gross 1978). In 1911, Peyton Rous in New York succeeded in demonstrating that sarcomas in chickens are also caused by filterable agents. Sarcomas were removed from chickens, homogenized, and filtered, and the filtrates were injected into healthy chickens. At the site of inoculation, a tumor developed that was similar to the original sarcoma used for the experiment (Hughes 1977).

Viruses infect bacteria: the discovery of the bacteriophages

These viruses were discovered independently by two investigators. F.W. Twort in 1915 described a phenomenon that he named *glassy transformation* in bacterial cultures (micrococcus) that were isolated from a preparation of smallpox vaccine. Twort suggested that the glassy transformation is caused by living protoplasma, an enzyme that can replicate, or possibly by a virus. In 1917, F. d'Herelle published an investigation into the properties of an agent antagonistic to dysentery bacilli isolated from convalescing patients, using Chamberland filters to retain the bacteria. The filtrates were able to lyse both cultures of dysentery bacteria and to cause the appearance of clear spots in bacterial cultures on agar. The clear spots contained bacteria that were unable to infect test animals. In his communication, d'Herelle used the term *bacteriophage* for the first time. His work was based on his previous study (performed in 1910 but not published) on the isolation of bacteria pathogenic to the locust. In cultures of the locust bacteria, clear spots appeared with a diameter of 2-3 mm, and the bacteria in the clear spot disappeared. The agent causing the clear-spot phenomenon was still active after it was passed through filters that retained bacteria (Brock 1961; Duckworth 1976).

Viral diseases in test animals

The search for a suitable host susceptible to different viruses led to the use of mice, rabbits, guinea pigs, ferrets, hamsters, horses, cattle, monkeys, and others as test animals. However, it became apparent that injection of tissue homogenates into test animals and the demonstration that they developed a disease could be misleading, since in certain circumstances a virus already latent in the test animal could be reactivated. In addition, animals to be used in virus research must be bred and kept under special conditions to protect them from infections by viruses or other agents, and large numbers of test animals are needed for quantitation of each infectious-virus preparation. Also the growing need for wild animals like monkeys might eventually threaten their existence. Thus the introduction of the chick embryo as a test animal during the 1930s was a great step forward in virus research. Fertilized eggs can be obtained in large numbers, and they provide a closed, sterile environment in which different viruses are able to develop, either in the embryo's tissues or in the different membranes surrounding the embryo. In the laboratory, the fertilized eggs are incubated for 10 or 11 days prior to inoculation with viruses, and the virus yield is harvested after a few more days of incubation.

Yellow fever virus attenuated by passage in chick embryos

The use of the chick embryo by Theiler for the cultivation of the virus of yellow fever led to a new breakthrough in the immunization of humans against virus diseases. Consecutive passages of yellow fever virus in the yolk sac of chick embryos led to the selection of a new virus strain that retained the antigenic properties of the original virus but lost its virulence for man. Thus the newly developed yellow fever virus, named *17D,* is the attenuated form of a highly pathogenic virus that can be safely administered to individuals who live or travel in areas where the yellow fever virus is endemic in local monkeys and mosquitoes. Theiler's investigation, therefore, added a new dimension to Jenner's concept that a safe, attenuated live-virus vaccine confers lifelong immunity. In the case of yellow fever, the virus was a laboratory-selected variant or mutant that developed when the wild-type virus was passed through an unnatural host like the chick embryo, which is a different species and has a higher developmental temperature than monkeys or mosquitoes, the natural hosts of the virus.

Viruses replicate in the infected host in selected tissues

Infection of test animals with disease-causing viruses led to the observation that different viruses infect different tissues. Pasteur's studies on rabid dogs led him to the understanding that the infectious agent resides in the dog's brain and saliva, and Pasteur's attempts to develop a heat-inactivated vaccine from brains of experimentally infected rabbits were based on these observations. In order to classify viruses precisely, it became acceptable to group the different viruses according to the tissues they affected. Thus viruses were classified as neurotropic if they affected the brain and viscerotropic if they affected the viscera. These observations on the tropism of viruses for specific tissues in the infected animals led to the realization that, in order to develop inactivated human virus vaccines, large numbers of animals would have to be infected to provide sufficient infected tissues for vaccine preparation. Thus preparation of an inactivated vaccine to protect against poliomyelitis, which appeared during the 1940s as a worldwide epidemic, was not feasible since poliovirus is a neurotropic virus capable of infecting monkey brains by direct inoculation. It became obvious that the use of monkeys for the preparation of such a vaccine was impractical, and scientists warned that injection of foreign neurological material into humans could lead to brain damage. In the face of an epidemic causing paralysis and death in large numbers of children and adults, a breakthrough was of the utmost urgency.

Animal viruses can replicate in isolated tissues in culture

Parallel with the development of virology, scientists made attempts to explant tissues from vertebrates in the hope of adapting them to growth under defined conditions in vitro. The technique was to remove tissues from animals or humans (such as skin explants) and to culture them under sterile conditions in a medium containing the essential nutrients. Enders, Weller, and Robbins (1949) used cultured human-embryonic tissues to determine whether poliomyelitis virus replicates in tissues other than those of neural origin. A small amount of poliovirus from a mouse-brain suspension was inoculated into human-embryonic tissue in culture, incubated at 37°C, and after a few days the virus content of the culture fluid was monitored in mice or monkeys. It was demonstrated that poliovirus replicates in embryonic muscle cells grown in culture. This finding opened up a new phase in virus research, since large quantities of poliomyelitis virus could now be obtained from cells cultivated under in vitro conditions.

Trypsinization of animal tissues was used to digest the connective tissue holding the cells together and to release the cells into a suspension that was placed in culture medium and seeded into suitable glass or plastic containers. The cells attach to the glass or plastic surfaces and develop into a monolayer. The use of antibiotics like penicillin and streptomycin put an end to the problems of bacterial contamination, and a new wave of activity was generated in all virus laboratories to isolate viruses in cell culture. Viruses were now characterized according to their ability to affect the cultured cells and cause a cytopathogenic effect (CPE) or a change in cell properties like cell transformation. The more practical impact of the introduction of cell cultures for virological research was their use for the production of human and animal vaccines. Shortly after the investigation by Enders, Weller, and Robbins, it was found that poliomyelitis can grow in monkey-kidney-cell cultures. Large quantities of poliovirus were produced in such cultures, and the formaldehyde-inactivated poliovirus vaccine was produced in the early 1950s by J. Salk.

It was thought that cultured cells are free of viruses, but subsequently it was discovered that monkey-kidney cells harbour a virus named *simian virus 40* (SV40) in a latent form. This virus was activated in the cultured monkey-kidney cells under in vitro conditions, and the poliovirus vaccine was indeed found to be contaminated with an undesirable virus that was shown to be capable of transforming cells in vitro. Subsequently, this virus was removed from the vaccine preparations. More rigorous control measures are now applied to the cell systems used in vaccine production. The large-scale vaccination of large numbers of people with the Salk vaccine had an immediate impact in reducing the severity of poliovirus epidemics throughout the world.

Plaque assay for animal viruses allowed quantitative analysis

Developments in culturing of cells made it possible to titrate animal viruses in cell monolayers. Dulbecco (1952) demonstrated that cells infected with an animal virus which were then covered with an agar overlay produced foci of infection called *plaques*. This plaque method of titration considerably enhanced the study of animal viruses.

Electron microscopy and the visualization of virus structure

Coincidentally with developments in adapting viruses to growth in cultured cells, the electron microscope as a tool in virus research became available. Until the introduction of the electron microscope, viruses had never been seen, with the exception of the poxviruses, which have a diameter large enough for resolution in the light microscope after staining of the virus particles. All other viruses were detected by their effects on animals or by their ability to cause pocks on the chorioallantoic membrane of the chick embryo. During the 1950s, the electron microscope became the eyes of the virologists, and for the first time it became apparent that viruses differed in their structure and morphology. Viruses were seen both outside and inside infected cells, in in vitro cultures as well as in situ in infected tissues. A new wave of research, which started with identifying all viruses morphologically, eventually led to the development of techniques for the visualization of the viral nucleic acids in the electron microscope.

Viruses and mouse leukemia

The role of viruses as causative agents of mouse leukemia was elucidated in 1951 by L. Gross, who was aware that spontaneous leukemia appeared at a high rate in mice from AK and C58 breedings, while C3H mice had a very low rate of leukemia. Gross injected into newborn C3H mice homogenates of tissues from normal female AK or C58 mice, after filtration to remove intact tissue cells, and showed the inoculated newborn C3H mice developed leukemia. Thus vertical transmission of leukemia by a filterable virus (namely introduction of the virus into the animal) is possible (Gross 1978).

Fungal viruses (mycorna)

Fungi, like plants and animals, are also infected by viruses. In 1950, J.W. Sinden and E. Hauser noted a disease in the fungus *Agasicus bisporus* and suggested that a virus was the causative agent. Electron microscopy of infected fungi revealed the presence of viral particles (Lemke and Nash 1974). These viruses of fungi were found to have a double-stranded ribonucleic acid (RNA) genome and belong to the family of Reoviridae (chapter 12).

MOLECULAR ASPECTS OF VIROLOGY: NUCLEIC ACIDS AS CARRIERS OF THE VIRAL GENES

Tobacco mosaic virus (TMV) genes in the viral RNA

The studies by D.I. Ivanovski and M.W. Beijerinck, which revealed that mosaic disease in plants is caused by a filterable agent, were continued by Stanley (1935), who noted that addition of ammonium or certain acids to the juice of infected plants resulted in the formation of crystals. These crystals were found to contain the tobacco mosaic infectious agent since inoculation of the dissolved crystals into tobacco plant leaves resulted in the development of the mosaic disease. Analysis of these crystals revealed that, in addition to a protein, they also contained a nucleic acid of the RNA type. Using available extraction procedures, Cohen and Stanley (1942) separated the RNA from the protein and analyzed its physical properties. Stanley subsequently inoculated tobacco plant leaves with the purified RNA and showed unequivocally that the typical disease symptoms occurred only in leaves infected with the viral RNA in which infectious virus particles were made. Stanley's conclusion that the viral RNA is the genetic element of the TMV (Stanley 1946) opened the way for the development of molecular virology and led to the understanding that TMV RNA molecules are the viral genomes that carry viral inheritance.

Experimental evidence that viral deoxyribonucleic acid (DNA) contains the viral genes

The studies by Rosalind Franklin and Maurice Wilkins on the x-ray crystallographic structure of the B form of DNA and the double-helix model of DNA provided by J. Watson and F. Crick in 1953 gave an insight into the three-dimensional structure of DNA (Olby 1974). Watson and Crick (1953*a,b*) analyzed the genetic implications of the double-helix model regarding DNA molecules present in virus particles and suggested that the viral DNA or RNA contains the genetic information necessary for expression of the virus in infected cells.

Hershey and his colleagues used T4 bacteriophage of *Escherichia coli* (*E. coli*), labeled with different radioisotopes in the DNA and proteins, to show that after attachment of the bacteriophage to the bacterial cell wall, the viral DNA is injected into the cell, while the viral protein coat is retained on the outside. Replication of the viral DNA in the bacterial cell leads to the formation of new viral particles that contain DNA coated with proteins. Thus Hershey concluded that the T4 DNA carries the viral genetic information that codes for the viral functions.

Viral DNA as template for messenger RNA in the synthesis of viral proteins

The central dogma in molecular biology, which states that the genetic information of an organism or virus is encoded in nucleotide sequences of the DNA and DNA molecules are made on DNA templates, came from the studies of Watson and Crick. However, the mechanisms that operate during the process of gene expression and that lead to the formation of the gene products—namely, the proteins—were not known. The concept of messenger RNA (mRNA) was first provided by Jacob and Monod (1961), when they proposed that the intermediate stage between the genes in the DNA and the final protein product represents short-lived RNA molecules that are transcribed from the DNA, keeping the base sequence of the genes unchanged. The first demonstration that mRNA molecules are transcribed from the DNA genome of a virus was presented by Becker and Joklik (1964), using cells infected with vaccinia virus. The mRNA molecules are then translated by the protein-synthesizing machinery of the cell—namely, the ribosomes that interact with the RNA and form polyribosomes. Penman, Scherrer, Becker, and Darnell (1963) were the first to show that RNA viruses can serve as mRNA in the infected cell on their own. It was reported that the viral RNA associates with the cellular ribosomes to form polyribosomes on which the viral-coat proteins are synthesized.

Studies by Kates and McAuslan (1967) and Munyon and associates (1967) revealed that vaccinia virus particles themselves contain an RNA polymerase, thus providing evidence that virus particles not only contain a nucleic acid and a protein coat, but also enzymes. A few years later, Temin and Mizutani (1970) and Baltimore (1970) independently discovered the RNA-dependent DNA polymerase (reverse transcriptase) inside retrovirus particles (chapter 21). This discovery revolutionized the central dogma that DNA can only be synthesized from other DNA molecules. The reverse transcriptase utilizes the RNA genome of a retrovirus as a template to synthesize a double-stranded DNA molecule which functions in infected cells and also integrates into the cellular DNA. Thus the genetic information of this group of viruses alternates between the RNA genome in the virion and the viral DNA genome in the infected cells. The RNA-dependent DNA polymerase is currently used to transcribe specific mRNA molecules from mammalian cells. Thus synthesis of mammalian genes became possible.

THE DEFINITION OF A VIRUS

A. Lwoff and S.E. Luria originally defined the virus as a particle containing a molecule (or molecules) of nucleic acid, either DNA or RNA, which carries all the genetic information needed for the virus to replicate in infected cells. Viruses are obligate parasites of these cells.

Subviruses as infectious agents

Viral agents that differ from most other viruses in their properties are called *nonconventional viruses,* since they do not have the ability to produce virus particles as the *conventional viruses* do. This group of agents includes viroids found in plants and certain pathogens of animals and humans (chapter 23). The viroids in plants are extremely small, circular RNA molecules; the agents in animals have not yet been clearly defined.

Genetic engineering technology in isolating and studying individual genes

The introduction of bacterial plasmids as cloning vehicles for DNA fragments has opened up a new horizon for the study of animal-virus genes. DNA viruses are digested by restriction enzymes and then cloned in bacterial plasmids (Bolivar et al. 1977; Sutcliffe et al. 1978; McKnight 1980). The genomes of RNA viruses can be transcribed into their DNA copy by the reverse transcriptase and thus can also be cloned in bacterial plasmids. This new technology allows individual viral genes to be analyzed and their nucleotide sequences to be determined.

Cloning of viral genes in bacterial plasmids may have practical application in the future—for example, in the use of the viral gene products in vaccine production.

BIBLIOGRAPHY

Baltimore, D. RNA-dependent DNA polymerase in virions of RNA tumour viruses. *Nature 226*:1209–1211, 1970.

Becker, Y., and Joklik, W.K. Messenger RNA in vaccinia infected cells. *Proc. Natl. Acad. Sci. USA 51*:577–581, 1964.

Bolivar, F.; Rodriguez, R.L.; Greene, P.J.; Betlach, M.C.; Heyneker, H.L.; Boyer, H.W.; Crosa, J.H.; and Falkow, S. Construction and characterization of new cloning vehicles. II. A multipurpose cloning system. *Gene 2*:95–113, 1977.

Brock, T.D. *Milestones in Microbiology*. American Society for Microbiology, Washington, D.C., 1961 (reprinted in 1975).

Cohen, S.S., and Stanley, W.M. The molecular size and shape of the nucleic acid of tobacco mosaic virus. *J. Biol. Chem. 144*:589–598, 1942.

Dulbecco, R. Production of plaques in monolayer tissue cultures by single particles of an animal virus. *Proc. Natl. Acad. Sci. USA 38*:747–752, 1952.

Duckworth, D.H. Who discovered the bacteriophage? *Bact. Rev. 40*:793–802, 1976.

Elgood, C. *A Medical History of Persia and the Eastern Caliphate*. Cambridge, The University Press, 1951.

Enders, J.H.; Weller, T.H.; and Robbins, F.C. Cultivation of the Lansing strain of polyomyelitis virus in cultures of various human embryonic tissues. *Science 109*:85–87, 1949.

Gross, L. The development of the concept of viral etiology of leukemia and related neoplastic diseases. Present status and prospects for the future. *In: Advances in Comparative Leukemia Research,* 1977, (P. Bentvelzen, J. Hilgers, and D.S. Yohn, eds.). Elsevier/North-Holland Biomedical Press, Amsterdam-New York, 1978, pp. 1–10.

Hughes, S.S. *The Virus: A History of the Concept*. Heinemann Educational Books, London, Science History Publications, New York, 1977.

Jacob, F., and Monod, J. Genetic regulatory mechanisms in the synthesis of proteins. *J. Mol. Biol. 3*:318–356, 1961.

Kates, J., and McAuslan, B.R. Poxvirus DNA-dependent RNA polymerase. *Proc. Natl. Acad. Sci. USA 58*:134–141, 1967.

Lechevalier, H. Dmitri Iosifovich Ivanovski (1864–1920). *Bact. Rev. 36*:135–145, 1972.

Lemke, P.A., and Nash, C.H. Fungal viruses. *Bact. Rev. 38*:29–56, 1974.

McKnight, S.L. The nucleotide sequence and transcript map of the herpes simplex virus thymidine kinase gene. *Nucleic Acids Res. 8*:5949–5964, 1980.

Munyon, W.; Paoletti, E.; and Grace, J.T., Jr. RNA polymerase activity in purified infectious vaccinia virus. *Proc. Natl. Acad. Sci. USA 58*:2280–2287, 1967.

Olby, R. *The Path to the Double Helix*. University of Washington Press, Seattle, 1974.

Penman, S.; Scherrer, K.; Becker, Y.; and Darnell, J.E. Polyribosomes in normal and poliovirus infected HeLa cells and their relationship to messenger RNA. *Proc. Natl. Acad. Sci. USA 49*:654, 1963.

Stanley, W.M. Isolation of a crystalline protein possessing the properties of tobacco mosaic virus. *Science 81*:644–645, 1935.

Stanley, W.M., 1946. The isolation and properties of crystalline tobacco mosaic virus. *In: Nobel Lectures, Chemistry, 1942–1962,* Amsterdam, 1964, pp. 137–157.

Sutcliffe, J.G. pBR322 restriction map derived from the DNA sequence: accurate DNA size markers up to 4361 nucleotide pairs long. *Nucleic Acids Res. 5*:2721–2728, 1978.

Temin, H., and Mizutani, S. RNA-dependent DNA polymerase in virions of Rous sarcoma virus. *Nature 226*:1211–1213, 1970.

Watson, J.D., and Crick, F.H.C. Molecular structure of nucleic acids. A structure for deoxyribose nucleic acid. *Nature 171*:737–738, 1953*a*.

Watson, J.D., and Crick, F.H.C. Genetic implications of the structure of deoxyribonucleic acid. *Nature 171*:964–967, 1953*b*.

RECOMMENDED READING

Chargaff, E.; Zamenhof, S.; and Green, C. Composition of human deoxypentose nucleic acid. *Nature 165*:756–757, 1950.

Franklin, R.E. Evidence for two chain helix in crystalline structure of sodium deoxyribonucleate. *Nature 172*:156–157, 1953.

Lechevalier, H.A., and Solotorovsky, M. *Three Centuries of Microbiology*. McGraw-Hill Book Co., New York, 1965.

Luria, S.E.; Darnell, J.E.; Baltimore, D.; and Campbell, A. *General Virology*. (Third edition). John Wiley & Sons, Inc., New York, 1978.

Lwoff, A., Horne, R., and Tournier, P. A system of viruses. *Cold Spring Harbor Symp. Quant. Biol. 27*:51–55, 1962.

2. CLASSIFICATION OF VIRUSES

With the accumulation of information on the morphology of virus particles (virions), their chemical composition, and the nature of their nucleic acid it became evident that viruses differ markedly from each other. Viruses with the same structure and morphology were isolated from a variety of hosts which included bacteria, fungi, plants, insects, and both lower and higher forms of vertebrates. Attempts to classify viruses were initially based on their morphology, although analyses of the nature of the viral genomes provided another method for classification based on the type of nucleic acid (DNA or RNA), the properties of the nucleic acid (double- or single-stranded), and molecular weight. To obtain a system of classification acceptable to all virologists, an International Committee on Nomenclature of Viruses (ICNV) (subsequently changed to International Committee on Taxonomy of Viruses [ICTV]) was elected. Expert subcommittees dealt with vertebrate viruses, invertebrate viruses, plant viruses, and bacterial viruses.

The committee for nomenclature decided not to use the Linnean system used for bacteria but instead opted for an internationally agreed-upon system to be applied universally to all viruses. The names of the viruses would be latinized; the genus should be a collection of various species of viruses with the name ending with ". . . virus" (e.g., herpesvirus). A family of viruses would contain a number of genera and the ending of the name would be ". . . viridae" (e.g., herpesviridae).

METHOD OF CLASSIFICATION

The viruses were divided into five different taxa, each containing different viruses according to the hosts they infect.

Taxon A: viruses that infect more than one host
Taxon B: viruses infecting vertebrates only
Taxon C: plant viruses
Taxon D: bacteriophages
Taxon E: viruses of invertebrates

The cryptogram

The cryptogram was designed to provide a short, precise description of the virus group, taking into consideration the biological and chemical properties of the viruses belonging to a particular family. The cryptogram contains four entries, each with two properties.

The first pair — the type of nucleic acid/strandedness of the nucleic acid; D stands for DNA and R for RNA; 1 means single-stranded and 2 means double-stranded.

The second pair — the molecular weight of the viral nucleic acid (in millions)/ percentage nucleic acid in the purified virions. When the virion nucleic acid is not in one molecule but in the form of a segmented genome, the Greek letter sigma (Σ) is added before the molecular weight.

The third pair — the morphology of virions, envelope, when present, the shape of the nucleocapsid. *S* is spherical; *E* is elongated with parallel sides and straight ends; *u* is elongated but with circular end(s); *X* is complex or otherwise different from previous shapes; *e* refers to envelope present in virions.

The fourth pair — the host/method of viral transmission/nature of vector. Hosts: A—alga; B—bacterium; F—fungus; I—invertebrate; S—seed plant; and V—vertebrate. Transmission: C—congenital; I—intestinal tract; O—contact with environment; R—respiratory tract; and Ve—invertebrate vector.

Vectors — Ac—ticks and mites (Acarina); AP—aphids; Au—leaf-, plant-, or tree-hoppers; Cl—beetles (Coleoptera); Di—flies and mosquitoes (Diptera); Si—fleas (Siphonaptera); Th—thrips (Thysanoptera); Fu—fungi; and Ne—nematodes.

In addition, an asterisk (★) symbolizes that the property of the virus is not fully known and parentheses ( ) are used when the information is doubtful or unconfirmed.

Virus classification

Taxon A: Viruses affecting a number of hosts

1. Family Iridoviridae (chapter 6)
   Cryptogram: C/2:130–160/12–16:S or Se/S:I, V/C, I, O, Ve/Ac
   Virus particles have a diameter ranging from 130–300 nm and are made up of about 1,500 capsomeres (the subunits of the viral capsid). The viral particle is made of a large number of proteins and enzymes; lipids are present in the envelope. These viruses replicate in the cytoplasm of the host cells.
2. Poxviridae (chapter 5)
   Cryptogram: D/2:130–240/5–7.5:X/★:I, V/O, R, Ve/Ac, Di, Si
   The virions are brick-shaped, measuring 300–450 × 170–260 nm with a lipid envelope containing more than 30 proteins, of which 10 are antigenic. The viruses replicate in the cytoplasm of the infected cell. Hemagglutinin is produced by orthopoxviruses.
3. Parvoviridae (chapter 11)
   Cryptogram: D/1: 1.5–2.2/19–32:S/S: I, V/C, I, O, R
   Nonenveloped virions with a diameter of 18–26 nm. The virions have a density of 1.38–1.46 g/ml. In some members the virions contain single-stranded DNA strands of opposite orientation; these are named plus (+) and minus (−) strands. In others, only the DNA(+) form is present in virions. There are two types of viruses: (1) self-replicating viruses and (2) viruses requiring adenovirus to provide helper functions to allow them to replicate. These viruses can be regarded as parasites of viruses.
4. Reoviridae (chapter 12)
   Cryptogram: R/2:Σ12–Σ19/15–30:S or So/S: I, S, V/I, O, Ve/Ac, Au, Di
   The virion has an isometric capsid with icosahedral symmetry and a diameter of 60–80 nm, usually without an envelope. The virion has a density of 1.31–1.38 g/ml, a sedimentation coefficient of 630 S. The capsomeres are arranged in two layers and an enzyme, the RNA-dependent RNA polymerase, is attached to the viral RNA in the core. These viruses replicate in the cytoplasm of infected cells. Members of the genus *Orbivirus* replicate in vertebrates and insects; plant reoviruses replicate in plants and insects.
5. Bunyaviridae (chapter 17)
   Cryptogram: R/1:Σ6–Σ7/★: Se/E:I, V/C, Ve/Ac, Di
   Enveloped virions with a diameter of 90–100 nm. The envelope contains at least one glycoprotein. A ribonucleoprotein complex with a diameter of 2–2.5 nm is present in the capsid. The viral genome is composed of three single-stranded RNA molecules of 0.8, 2, and 4 × 10^6 daltons, respec-

tively. The virus replicates in the cytoplasm of the infected cell and the nucleocapsids bud through the cytoplasmic membrane to acquire their envelopes.

6. Picornaviridae (chapter 18)
   Cryptogram: R/1:2.3–2.8/30:S/S:I, V/I, O, R
   Virions are not enveloped, the capsomeres are arranged in the capsid in an icosahedral shape, the diameter of the virion is 20–30 nm, and the virus replicates in the cytoplasm. The viral RNA is monocistronic, infectious, and acts as messenger RNA for the synthesis of viral peptides which are cleaved from a precursor into the functional peptides, the viral capsid protein, and the viral replicase (RNA-dependent RNA polymerase).
7. Rhabdoviridae (chapter 13)
   Cryptogram: R/1:3.5–4.6/2–3:Ue/E:I, S, V/C, O, Ve/Ap, Au, Di
   The virions are bullet-shaped, 130–300 nm in length and 70 nm in width, with an envelope covering an elongated nucleoprotein with helical symmetry. The virions have a density of 1.20 g/ml and contain five major proteins, including the RNA-dependent RNA polymerase. They replicate in arthropods, insects, vertebrates, and higher plants.
8. Togaviridae (chapter 19)
   Cryptogram: R/1:3.5–4/5–8:Se/S:I, V/C, I, O, R, Ve/Ac, Di
   The virions are spherical with a diameter of 40–70 nm with a lipoprotein envelope attached to a capsid with icosahedral symmetry. The density of the virions is 1.25 g/ml, and they replicate in arthropods and vertebrates.

Taxon B: Vertebrate viruses

This taxon includes eight virus families that replicate only in vertebrates.

1. Adenoviridae (chapter 8)
   Cryptogram: D/2:20–30/12–17:S/S:V/I, O, R
   The virions lack an envelope; the 252 capsomeres are arranged in icosahedral symmetry in the capsid which has a diameter of 70–90 nm. Capsomeres with filaments attached are antigenically distinct from others. The density of virions is 1.33–1.35 g/ml and the sedimentation coefficient is 795 S. The G + C content of the viral double-stranded DNA is 48 to 61% in different strains. The virus replicates in the nucleus and several strains have the ability to transform cells. Certain strains are capable of agglutinating cells of different species.
2. Herpesviridae (chapter 7)
   Cryptogram: C/2:92–102/8.5:Se/S:(F), (I), V/C, O, R
   The enveloped virions contain a nucleocapsid of 162 capsomeres arranged in icosahedral symmetry; the diameter is 120–150 nm. The virions have a density of 1.27–1.29 g/ml and contain approximately 33 proteins with a total molecular weight of 290,000 daltons. The G + C contents of the viral DNA range from 33–74% in different strains. The virus replicates in the

nucleus of the host cell. Herpes-type viruses that are also found in fungi and in oysters, occur in lower and higher vertebrates.

3. Papovaviridae (chapter 9)
   Cryptogram: D/2:3–5/12:S/S:V/O, Ve/Ac, Si
   Unenveloped spherical virions with a diameter of 45–55 nm are composed of 72 capsomeres. Elongated virions were also described. The G + C content of the circular, double-stranded DNA molecules is 41–49%; density is 1.34 g/ml. The virus replicates in the nucleus and most strains are capable of cell transformation.
4. Arenaviridae (chapter 16)
   Cryptogram: R/1: Σ3.2–Σ5.6/*:Se/*:V/C, O
   Virions are spherical or pleomorphic with a diameter of 50–300 nm, have envelopes and contain particles similar to cellular ribosomes (two glycoproteins and two polypeptides). The genome is composed of 1–3 short and four longer single-stranded RNA molecules of 0.03; 0.7; 1.7; 1.1, and 2.1×10^6 daltons. The virions contain an RNA-dependent RNA polymerase. Most viruses infect a particular rodent and cause a persistent infection accompanied by viruria (excretion of virus in the urine). Infection of other rodents or of humans is infrequent.
5. Coronaviridae (chapter 20)
   Cryptogram: R/1:9/*:Se/E:V/I, R
   Pleomorphic-enveloped virions with a diameter of 100 nm. Peplomers are found on the surface of the envelope. Single-stranded RNA genome of 9×10^6 daltons. These viruses replicate in the cytoplasm of infected cells.
6. Orthomyxoviridae (chapter 15)
   Cryptogram: R/1: Σ4/1:Se/E:V/R
   Spherical or elongated enveloped virions with a diameter of 80–120 nm. On the envelope there are glycoprotein peplomers 10–14 nm in length and 4 nm in diameter, consisting of the hemagglutinin (HA) and the enzyme neuraminidase (NA). The virions contain an RNA genome composed of eight single-stranded RNA molecules and the RNA-dependent RNA polymerase.
7. Paramyxoviridae (chapter 14)
   Cryptogram: R/1:5–8/1:Se/E:V/O, R
   Pleomorphic, usually spherical, enveloped virions with a diameter of 150 nm or more. Long filamentous forms also occur. The envelope contains an HA and an NA. The genome is unsegmented, single-stranded, and linear RNA ($5–8 \times 10^6$ daltons) and forms a ribonucleoprotein complex in the virion containing an RNA-dependent RNA polymerase.
8. Retroviridae (chapter 21)
   Cryptogram: R/1:7–10/2:Se/*:V/I, C, O, R
   Spherical-enveloped virions with a diameter of 100 nm with an icosahedral core containing single-stranded RNA of $7–10 \times 10^6$ daltons which dissociates into two or three pieces. The virion contains RNA-dependent

DNA polymerase (reverse transcriptase), which is responsible for the synthesis of the double-stranded DNA genome of the virus that can integrate into the host nuclear DNA. This integrated viral DNA is also the proviral DNA.

Taxon C: Plant viruses

Twenty families of viruses, all of which (except one) have an RNA genome.

Taxon D: Bacteriophages

1. Corticoviridae
   Cryptogram: D/2:5/12:S/S:B/O (includes phage PM2 with a lipid envelope)
2. Myoviridae
   Cryptogram: D/2:21–190/4–49:X/X:B/C, O (T even phage)
3. Pedoviridae
   Cryptogram: D/2:12–73/44–51:X/S:B/C, O (T uneven phage, like T7)
4. Styloviridae
   Cryptogram: D/2:25–79/44–62:X/X:B/C, O (Lambda (X) phage)
5. Inoviridae
   Cryptogram: D/1:1.9–2.7/12:E/E:B/C, O (fd phage)
6. Microviridae
   Cryptogram: D/1:1.7/26:S/S:B/O (ØX phage)
7. Cystoviridae
   Cryptogram: R/2:Σ13/10:Se/S:B/O (Ø6 phage)
8. Leviviridae
   Cryptogram: R/1:1/31:S/S:B/O (Ribophage like f2, OB)

Classification of viruses as a continuing process

The classification of viruses is a continuing effort that is updated and augmented annually. Different vertebrate virus families will be presented in this book according to the properties of their nucleic acid, starting with the viruses containing the largest DNA genome (poxviridae) and ending with the unclassified subviruses.

BIBLIOGRAPHY

Matthews, R.E.F. Classification and nomenclature of viruses. Third report on the International Committee on Taxonomy of Viruses. S. Karger, Basel, *Intervirology 12,*No. 3–5, 1979.

Melnick, J.L. Taxonomy of viruses, 1980. *Prog. Med. Virol. 26,*214–232, 1979.

3. MOLECULAR CONSIDERATIONS OF VIRUS REPLICATION AND VIRUS–CELL INTERACTIONS

Virions were defined by A. Lwoff and S.E. Luria as infectious particles containing the viral genetic material in the form of a DNA or RNA molecule covered by a protein coat. These particles constitute the stable infectious form of the virus since during the growth cycle in infected cells the virions disappear, and after an eclipse period, new progeny appear. The viral nucleic acid is packaged inside the virions in a form that protects it from damaging agents outside the living host cell; the virions can, therefore, be regarded as the stable or dormant phase in the virus life cycle. In certain virus families, the virions themselves carry enzymes essential for virus replication after entry into the new host cell.

Viruses as obligate parasites of cells

The viral nucleic acids vary in size and molecular weight ranging from the largest viral DNA of vaccinia virus, which has a molecular weight of about 150×10^6, to the smallest viral RNA or DNA with a molecular weight of $1\text{–}2 \times 10^6$ and containing only a few genes (e.g., picornaviruses). In comparison, *Chlamydia trachomatis,* a prokaryotic obligate parasite of mammalian cells, contains a DNA genome of 660×10^6 daltons with several hundred genes. In order to replicate, the viral nucleic acid requires the cellular energy-producing systems, protein-synthesizing systems, as well as specific cellular enzymes and proteins that the cell normally utilizes for its own metabolism. Viruses that encapsidate replication enzymes in their virions can be studied under in vitro conditions after removal of the virion envelope and incubation of the exposed viral enzymes and the viral nucleic acid in a suitable reaction mixture.

The entry of virions into host cells is followed by the uncoating process—namely, the release of the viral nucleic acid from its protein coat. A series of processes is then set in motion which leads to the synthesis of viral proteins, replication of the viral genome, and finally the formation of new virus progeny, identical to those that initiated the infectious process.

In certain instances, cell–virus interactions can lead to the suppression of the virus replicative cycle. In such nonpermissive cells, expression of the viral genetic information is suppressed, and the viral genomes can be retained in the cells for prolonged periods. Only when certain changes take place in these cells can the virus escape from the control of the cell and replicate itself.

Virus genes code for different groups of functional proteins

In the different virus families (except viroids and viruses, which are parasites of other viruses), the genome is essentially made up of three types of genes: (1) genes for enzymes that participate in the replication of the viral nucleic acid; (2) genes for proteins that are involved in regulatory processes; and (3) genes coding for the structural viral proteins—namely, the proteins that are needed to form the viral capsids and envelopes. A number of virus structural peptides form the capsomere—the subunits that assemble into the viral capsid. The number of capsomeres in the virion capsid differs according to the size of viral nucleic acid. The formation of a capsid requires that the capsomeres be arranged into stable structures.

Viral capsomeres assemble into capsids

The internal organization of the capsomeres in the viral capsid was examined in the electron microscope and also by x-ray diffraction methods (summarized by Caspar and Klug 1962). Three types of cubic symmetry are possible in the viral capsid: tetrahedral (2:3), octahedral (4:3:2), and icosahedral (5:3:2). This means that these structures contain 12, 24, or 60 capsomeres, respectively, arranged on the surface of a sphere. Crick and Watson (1956) had originally suggested that cubic symmetry was the most fitting form for viruses. With the accumulation of evidence on viral structure, it became obvious that a number of viruses have the shape of a regular icosahedron. This was first shown by Caspar (1956) in x-ray diffraction studies on tomato bushy stunt virus, and shortly afterward by Klug, Finch, and Franklin (1957) with turnip yellow mosaic virus. The most conclusive evidence of icosahedral symmetry was presented by Williams and Smith (1958) with Tipula iridescent virus, and subsequently by Finch and Klug (1959) with poliovirus. The theory of Caspar and Klug (1962) for the icosahedral structure of viruses was based on the structure of a geodesic dome, which is composed of triangular subunits arranged in groups of five. Icosahedral symmetry appeared to be the most efficient form of packing subunits into a sphere, but further analysis showed this to be only one aspect of the more comprehensive idea of the optimum design of a shell.

Tobacco mosaic virus (TMV) has its capsids arranged in the form of helical symmetry. The helical framework is most suited for packaging of a long, single-stranded RNA molecule since it allows optimal interaction between the viral structural coat protein and nucleic acid. The capsid subunits might also be assembled in the absence of the viral ribonucleic acid.

Virion formation and capsid–nucleic acid interactions

In the final stages of virion assembly, the coiling of the viral nucleic acid into a structure that can fit into the capsid requires specialized proteins that vary according to the size and properties of the viral nucleic acid and sometimes also polyamines. In the family Herpesviridae (chapter 7), the viral DNA is large (100×10^6 daltons) and is coiled into the form of a torus, whereas in small DNA viruses (e.g., Papoviridae), cellular histones attach to the viral DNA which attains the form of a minichromosome.

In other viruses (e.g., Orthomyxoviridae), the viral RNA forms a ribonucleoprotein complex due to the interaction with specific viral proteins. Such complexes are incorporated into a membrane that contains the viral glycoprotein and buds from the cell membrane. The mechanism of insertion of the viral nucleic acid into a capsid differs in the various virus families; some (e.g., Herpesviridae) acquire a lipid envelope when passing through the membranes of the infected cell which contain the virus-coded glycoprotein. Thus viral structural proteins (glycoproteins) form part of the virion envelope and provide the virus specificity of the membrane.

Cells transfected with naked viral nucleic acids

Naked viral nucleic acids (DNA or RNA) might retain their infectivity for cells under in vitro conditions, provided special conditions of infection are used. Double-stranded viral DNA molecules are made to enter host cells by forming a calcium phosphate precipitate on the cell membrane which also contains the viral nucleic acid. The precipitate is taken up by the cells, and the viral DNA can function in the cell. Viral RNA molecules were introduced into cells by adding DEAE dextran to the medium of the cells infected with purified viral RNA.

Virions released to initiate new infections

A number of different mechanisms are used for the release of virions from infected cells. At the end of the replicative cycle, icosahedral viruses that lack an envelope and replicate in the nucleus (e.g., adenoviruses) or in the cytoplasm (e.g., picornaviruses) are neatly arranged in crystal-like arrays and are released from the infected cell after it disintegrates.

Enveloped virions of different virus families (e.g., Herpesviridae, Togaviridae, Retroviridae), on the other hand, utilize an active mechanism of release. The virions of the Herpesviridae obtain part of their envelope when passing through the nuclear membrane and an additional part of their envelope during egress through the cytoplasmic vacuoles. Other virus families (e.g., Myxoviridae, Retroviridae) obtain their envelopes during passage of the viral nucleocapsids through the cellular membrane. The nucleocapsid is enveloped with a cellular membrane that contains inserted viral glycoproteins.

Transfer of viruses from one host to another

Many viruses spread from host to host via the respiratory and alimentary tracts. Thus viruses that are excreted in the feces (e.g., poliovirus or Coxsackie virus) or urine (e.g., hemorrhagic fever virus) or via the respiratory tract (e.g., influenza virus or the common cold viruses) can be disseminated with ease.

Another mode of transfer of viruses involves introduction of the virus directly into the recipient tissues. Rabies virus is spread by rabid animals which secrete the virus in their saliva. An animal or person bitten by a rabid animal becomes infected. Viruses that belong to Taxon A (capable of replicating in more than one host) can infect blood-sucking insects. When such an insect feeds on an infected animal at the stage when the virus is in the blood (the viremic state), the insect is also infected and can transmit the virus to a healthy host.

Insects also transmit viruses mechanically by transferring the virions present on the proboscis to a new host during biting, without the virus actually replicating in the insect. Viruses that are able to infect via the alimentary tract must overcome the acid barrier of the stomach. Indeed, it was found that such viruses (like poliovirus) have a capsid that can withstand the acidity in the stomach, and thus the virions can reach the susceptible cells in the intestine and initiate an infection. Enveloped virions that are destroyed by acid conditions infect mucous membranes of the respiratory tract (e.g., poxviruses and myxoviruses) or the mouth and genitalia (e.g., herpesviruses). Viruses that are transferred by insects directly into the bloodstream of the host are also enveloped.

The structure of the virion and the properties of the envelope and internal capsids play an important role in stability of viruses. In some instances, the virions can reside in dust or in organic debris for extended periods prior to infection. Smallpox virus (chapter 5), for example, is very stable and can withstand dryness and temperatures of up to 45°C for prolonged periods.

Infection by attachment of virions to receptors in the cell membrane

For the infectious process to start, the viral genome must be able to reach the susceptible cell. In an immunized host who has specific antibodies, the invading virions interact with the antibodies in the bloodstream, on the surface of T cells, or elsewhere in the body, and virions to which antibodies are attached cannot adsorb to the receptors on the cell membrane. Such virion–antibody complexes may enter into a cell, but virus replication is prevented since the viral nucleic acid cannot be released from its protein coat.

The first step in the infectious process—namely, attachment of virions to the outer-cell membrane—is electrostatic in nature. Usually, the outer surface of the virion is negatively charged, and attachment takes place in areas of the cell membrane that are positively charged. This interaction is not a stable one, and virions can be released from the membrane if the electrostatic charge in the area of virion attachment is changed (by addition of heparin, for example). However, if the virions attach to the correct receptor on the cell membrane, the second step of attachment can take place. This step is irreversible, and the virions can no longer be released.

The cellular receptors vary for the different virus families and are controlled by the host-cell genome. In the absence of such a gene, or in the absence of the receptors, the cells will be unable to adsorb the particular virions. It was demonstrated that the gene for the poliovirus receptor is localized in chromosome 19 of the human cell (chapter 4). Mouse cells do not have a gene for a receptor for poliovirions and are therefore resistant to infection with this virus. When human cells that have a receptor for poliovirus are fused with mouse cells, human-mouse hybrid cell lines can be isolated that have the receptor for poliovirus adsorption. The mouse cell synthesizes a receptor protein for poliovirus only when the human chromosome number 19 is present. Hybrid cell lines containing other human chromosomes cannot produce the receptor protein. Other genes in human cells that affect the susceptibility of cells to virus infection are described in chapter 4.

Enveloped virions that are released from infected cells by a process of budding through the membrane (e.g., Myxoviridae, Togaviridae) enter new cells by a cellular response mechanism called *viropexis.* Since the virion envelope and outer membrane of the cell are very similar (apart from the presence of the viral proteins) the two fuse to form one continuous membrane, and the viral nucleocapsid is released directly into the cytoplasm of the infected cell. Virions lacking an envelope (e.g., Adenoviridae, Picornaviridae) or with envelopes that do not resemble the outer membrane of the cell (e.g., Poxviridae) are introduced into the cells by a mechanism of pinocytosis. The cell membrane responds to the attachment of virions by invagination into the cytoplasm, the upper part of the invaginated membrane fuses, and a vacuole containing the virions is formed in the cytoplasm. Lysosomes attach to the vacuole and release proteolytic enzymes in an attempt to degrade the foreign particles. As a result, the protein coat of the virus is digested and the nucleic acid molecules are released. After uncoating, the viral nucleic acid molecules are able to express their genetic information in the cytoplasm or nucleus. Cellular transport mechanisms are used to move molecules between the different compartments of the infected cell.

RNA genomes as plus or minus molecules

In some virus families, the RNA genome can serve as messenger RNA for the synthesis of viral proteins (e.g., poliovirus). These are RNA^+ viruses. In other virus families, the nucleic acid of the virion already has certain enzymes attached for the initiation of viral RNA synthesis. These are RNA^- viruses.

Replication of the virus in the infected cell: one-step growth cycle

In most instances, only some of the virions that attack a cell irreversibly attach to the cell membrane. Not all the attached virions are transferred into the cytoplasm and not all the virions are uncoated. Thus in order to infect a cell, more than one virion is necessary. Essentially, as in the case of poliovirus, 100 virions are considered as one plaque forming unit (PFU), which means that at least 100 virions are needed to infect one cell. Herpes simplex virus requires about 36 virions to successfully infect one cell. This is because only 50% of the virions enter the cell, of which only 50% are uncoated, and finally only about 4 viral genomes make their way into the nucleus, the site of virus replication.

The technique of plaque formation was developed by R. Dulbecco for the quantitation of viruses. The infected cells in a monolayer are covered with a layer of agar to restrict the spread of virus progeny. Beneath the agar, the virus can infect and destroy only neighboring cells, and staining of the infected cell monolayers with a stain that enters only living cells results in the appearance of clear areas, or plaques. One PFU is due to the successful infection of a single cell by a number of virus particles.

In nature, the infection of an animal or man is initiated by a relatively small amount of virus infecting a limited number of cells. The virus progeny spreads to neighboring cells, infecting more and more cells, and even though the virus infection stimulates the defense mechanism of the host (like the production of interferon), the infection continues. To investigate the molecular processes in the replication of a virus, experimental cell–virus systems were developed in which all the cells were infected simultaneously. Such a synchronous infection could be measured at different times after infection by titration of the virus progeny by the plaque method. In addition, virus-induced molecular processes can be studied, using analytical chemical procedures. It was noted that after adsorption of virions to the cells, the infectivity disappeared for a certain period (called the eclipse phase), and new progeny of infectious virions appeared afterward. The time between infection and synthesis of new infectious progeny differs in various virus families and can also be affected by the host cell.

Enzymes in virions of some virus families

Viruses, being obligate parasites of cells, utilize numerous cellular biochemical processes like the energy-producing systems, the protein-synthesizing mechanisms, the enzymes involved in phosphorylation of proteins, the enzymes involved in transcription of mRNA from DNA genomes, as well as enzymes required for the synthesis of the cap and poly A sequence attached to the viral mRNA. The cellular systems required for lipid and membrane synthesis are also utilized for viral synthesis. The viral genome contains the information needed for the synthesis of specific viral enzymes that replicate the viral nucleic acid and structural proteins, while the rest of the enzymatic processes are contributed by the host cells. However, for the initiation of their replication cycle, some viruses require specific enzymes that are absent in the host cells.

Such viruses carry the specific enzyme to ensure their replication in the infected host. Three specialized enzymes were found to be carried by virions:

1. *DNA-dependent RNA polymerase.* This enzyme, which is responsible for transcribing 14% of the viral genome, is found in the Poxviridae.
2. *RNA-dependent RNA polymerase.* Viruses that have an RNA genome that cannot serve as messenger RNA (designated RNA^-) produce copies of the genome to serve as mRNA for the production of the viral proteins. (These mRNA molecules are designated RNA^+.) Since the cells do not have an enzyme that synthesizes RNA molecules on an RNA template, the virions were found to contain an RNA-dependent RNA polymerase which produces mRNA (regarded as RNA^+) from the viral RNA genome (regarded as RNA^-). Virions of the families Rhabdoviridae, Paramyxoviridae, Orthomyxoviridae, Arenaviridae, and Bunyaviridae contain RNA-dependent RNA polymerase.
3. *RNA-dependent DNA polymerase (reverse transcriptase).* This enzyme, found in virions of the Retroviridae, utilizes the viral genome as a template for the synthesis of a complementary DNA molecule. The same enzyme synthesizes a second DNA strand, complementary to the first, to form a double-stranded DNA molecule. Such DNA molecules interact with the host cell genome, causing its transformation into a cancer cell.

The mechanisms utilized for virus replication in infected cells depend on the viral nucleic acid

The organization of the viral genome into single-stranded or double-stranded DNA or RNA determines the molecular processes required for its replication:

A. Double-stranded DNA viruses:
   1. Uncoated viral double-stranded DNA + cellular or viral RNA polymerase → synthesis of early mRNA.
   2. Early mRNA + cellular ribosomes → early viral proteins, including enzymes and proteins for the replication of the viral nucleic acid.
   3. Parental viral DNA + replication proteins → DNA replication complex involved in the synthesis of progeny viral DNA.
   4. Progeny viral DNA + cellular and/or viral RNA polymerase → synthesis of late mRNA from the late viral genes.
   5. Late mRNA + cellular ribosomes → structural viral proteins.
   6. Structural proteins + progeny viral double-stranded DNA → nucleocapsids.
   7. Nucleocapsids + envelopes and cellular membranes → infectious virions.

B. Single-stranded DNA viruses:
   1. Single-stranded viral DNA + cellular DNA polymerase → double-stranded DNA

2. Subsequent steps are similar to A.1. to 3., but the progeny viral DNA is single stranded.
3. The single-stranded DNA genome + structural proteins → virions.

C. Double-stranded RNA viruses:
1. Virions contain 10–12 fragments of parental double-stranded RNA molecules + RNA polymerase present in the virions → synthesis of mRNA molecules from each individual fragment.
2. Viral mRNA + ribosomes → viral structural proteins and enzymes.
3. Viral nonstructural (RNA-dependent RNA polymerase) and structural proteins + viral mRNA molecules → synthesis of double-stranded RNA in the nucleocapsids.
4. Nucleocapsid + capsomeres → virions with two layers of capsomeres.

D. Viruses with negative single-stranded RNA genome (RNA^-).
1. Single-stranded RNA^- + RNA-dependent RNA polymerase attached to the viral genome → viral mRNA (RNA^+).
2. Viral mRNA + ribosomes → viral proteins.
3. Viral mRNA (RNA^+) + RNA-dependent RNA polymerase → synthesis of progeny RNA^- molecules.
4. Progeny RNA^- molecules + viral proteins (including RNA polymerase) → nucleocapsids.
5. Viral proteins + cellular enzymes → viral glycoproteins.
6. Viral glycoproteins + cellular membranes → regions in the cell membranes containing viral glycoproteins.
7. Viral nucleocapsid attaches to the envelope at the site of viral glycoproteins → budding to form virions.

E. Viruses with single-stranded RNA^+ genomes.
1. Virion RNA^+ molecules + cellular ribosomes → one high molecular weight viral precursor polypeptide (from monocistronic mRNA).
2. Viral precursor polypeptide + proteases → viral polypeptides, structural and functional (RNA-dependent RNA polymerase).
3. Viral RNA-dependent RNA polymerase + parental RNA^+ → synthesis of RNA^- molecules which serve as templates for the synthesis of progeny RNA^+ molecules.
4. Progeny viral RNA + RNA dependent-RNA polymerase → synthesis of subgenomic (part of the parental RNA^+) RNA molecules serving as mRNA.
5. Progeny viral RNA^+ + capsid (made of a number of capsomeres) → virions.

F. Viruses with single-stranded RNA^+ genome but with a double-stranded DNA phase.
1. Parental RNA^+ + RNA-dependent DNA polymerase (in the viral capsid) → synthesis of RNA-DNA hybrid → synthesis of double-stranded DNA.

2. Linear progeny double-stranded DNA → circular double-stranded DNA.
3. Circular double-stranded DNA + chromosomal DNA → integration into cellular DNA. Some viral double-stranded DNA genomes are retained in the nuclei of infected cells as episomes.
4. Viral DNA in the chromosomal DNA + cellular DNA-dependent RNA polymerase II of the host cells → viral mRNA.
5. Viral mRNA + ribosomes → viral proteins.
6. Viral proteins + viral RNA^+ molecules → viral nucleocapsids.
7. Nucleocapsids + cellular membranes (containing viral glycoproteins) → budding of virions.

Cell response to the virus infection

Certain viruses are cytocidal, while other viruses can transform cells, causing them to proliferate indefinitely. Several cell–virus relations can be noted:

1. The cell dies at the end of the virus replicative cycle, and the virus progeny is released.
2. The infected cell remains alive but continues to synthesize virions.
3. The infected cell is transformed by the virus but continues to produce virions.
4. The infected cell is transformed by the virus, but virions are not synthesized.
5. The cell is infected by the virus, but the virus cannot replicate (latent infection); the cells remain alive, and the virus can be activated.

BIBLIOGRAPHY

Caspar, D.L.D. Structure of tomato bushy stunt virus. *Nature 177*:476–477, 1956.

Caspar, D.L.D., and Klug, A. Physical principles in the construction of regular viruses. *Cold Spring Harbor Symp. Quant. Biol. 27*:1–24, 1962.

Crick, F.H.C., and Watson, J.D. The structure of small viruses. *Nature 177*:473–475, 1956.

Finch, J.T., and Klug, A. Structure of poliomyelitis virus. *Nature 183*:1709–1714, 1959.

Klug, A.; Finch, J.T.; and Franklin, R.E. The structure of turnip yellow mosaic virus: X-ray diffraction studies. *Biochim. Biophys. Acta 25*:242–252, 1957.

Williams, R.C., and Smith, K.M. The polyhedral form of the Tipula iridescent virus. *Biochim. Biophys. Acta 28*:464–469, 1958.

RECOMMENDED READING

Bachrach, H.L. Comparative strategies of animal virus replication. *Adv. Virus Res. 22*:163–186, 1978.

Baltimore, D. Expression of animal virus genomes. *Bact. Rev. 35*:235–241, 1971.

Bishop, D.H. Virion polymerases. *Comprehensive Virology* (H. Fraenkel-Conrat and R.R. Wagner, eds.) Plenum Press, New York, 1977, Vol. 9, pp. 117–278.

Butterworth, B.E. Proteolytic processing of animal virus proteins. *Curr. Top. Microbiol. Immunol.* 77:1–41, 1977.

Dales, S. Early events in cell-animal virus interactions. *Bact. Rev. 37*:103–135, 1973.

Doerfler, W. Animal virus-host genome interaction. *In: Comprehensive Virology* (H. Fraenkel-Conrat and R.R. Wagner, eds.) Plenum Press, New York, 1977, Vol. 9, pp. 279–400.

Huang, A.S., and Baltimore, D. Defective interfering animal viruses. *In: Comprehensive Virology* (H. Fraenkel-Conrat and R.R. Wagner, eds.) Plenum Press, New York, 1977, Vol. 9, pp. 73–116.

Naha, P.M. Early functional mutants of mammalian cells. *Nature New Biol. 241*:13–14, 1973.

Raghow, R., and Kingsbury, D.W. Endogenous viral enzymes involved in messenger RNA production. *Annu. Rev. Microbiol. 30*:21–39, 1976.

Russell, W.C., and Winters, W.D. Assembly of viruses. *Prog. Med. Virol. 19*:1–39, 1975.

Shatkin, A.J.; Banerjee, A.K.; and Bott, G.W. Translation of animal virus mRNA in-vitro. *In: Comprehensive Virology* (H. Fraenkel-Conrat and R.R. Wagner, eds.) Plenum Press, New York, 1977, Vol. 9, pp. 1–72.

Simchen, G. Cell cycle mutants. *Annu. Rev. Genet. 12*:161–191, 1978.

4. GENES IN HUMAN CELLS DETERMINING VIRUS SUSCEPTIBILITY

The use of somatic cell hybrids formed between human and mouse cells led to the discovery that certain genes in human cells determine cell sensitivity or resistance to virus infections. The technique is based on the property of human–mouse somatic cell hybrids to lose most of the human chromosomes during cultivation under in vitro conditions. The human and mouse cells are fused by a virus (Sendai virus) or by the chemicals lysolecithin or polyethyleneglycol. The hybrid cells are propagated in culture, and the human and mouse chromosomes are defined by selective staining. With this technique, the following genes in human cells were defined:

Chromosome 1 is an integration site for adenovirus 12 DNA (adenovirus 12 chromosome modification sites 1q and 2q3) (McKusick 1975; 1978).

Chromosomes 2 and 5 contain genes (If1 and If2) for the production of interferon. (Animal cells respond to virus infection with the production of protein with a molecular weight ranging from 16,500 to 23,000 that has antiviral activity) (Tan et al. 1974). The interferon-1 gene (alpha-leukocyte interferon) has three loci. Another gene (interferon-2), responsible for the antiviral state, is found in the short arm of chromosome 5. Chromosome 9 contains the gene for interferon-3 (fibroblast interferon).

Chromosome 3 contains a gene for sensitivity to herpes simplex virus type 1 (HSV-1). This gene determines the ability of HSV-1 to replicate in the host cell. The function of this gene is not yet known, but a gene in chromosome

3 is responsible for the ability of the cell to pass from the G_1 stage of the cell cycle to the S phase, the phase of DNA synthesis. A cell mutant was isolated that could not synthesize chromosomal DNA at a raised temperature (referred to as temperature sensitive [ts]), and thus remained in the G_1 phase. In such cells, at the nonpermissive temperature, HSV-1 cannot replicate (Yanagi et al. 1978). The relationship between the gene function required for HSV-1 replication and the gene function required for the S phase is not known (Carritt and Goldfarb 1976; Smiley et al. 1978).

Chromosome 6 contains an integration site for the proviral DNA of a baboon retrovirus. The locus designated Bevi contains an integration site for the proviral DNA of the baboon type C virus (Lemons et al. 1977).

Chromosome 7 contains an integration site for SV40 virus DNA (site 1) (Croce and Koprowski 1975).

Chromosome 11 contains xenotropic BALB virus induction gene designated BVIX.

Chromosome 15 contains BALB virus induction gene (N tropic) designated BVIN.

Chromosome 17 contains the second integration site of SV40 DNA (site 2) (Kelly and Nathans 1977).

Adenovirus 12 chromosome modification site 17: this virus causes gaps and breaks at a site on the long arm of chromosome 17 (McDougall 1971).

Chromosome 19 contains the gene for poliovirus receptor and a gene that determines Echo virus 11 sensitivity and Coxsackie B virus sensitivity. This gene is also responsible for infection by baboon M7 endogenous virus (Miller et al. 1974).

Chromosome 21 contains the gene for the receptor for interferon in the cell membrane. The interferon molecules synthesized by the genes in chromosomes 2 and 5 are released from the stimulated cells. In order to protect cells, the interferon molecules must bind to a receptor protein molecule present on the outer surface of the cell membrane. The gene for such a receptor is present in chromosome 21 (Tan, 1975; Epstein and Epstein, 1976; Wiranskowa-Stewart and Stewart, 1977; Slate et al., 1978). After attachment of interferon molecules to the receptor, the cell is induced to activate a gene (its location is not yet known) to synthesize mRNA for a new protein molecule which is responsible for the antiviral state of the cell. In interferon-treated cells, this protein prevents the replication of the invading virus (Chapter 26).

Further information on the mapping of genes in chromosomes of human cells is given in the journal entitled *Cytogenetics and Cell Genetics* (see in particular, McKusick, 1982).

BIBLIOGRAPHY

Carritt, B., and Goldfarb, P. A human chromosomal determinant for susceptibility to herpes simplex virus. *Nature 264*:556–558, 1976.

Croce, C.M., and Koprowski, H. Assignment of gene(s) for cell transformation to human chromosome 7 carrying the simian virus 40 genome. *Proc. Natl. Acad. Sci. USA* 72:1658–1660, 1975.

Epstein, L.B., and Epstein, C.J. Localization of the gene AVG for the antiviral expression of immune and classical interferon to the distal portion of the long arm of chromosome 21. *J. Infect. Dis. 133* (Suppl.):A56–A62, 1976.

Kelly, T.J., and Nathans, D. The genome of simian virus 40. *Adv. Virus Res. 21*:85–173, 1977.

Lemons, R.S.; O'Brien, S.J.; and Sherr, C.J. A new genetic locus, Bevi, on human chromosome 6 which controls the replication of baboon type C virus in human cells. *Cell 12*:251–262, 1977.

McDougall, J.K. Adenovirus-induced chromosome aberrations in human cells. *J. Gen. Virol. 12*:43–51, 1971.

McKusick, V.A. *Mendelian Inheritance in Man.* (Fourth edition). The Johns Hopkins University Press, Baltimore and London, 4th edition, 1975; 5th edition, 1978.

McKusick, V.A. The human genome through the eyes of a clinical geneticist. *Cytogen. Cell Genet. 32*:7–23, 1982.

Miller, D.A.; Miller, O.J.; Dev, V.G.; Hashmi, S.; Tantravaki, R.; Medrano, L.; and Green, H. Human chromosome 19 carries a poliovirus receptor gene. *Cell 1*:167–174, 1974.

Slate, D.L.; Shulman, L.; Lawrence, J.B.; Revel, M.; and Ruddle, F.H. Presence of human chromosome 21 alone is sufficient for hybrid cell sensitivity to human interferon. *J. Virol. 25*:319–325, 1978.

Smiley, J.R.; Steege, D.A.; Juricek, D.K.; Summers, W.P.; and Ruddle, F.H. A herpes simplex virus 1 integration site in the mouse genome defined by somatic cell genetic analysis. *Cell 15*:455–468, 1978.

Tan, Y.H. Chromosome-21-dosage effect on inducibility of anti-viral gene(s). *Nature 253*:280–282, 1975.

Tan, Y.H.; Creagan, R.P.; and Ruddle, F.H. The somatic cell genetics of human interferon: assignment of human interferon loci to chromosomes 2 and 5. *Proc. Natl. Acad. Sci. USA 71*:2251–2255, 1974.

Wiranskowa-Stewart, M., and Stewart, W.E. The role of human chromosome 21 in sensitivity to interferons. *J. Gen. Virol. 37*:629–633, 1977.

Yanagi, K.; Talavera, A.; Nishimoto, T.; and Rush, M.G. Inhibition of herpes simplex virus type 1 replication in temperature-sensitive cell cycle mutants. *J. Virol. 25*:42–50, 1978.

II. VIRUS FAMILIES

A. DOUBLE-STRANDED DNA VIRUSES

5. POXVIRUSES

POXVIRIDAE: THE POXVIRUS FAMILY

Poxviruses are the largest viruses among the members of Taxa A and B. The virions that replicate in the host cell cytoplasm contain a double-stranded DNA genome of 130–240 × 10^6 daltons and more than 30 proteins, including an enzyme, a DNA-dependent RNA polymerase. The virion proteins elicit antibodies to at least 10 antigens. One antigen is common to all the members of this virus family.

Members of the poxviridae

Vaccinia subgroup: *Orthopoxvirus*

- Vaccinia virus in man: used for vaccination against smallpox.
- Cowpox in cattle and man.
- Variola in man, responsible for smallpox. (Two forms of the disease are known: variola major and variola minor.)
- Monkeypox in monkeys.
- Rabbitpox in rabbits.
- Ectromelia in mice.
- Also poxviruses of buffalo and camels.

Fowlpox subgroup: *Avipoxvirus*
Viruses of birds, like canarypox virus, turkeypox virus, pigeonpox virus, quailpox virus.

Sheeppox subgroup: *Capripoxvirus*
Goatpox virus and sheeppox virus.

Myxoma subgroup: *Leporipoxvirus*
Myxomavirus, rabbit fibroma virus (Shope), squirrel fibroma virus, hare fibroma virus.

Orf subgroup: *Parapoxvirus*
These viruses can infect man: milker's node virus, bovine pustular stomatitis virus.

Insect subgroup: *Entomopoxvirus*
These are insect viruses that most probably do not infect vertebrates and have no antigenic relationship with the other poxviruses.

There are still some unclassified poxviruses: Tanapoxvirus and molluscum contagiosum virus of man, Yaba monkey tumor poxvirus, and swinepox virus.

Relatedness between viruses by DNA–DNA reassociation techniques

The relatedness of the different poxviruses is determined by the DNA–DNA hybridization technique. In this procedure, the DNA molecules from two viruses are extracted and denatured to the single-stranded form of the virion DNA. The two DNA preparations, one of which is radioactively labeled, are mixed in equivalent amounts in a suitable salt solution at the proper temperature, and the molecules are allowed to regain their double-stranded form (annealing or reassociation). The extent of reassociation (hybridization) of DNA is determined after digestion of the residual single-stranded molecules by a DNase that specifically digests DNA (S_1 DNase), and the percentage of radioactive DNA that reassociates into double-stranded DNA is calculated.

Such studies showed that variola virus (the cause of smallpox) has 30% homology with vaccinia virus which is used for human vaccination. Vaccinia virus DNA is completely homologous with rabbit poxviruses.

Since variola virus is a human pathogen and vaccinia virus is the vaccine strain that protects man against smallpox, most of this chapter will focus on these two viruses.

Orthopoxvirus subgroup

The virions have a specific brick shape (figure 3), as seen in the electron microscope. Thin sections through the virions revealed a central core covered with a lipid envelope. Two lateral bodies are situated between the envelope and the core. The outer surface of the virion envelope contains protein projections (peplomeres). The composition of virions purified through a sucrose

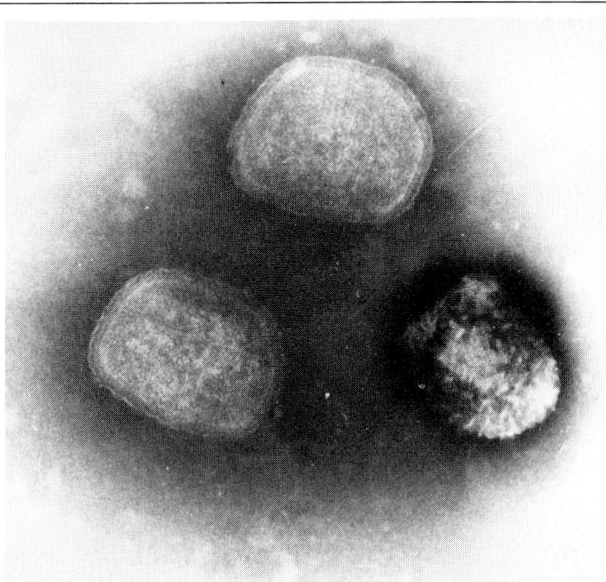

Figure 3. Electron micrograph of virions of vaccinia virus.
(Department of Molecular Virology, Hebrew University-Hadassah Medical School, Jerusalem, Israel.)

gradient (Joklik 1962), can be analyzed following the disruption of their coat by treatment with detergents.

The DNA genome contains more than 150 genes

The DNA of vaccinia virus is linear with a molecular weight of 153 × 10^6 and length, as determined by electron microscopy, of 75–80 nm (Becker and Sarov 1968). The sedimentation profile of the viral DNA in a sucrose gradient is presented in figure 4A. Digestion of the viral DNA with restriction enzymes allowed the various virus species to be classified according to the digestion product pattern after agarose gel electrophoresis (Esposito et al. 1978) (figure 4Ba).

Restriction endonuclease analysis has shown that the genomes of orthopoxviruses vary considerably in size. Rabbitpox, vaccinia, and variola viruses were found to have similarly sized genomes with an average molecular weight of 120, 124, and 121 × 10^6, respectively. The genomes of monkeypox, cowpox, or ectromelia viruses are significantly larger, with molecular weights of 128, 145, and 136 × 10^6, respectively (Machett and Archard 1979).

A unique property of poxvirus DNA is the presence of covalent cross-links between several nucleotides present on opposite DNA strands, usually at the ends of the molecule. The long inverted terminal repetition (molecular weight 6.8 × 10^6) of vaccinia virus DNA was found to contain a fragment that

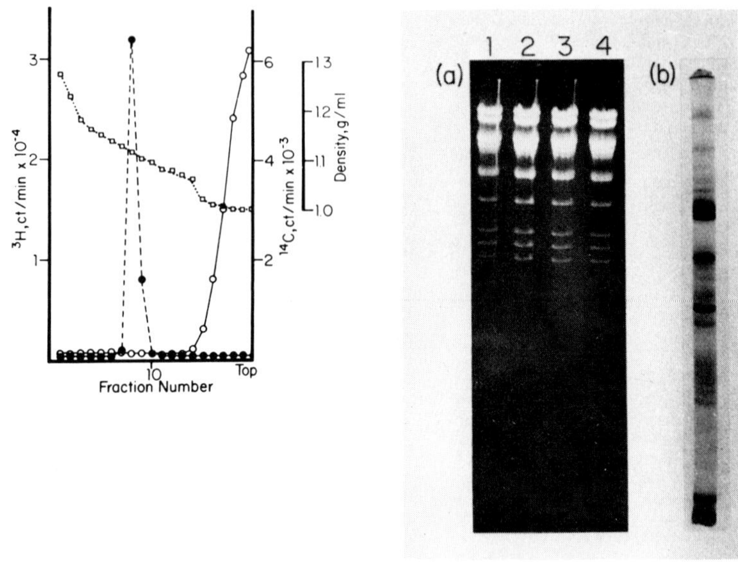

Figure 4. Properties of vaccinia virus DNA

A. Equilibrium centrifugation of vaccinia virus DNA in sodium diatrizoate density gradients (□ —— □ density; ● —— ● [^{14}C] thymidine; ○ —— ○ [^3H] leucine-labeled virions.

B. (a) Analysis of vaccinia virus DNA (from A) after cleavage with Hind III and electrophoresis in agarose gels.

(b) Structural proteins of vaccinia virions in SDS-discontinuous gel electrophoresis.

(Esposito, et al. 1978. Reprinted by permission from *Virology*, Vol. 89, Figs. 3 & 4, p. 59.)

encodes early mRNAs (Witteck et al. 1980). The purified viral DNA molecules are not infectious and cannot replicate when introduced into cells.

Structural proteins and enzymes in poxvirions

Proteins released from the virions can be analyzed by electrophoresis in polyacrylamide gels. Under such conditions, the faster migrating polypeptides have the lower molecular weights. The proteins in the gels can be stained by the dye Coomassie blue or detected by radioautography after labeling with radioactive amino acids. A typical analysis of viral proteins under denaturing and reducing conditions is presented in figure 4B(b). About 30 polypeptides ranging from 200,000–8,000 daltons were detected in vaccinia virions. The overall calculated molecular weight of the 30 virion peptides is 2×10^6. The relative amounts of the different virion peptides were also determined: four peptides constituted 10.5%, five other peptides constituted 6.5% to 9.1%, and the rest of the peptides each were 0.3 to 3.4% of the total peptide content. This does not rule out the possibility that some cellular proteins may also be present in the virions.

In addition, a glycoprotein of 38,000 daltons and a phosphoprotein of 11,000 daltons were identified.

The enzymes in the poxvirion

1. The discovery by Kates and McAuslan (1967) that vaccinia virions contain the virus coded enzyme DNA-dependent RNA polymerase was a turning point in the understanding of the molecular biology of viruses, since it showed that poxvirions carry with them a specific enzyme not provided by the host cell for the transcription of the viral DNA. It was found that this DNA-dependent RNA polymerase is indeed coded for by the viral DNA and is responsible for the transcription of 14% of the viral genome. The mRNA molecules transcribed by this viral enzyme have a sedimentation coefficient of 8–10 S and represent the early mRNA of the virus—namely, the mRNA molecules which are transcribed from the parental DNA prior to the synthesis of viral progeny DNA.
2. An additional enzyme that was found in the poxvirions polymerizes adenosine monophosphates into a polymer. This enzyme was named poly(A) polymerase; it is made up of two polypeptides of 35,000 and 50,000 daltons and is capable of synthesizing a polyA stretch of 200 bases which is covalently attached to the 3′ end of the viral mRNA molecule.
3. Two enzymes, each of 65,000 daltons, were also described. One hydrolyses ATP or dATP and the other hydrolyses ribo- and deoxyribonucleoside triphosphates. The role of these enzymes is not yet known.
4. A deoxyribonuclease (DNase) of 100,000 daltons was also isolated that has a pH optimum of 4.4 and degrades single-stranded DNA with both endo- and exo-nucleolytic activity.
5. The enzyme protein kinase capable of phosphorylation of proteins was also found.
6. DNA nick closing enzyme (made of two subunits of 35,000 and 24,000 daltons).

Cell infection with a poxvirus as a multistage process

Poxvirions phagocytized by infected cells

Poxvirions attach to the outer membrane of a mammalian cell and are engulfed into a cytoplasmic vacuole called a *phagosome*. The process of phagocytosis of virions is inhibited if the cells are treated with NaF and the attached virions are retained on the surface of the cell.

The phagosome is the cellular organelle used to degrade foreign bodies with the aid of lysosomal enzymes. Indeed, inside the phagosomes the viral cores are released after digestion of the outer virion envelope.

Poxvirions treated with specific antibodies are able to attach to the cell membrane and enter into the phagosomes, but the cellular proteolytic enzymes cannot remove the virion envelope, making release of the core impossible. Similarly, heat-inactivated virus in a phagosome cannot be digested by the cellular enzymes. Reactivation of the heat-inactivated virus can be achieved by infecting the same cell with active virus that is able to bring about the release of the cores from the inactive virion. Virions inactivated by ultraviolet (uv) treatment cannot infect cells since the viral cores are not released in the phagosomes.

Uncoating of the virions leads to release of viral cores

The cores are released from the phagosomes into the cytoplasm where the availability of the four ribonucleoside triphosphates (ATP, UTP, GTP, and CTP) stimulates the virion DNA-dependent RNA polymerase to synthesize the early viral mRNA molecules. Inactivation of poxvirions by heat treatment may be due to the inability of the DNA-dependent RNA polymerase in the core to function. Under suitable conditions, one virion is able to infect a cell and produce virus progeny, but it is generally accepted that one to ten virions constitute one PFU, the smallest number of virions able to infect one cell.

To relate the molecular processes that take place in an infected cell with the fate of the parental and progeny virus, it is necessary to synchronize virus infection. This can be done by infecting cells in suspension (e.g., L cells) at a high multiplicity of infection (a large number of PFUs per cell), allowing the adsorption of the virions to be completed within 15 min of incubation. The amount of infectious virus in the cells rapidly decreases due to uncoating of the virions, and the first new progeny appears only 8 hr after the initiation of the infection, gradually increasing to a maximal virus yield at 24 hr postinfection. The virus titer is determined by sampling infected cells at different time intervals and determining the virus content by the plaque assay.

Effect of virus infection on nuclear processes

The synthesis of the nuclear RNA in the infected cells was found to continue for three hr after infection and then to stop gradually. At seven hr after infection, more than 85% of nuclear RNA synthesis is inhibited. The synthesis of nuclear 45 S RNA ribosomal precursor molecules is inhibited, indicating that the infected cell is unable to synthesize new ribosomes.

Synthesis of cellular DNA and proteins in the infected cell is also affected by an unknown mechanism. The virus infection thus leads to paralysis of the activities of the host cell and to a takeover by the virus.

Expression of early viral genes leads to the synthesis of early viral mRNA molecules

The DNA-dependent DNA polymerase present in the virion core is responsible for the transcription of 14% of the parental viral DNA genome in the cytoplasm and also when the virions are incubated under in vitro conditions. The mRNA species produced under the in vitro conditions resemble the mRNA synthesized in infected cells prior to the replication of viral DNA. The mRNA species synthesized at different stages after infection are shown in

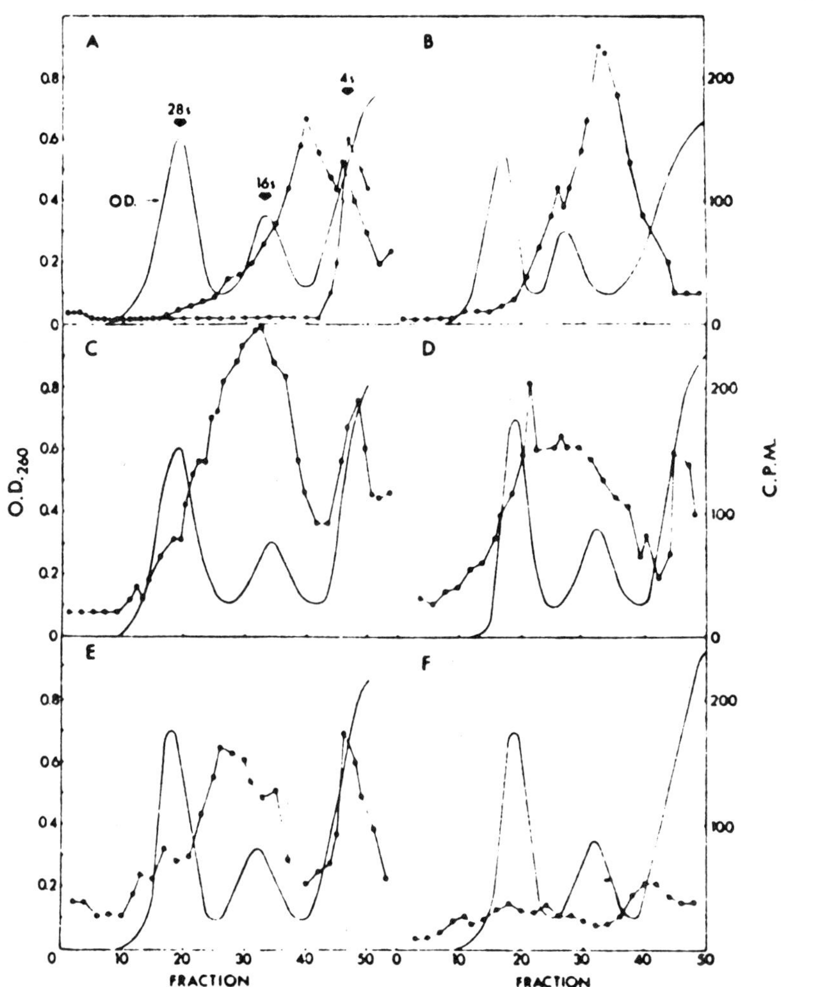

Figure 5. Size distribution of the rapidly labeled cytoplasmic RNA. HeLa cells were infected and samples of 6 × 10^7 cells were pulse labeled for 10 min with uridine-2-^{14}C at ½ hr (A), 1 hr (B), 3 hr (C), 5 hr (D), 7 hr (E), and 9 hr (F) after infection. 6 × 10^7 uninfected cells were similarly treated (A). Radioactive material insoluble in 18% TCA was measured. Closed circles: infected cells. Open circles: uninfected cells. Line: optical density.

(Becker and Joklik 1964. Reprinted by permission from *Proc. Natl. Acad. Sci. USA* 51, Fig. 4, p. 581.)

figure 5 and include the 10 S mRNA species (Becker and Joklik 1964). This was the first demonstration that a viral double-stranded DNA is responsible for the synthesis of mRNA to produce viral proteins. Short labeling periods with ^{3}H-uridine were utilized to label the viral mRNA in the cytoplasm in order to circumvent the release of cellular labeled RNA from the nuclei.

Early mRNA molecules which have a sedimentation coefficient of 10 S with

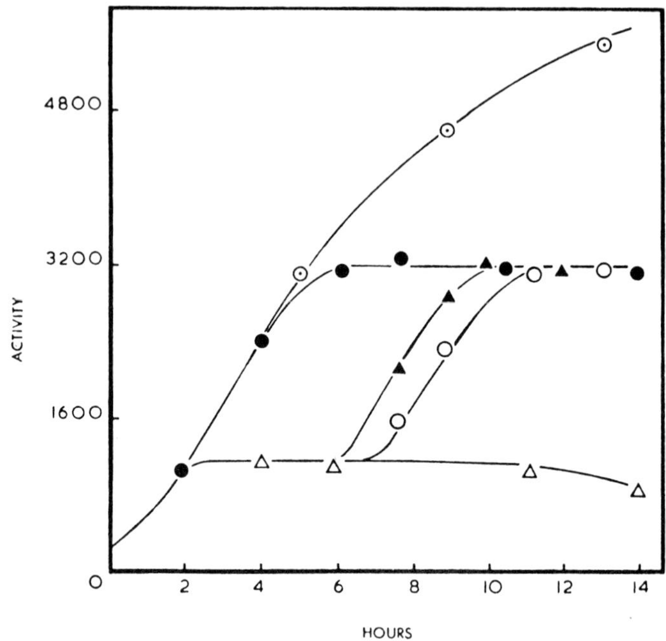

Figure 6. The effect of actinomycin D and puromycin hydrochloride on the establishment of repression of thymidine kinase synthesis. ● thymidine kinase activity in CP-infected cells; ⊙ thymidine kinase activity after addition of actinomycin at 2 hr; △ thymidine kinase activity after addition of puromycin hydrochloride at 2 hr; ▲ thymidine kinase activity after removal of puromycin inhibition at 5½ hr; ○ thymidine kinase activity after addition of actinomycin 30 min before removal of puromycin hydrochloride.
(McAuslan 1963. Reprinted by permission from *Virology* 21, Fig. 5, p. 388.)

a tail of 100–200 adenosines (poly-adenosine) attached at the 3′ end are synthesized by the viral poly(A) polymerase. At the 5′-terminus of vaccinia virus mRNA, there is a cap structure $m^7G^{5'}$ pppAm- or $m^7G^{5'}$pppGm- that is synthesized by a series of RNA polymerase and capping enzymes contained in the virus particle (Urushibara et al. 1981).

Analysis of the tRNA present in the cytoplasm of infected cells revealed a mixture of cellular and viral tRNA molecules, the latter probably synthesized on the viral DNA template.

The mRNA molecules transcribed from the viral DNA attach to cytoplasmic ribosomes present around the site of virus replication. The viral proteins synthesized by the cellular ribosomes according to the nucleotide sequences in the viral mRNA include the enzymes required for viral DNA biosynthesis—namely, DNA-dependent DNA polymerase, thymidine kinase, alkaline deoxyribonuclease, and polynucleotide ligase.

The biosynthesis of viral enzymes like the thymidine kinase (TK) is controlled by the viral DNA. McAuslan (1963) demonstrated (figure 6) that the cessation of synthesis of the viral TK enzyme is naturally regulated and stops 6 hr after infection. However, treatment of the infected cells with the antibiotic actinomycin D, which inhibits the synthesis of DNA-dependent RNA and DNA replication, resulted in the continued synthesis of the TK enzyme. This finding revealed that the synthesis of new viral mRNA is needed to switch off the synthesis of the viral TK.

The biosynthesis of viral DNA in discrete cytoplasmic sites

The viral DNA is synthesized in discrete cytoplasmic sites (or so-called factories), designated viroplasm, in which the viral proteins accumulate. The viral mRNA molecules synthesized on the viral DNA template associate with ribosomes present at the periphery of the viroplasm, and the viral proteins are produced close by.

Labeling of infected cells with ^{3}H-thymidine for short intervals (pulse-labeling) revealed that viral DNA synthesis in the cytoplasm is initiated 90 min after infection. Four and a half hours after infection, the synthesis of viral DNA is terminated. Electron microscopy of replicating viral DNA molecules suggested that the replication fork of the newly synthesized DNA strand moves in one direction on the viral double-stranded DNA template. Synthesis is initiated on one DNA strand close to the end of the molecule, which opens up to form a circle. This is possible because the viral DNA contains cross-links close to the end of the molecule. After synthesis of the DNA strand by the viral DNA polymerase in one direction on one template DNA strand, the viral DNA polymerase continues to synthesize DNA on the opposite strand in the opposite direction. At the end of the DNA replication process, the ends of the molecule are cleaved, and two viral double-stranded DNA molecules become available.

Enzymes involved in DNA biosynthesis include the viral DNA polymerase, which was found to differ from the cellular DNA polymerases α, β, and γ. A viral polynucleotide ligase is capable of ligating nicks present in DNA. Three viral DNases were identified, each with a different pH optimum for their nucleolytic activity.

Expression of late viral genes in the infected cell after viral DNA synthesis.

Late mRNA is transcribed from progeny viral DNA only. The viral DNA-dependent RNA polymerase responsible for transcription of late viral genes differs from the viral RNA polymerase that transcribes early genes. The profile of the late viral mRNA species is presented in figure 5.

RNA–DNA hybridization experiments revealed that during the early phase of virus replication (prior to the synthesis of viral DNA), certain mRNA species are transcribed that await translation until after viral DNA biosynthesis and thus constitute a portion of the late viral mRNA. The mechanism that controls this process is not known. The majority of the late mRNA species that are transcribed from progeny viral DNA are not synthesized when: (1)

viral DNA synthesis is inhibited; (2) the infected cells are incubated in an arginine-deficient medium; or (3) infected cells are treated with protein inhibitors early in infection. The late viral mRNA molecules are polyadenylated (contain a poly A stretch of 100–200 adenines in the 3′ end) and have a methylguanine cap in the 5′ end.

Most of the proteins synthesized by translation of the late mRNA molecules are structural proteins needed for the synthesis of new virions. However, the DNA-dependent RNA polymerase that is inserted into the virions for the early transcription of the viral DNA in the new host is also a late protein as well as being a single-stranded DNase.

A number of structural proteins are modified posttranslationally to enable them to function in the assembly of the virions. At least three structural proteins were found to arise from one viral polypeptide. The cleavage of such precursor polypeptides takes place after their incorporation into the viral envelope. The antibiotic rifampicin, which inhibits cleavage of the precursor polypeptide, prevents the formation of the viral envelope and thus inhibits the production of infectious virions.

Some late polypeptides are glycosylated posttranslationally to make them suitable for insertion into the viral envelope, and some other viral polypeptides are phosphorylated.

Morphogenesis of the poxvirions

Studies on the assembly of poxvirions by electron microscopy showed that the initial capsids formed in the cytoplasmic viroplasms are vesicle-shaped. The viral DNA molecule is introduced into these capsid-like structures. Subsequently, the capsids are completed and the cores are formed with a lipid envelope. Since the poxviruses are cytocidal, the infected cells disintegrate and the virions are released into the environment.

Antiviral agents specific for poxviruses

Isatin-β-thiosemicarbazone (IBT, Marboran)

The thiosemicarbazone side chain that is bound to isatin in the β position (relative to the NH) has specific antiviral activity against poxviruses. The position of the side chain in the isatin molecule is important, since a derivative of isatin, in which the side chain is in the α position, has no antiviral activity. This antiviral compound affects the late viral mRNA (Cooper et al. 1979). In infected, untreated cells, these mRNA molecules have a sedimentation coefficient of 16 S, as opposed to 8 S in IBT-treated cells. Vaccinia virus mutants that are either resistant to, or dependent on, IBT develop at high frequency in the virus population (Ghendon and Chernos 1972).

MARBORAN AS A PROPHYLACTIC DRUG AGAINST SMALLPOX. Marboran was tested in countries where smallpox was endemic. It was found that, when given to smallpox patients and to healthy attendants, Marboran had no beneficial effect on the smallpox victims but markedly reduced the extent of disease incidence in the healthy attendants (chapter 26).

RIFAMPICIN AS AN INHIBITOR OF POXVIRUSES. Rifampicin is a semisynthetic compound: rifamycin SV is a fermentation product of the mold *Streptomyces mediterranei*. Addition of the side chain —CH=N—$(CH_2)_2N(CH_3)CH_2CH_2$ to rifamycin SV by a chemical reaction yields the compound rifampicin. The hydrazone side chain of rifampicin enhances the ability of the drug to inhibit the replication of poxviruses, and modifications in the side chain lead to loss of antiviral activity. Rifampicin inhibits the cleavage of a structural viral polypeptide needed for the virion envelope, and therefore the formation of complete infectious virions is inhibited (chapter 26).

INTERFERON. Interferon, a cell-made polypeptide, is released from the induced cell, attaches to a specific receptor on a noninfected cell, and induces the synthesis of a 2′,5′-oligoadenylate synthetase. The 2′,5′oligoadenine induces a nuclease that cleaves the viral mRNA and thus prevents virus replication (chapter 26). Interferon is capable of preventing vaccinia virus replication in cultured cells in vitro.

DISTAMYCIN A. Distamycin A is an antiviral substance that is synthesized by the mold *Streptomyces distallicum* and is capable of binding to A-T rich regions in the DNA. In the presence of the antiviral substance, synthesis of vaccinia virus DNA is inhibited.

Smallpox: past, present, and future

This much-feared disease—smallpox—spread from the Middle East and North Africa to Europe, and from Spain to Central America during the Spanish conquest in the sixteenth century. With the advent of slavery, the virus migrated from Central Africa to America. Until recently, smallpox was still endemic in large parts of Africa and Latin America, as well as in India, Sumatra, and Borneo. As a result of the comprehensive smallpox-eradication program undertaken by the World Health Organization (WHO) in 1967, all persons in endemic areas were vaccinated, and the incidence of the disease markedly decreased. By 1978, isolated cases of smallpox were found only in the horn of Africa: Somalia, Ethiopia, and Sudan. The child with smallpox shown in figure 7 is one of the last smallpox victims in Africa. The success of the immunization program, which could mean an end to smallpox, has led to the declaration by WHO that the disease has been eradicated. Nonetheless, surveillance will continue since the smallpox virus is very stable, and it could reappear after relaxation of the immunization program.

The infectious process in humans

The smallpox virus (variola) is present in the vesicles on the skin of smallpox patients (figure 7). The virus in skin debris is spread in the air, and the portal of entry in healthy individuals is through the respiratory tract. After an initial sequence of virus replication, the virions enter the bloodstream (viremia) and

Figure 7. Last recorded case of smallpox in Kenya during an outbreak that was brought under control in February 1977.
(WHO/17877, Afro, Smallpox, SM 4, 1978. Reprinted by permission from World Health Organization, Geneva, Switzerland.)

spread into the capillaries in the skin, internal organs, and mucous membranes. During the viremia stage, which lasts 4–6 days, the patient's body temperature increases, followed by extreme headache which subsides when vesicles appear on the skin (first on the face and hands and later on the body and legs). After 14–16 days, the vesicles begin to dry, and within a week they disappear, leaving pitted scars.

Some patients die within one to two weeks of the appearance of the symptoms. In these patients, bleeding from the mucous membranes of the nose and mouth is common. Death is attributed to the appearance of toxic substances in the blood. The pathology also shows damage to the heart.

Smallpox appears in two forms: variola major (the severe form) and variola minor (a milder disease). In pregnant women, infection causes intrauterine bleeding and abortion. The embryo is infected due to the viremia in the mother.

Smallpox patients must be isolated, and the medical staff must be protected by immunization and prophylactically treated with Marboran.

Importance of differential diagnosis of smallpox.

The most rapid method for identification of the virus present in skin vesicles is by observation in the electron microscope. Fluid is withdrawn from skin vesicles, stained with phosphotungstate (negative staining), and viewed in the electron microscope. Particles can easily be diagnosed as poxvirions with typical morphology (see figure 3).

It is possible to isolate smallpox virus by infecting the chorioallantoic membrane of embryonated eggs. This results in the formation of typical pocks on the surface of the membrane. Cultured cells infected under in vitro conditions are incubated and stained with immunofluorescent antibodies specific for smallpox virus for identification of the virus.

Virulent smallpox viruses in research laboratories as a worldwide danger

With the disappearance of smallpox, the only places where the virus can still be found are in research laboratories. Although the virus is extremely virulent, no special precautions are taken apart from those usually in practice in virological laboratories. This is in contrast to viruses like Marburg disease and Ebola (chapter 22), which are studied in special security laboratories completely sealed off from their surroundings by special treatment of the air and refuse coming out of the laboratory.

Two recent incidents of smallpox infections in humans reported in England demonstrated that smallpox virus can leak from a regular virology laboratory. In the first instance, a technician who made unauthorized use of a balance in a smallpox-virus laboratory developed a respiratory infection that was not diagnosed as smallpox. During hospitalization, she infected two visitors who subsequently died of smallpox. In the second incident, the virus leaked from the virology laboratory through the air-conditioning system and infected a woman (who subsequently died of the disease) working on another floor of the same building.

Laboratories in China, West Germany, Japan, Peru, South Africa, Great Britain, the United States, and USSR still preserve smallpox (variola) virus; there are more than one laboratory in China and 15 laboratories in the other eight countries. It is the aim of WHO to limit variola virus maintenance to only four laboratories in the world that would then be potential collaborators on diagnostic studies of poxviruses of universal importance.

Humans and monkeypox virus

In Africa, in the Democratic Republic of Congo, Liberia, and Sierra Leone, five children were diagnosed as suffering from smallpox. From these children, a monkeypox virus was isolated. Another smallpox patient, a 24-year-old man, had handled internal organs of a monkey three–four weeks prior to developing the disease. All these patients recovered from their illness. Examination of captured monkeys in Liberia revealed that they had antibodies to monkeypox virus. Until now, 29 cases of monkeypox were known (21 of them in Zaire) but this virus does not seem to be a serious public-health problem.

Another poxvirus isolated from monkeys was termed *whitepox* because of the white pocks it forms on the chorioallantoic membrane of embryonated chicken eggs. This virus has not been associated with human disease. Monkeypox mutants arising spontaneously or after serial high multiplicity passage resemble whitepox and variola viruses in several markers tested, but all are distinguishable phenotypically from these. None resembles whitepox viruses in genome structure (Dumbell and Archard 1980).

Animal poxviruses

Molluscum contagiosum is a poxvirus disease noted in cynomolgus monkeys by von Magnus in Copenhagen. About 0.5% of the monkeys die from infection during shipment. The virus can be isolated in cultured cells.

Yaba virus of monkeys

This virus is associated with localized tumors in the skin. The properties of the virus are not known.

Myxomatosis in rabbits

European rabbits brought to Australia by early settlers proliferated to enormous numbers in the absence of natural enemies and constituted a danger to agriculture. These rabbits were found to be highly susceptible to the myxomavirus that causes myxomatosis in South American rabbits and is transferred from one rabbit to another by insects. F. Fenner released a number of infected Australian rabbits and thus introduced a highly virulent myxomavirus into the wild rabbit population of Australia. Most of the rabbits (except those that developed immunity) were killed by the virus. Myxomavirus remained in the wild rabbit population, but with changed pathogenicity, and a stable relationship or biological balance eventually developed between the virus and the rabbits. Thus during periods when natural conditions allow the rabbits to multiply, the virus becomes more virulent and reduces the rabbit population.

BIBLIOGRAPHY

Becker, Y., and Joklik, W.K. Messenger RNA in vaccinia infected cells. *Proc. Natl. Acad. Sci. USA 51*:577–585, 1964.

Becker, Y., and Sarov, I. Electron microscopy of vaccinia virus DNA. *J. Mol. Biol. 34*:655–660, 1968.

Cooper, J.A.; Moss, B.; and Katz, E. Inhibition of vaccinia virus late protein synthesis by isatin-β-thiosemicarbazone: characterization and in-vitro translation of viral mRNA. *Virology 96*:381–392, 1979.

Dumbell, K.R., and Archard, L.C. Comparison of white pock (h) mutants of monkeypox virus with parental monkeypox and with variola-like viruses isolated from animals. *Nature 286*:29–32, 1980.

Esposito, J.J.; Obijeski, J.F.; and Nakano, J.H. Orthopoxvirus DNA: strain differentiation by electrophoresis of restriction endonuclease fragmented virion DNA. *Virology 89*:53–66, 1978.

Ghendon, Y.Z., and Chernos, V.I. Mutants of poxviruses resistant to acetone and N-methyl-isatin-beta-thiosemicarbazone. *Acta Virol. 16*:308–312, 1972.

Joklik, W.K. The multiplication of poxvirus DNA. *Cold Spring Harbor Symp. Quant. Biol. 27*:199–208, 1962.

Kates, J., and McAuslan, B.R. Poxvirus DNA-dependent RNA polymerase. *Proc. Natl. Acad. Sci. USA 58*:134–141, 1967.

Machett, M., and Archard, L.C. Conservation and variation in orthopoxvirus genome structure. *J. Gen. Virol. 45*:683–701, 1979.

McAuslan, B.R. The induction and repression of thymidine kinase in the poxvirus infected HeLa cell. *Virology 21*:383–389, 1963.

Urushibara, T.; Nishimura, C.; and Miura, K.-I. Process of cap formation of messenger RNA by vaccinia virus particles carrying an organized enzyme system. *J. Gen. Virol. 52*:49–59, 1981.

Wittek, R.; Barbosa, E.; Cooper, J.A.; Garon, C.F.; Chan, H.; and Moss, B. Inverted terminal repetition in vaccinia virus DNA encodes early mRNAs. *Nature 285*:21–25, 1980.

RECOMMENDED READING

Bauer, D.J. Thiosemicarbazone. International Encyclopedia of Pharmacology and Therapeutics. Section 61. *Chemotherapy of Virus Diseases.* Vol. 1. New York, Pergamon Press, 1972, pp. 35–113.

Baxby, D. The origins of vaccinia virus. *J. Infect. Dis. 136*:453–455, 1977.

Cho, C.T., and Wenner, H.A. Monkeypox virus. *Bact. Rev. 37*:1–18, 1973.

Downie, A.W. Poxvirus group. *In: Viral and Rickettsial Infections of Man.* (F.L. Horsfall and I. Tamm, eds.), pp. 932–964. J.B. Lippincott Co., 1965.

Fenner, F. The eradication of smallpox. *Prog. Med. Virol. 23*:1–21, 1977.

Follett, E.A.C., and Pennington, T.H. The mode of action of rifamycins and related compounds on poxvirus. *Adv. Virus Res. 18*:105–142, 1973.

Moss, B. Reproduction of poxviruses. *In: Comprehensive Virology* (H. Fraenkel-Conrat and R.R. Wagner, eds.), Plenum Press, New York, 1974, Vol. 3, pp. 405–474.

Wilcox, W.C., and Cohen, G.H. The poxvirus antigens. *Curr. Top. Microbiol. Immunol. 47*:1–19, 1969.

6. IRIDOVIRUSES

FAMILY IRIDOVIRIDAE

Iridoviruses, which replicate in the cytoplasm, have a double-stranded DNA genome of 100–130 × 10^6 daltons. The virions, which range from 130–300 nm, are icosahedral with a capsid of 1500 capsomeres. Many proteins, including several enzymes, are present in the virus particles.

Genus iridovirus.

These are arthropod viruses that are unenveloped. This iridescent group of viruses consists of:

African swine fever virus (ASFV) (figure 8), which can be found in ticks, is a virus of pigs.
Frog virus types 1-3, 5-24, L2, L4, L5 from *Rana pipiens;* LT1-4, and T6-20 from newts; T21 from *Xenopus laevis.*
Fish virus: lymphocystis virus.
Gecko virus.
Insect virus: Tipula iridescent virus

African swine fever virus (ASFV)

ASFV resembles other iridoviruses in the structure of the virions. The viral DNA replicates in the nuclei of infected cells, while the assembly of the virions takes place in the cytoplasm (Ortin and Vinuela 1977; Tabares and Sanchez-

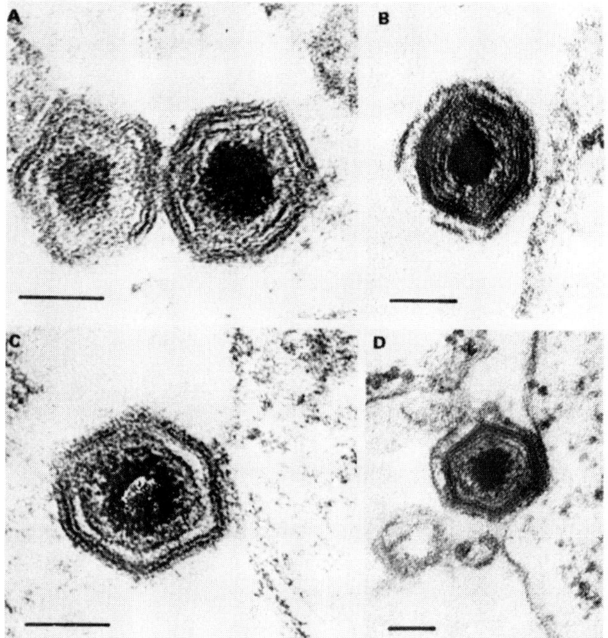

Figure 8. Thin sections of ASFV particles in the process of penetration into cells. A. Free virions. B. Virion attaching to the cell membrane. C. Virion in contact with the cell membrane. D. Virion in the process of penetration.
(EUR 5626e, 1977, Plate XI, p. 64. Reprinted by permission from the Commission of the European Communities, Brussels, Belgium.)

Botija 1979). This virus, which infects wild and domestic swine and occurs in wild boars in central Africa, also spread to Iberia, most probably in pork. The virus, originally isolated from infected domestic pigs, infects pig lymphocytes and can be grown in other cell types in culture. The virions are icosahedral and were found to contain a double-stranded DNA genome of 100×10^6 daltons (figure 9; Adlinger, et al. 1966; Sanchez-Botija, et al. 1977; Enjuanes, et al. 1976). The virion protein coat is made up of 12 different peptides, of which 3 or 4 are antigenic (Tabares, et al. 1980a *b*). The capsid is enveloped with a lipid envelope that contains the viral glycoproteins. The envelope is essential for the adsorption of the virions to cells. Removal of the envelope by treatment with detergent causes the loss of the ability of the virions to infect cells.

Since the virions do not contain the transcription enzyme RNA polymerase, it is assumed that the viral DNA utilizes the nuclear RNA polymerase II of the host cell. The viral DNA is transcribed and replicated in the nuclei. Later the progeny viral DNA is transported to the cytoplasm and is incorporated into the viral capsid. Assembly of virions takes place when the particles pass through the cell membrane (figure 8).

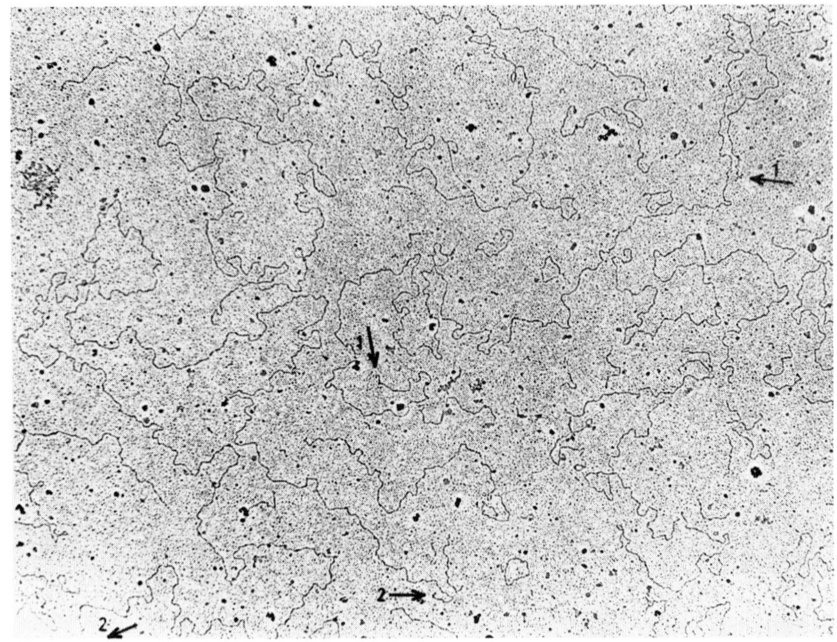

Figure 9. Electron micrograph of ASFV DNA. The molecular length is close to 50 nm. (EUR 5626e, 1977, Plate X, p. 63. Reprinted by permission from the Commission of the European Communities, Brussels, Belgium.)

ASFV in pigs

The disease in the pig is recognized by the appearance of hemorrhages in the skin. When introduced into pigs for the first time, the virus is highly virulent and many of the pigs die; after the virus becomes established in the swine population, however, its virulence is reduced. The virus is transmitted as a respiratory infection. African swine fever is endemic to the Iberian peninsula. Ticks, after feeding on an infected pig, can be maintained for up to ten years and still be able to transfer the virus to a healthy pig. The virus is maintained in wild boars and ticks in addition to domestic pigs. Constant surveillance on movement of animals in Europe is needed to limit the virus from spreading into countries free of the disease. Trading in pigs can spread the virus from one country to another and, indeed, such trade is prohibited. The virus is very resistant to heat and can also be transmitted via food wastes given to pigs. In this way, the virus can spread from herd to herd. To date, no vaccine against ASFV is available, and the only way to prevent spread of the virus is to eliminate the pigs in an infected area. The animals are killed and then buried.

FROG VIRUSES

Frog virus 3 (FV3)—Structure and organization of the virions

FV3 has a double-stranded DNA genome of 100 × 10^6 daltons with a G + C content of 56%. Other frog viruses have a G + C content of 30–40% and double-stranded DNA genomes of 130–160 × 10^6 daltons.

Analysis of virion proteins revealed 20 polypeptides ranging from 10,000–200,000 daltons. The major peptide is 65,000 daltons (Tan and McAuslan 1971; Goorha and Granhoff 1974). Five enzymes are present in the virions: nucleotide phosphorylase (ATPase) inside the core, a DNase (active at pH 5.0) attached to the viral DNA, a DNase (active at pH 7.5), an RNase capable of digesting double-stranded RNA, protein kinase, protein phosphatase, and the RNA polymerase (Vilagines and McAuslan 1971; Kang and McAuslan 1972).

Biosynthesis of virions

FV3 virus replicates in cells from amphibia, birds, and mammals. The optimal temperature for virus replication is 28°C and replication is inhibited at 37°C. After infection, the virus inhibits the cellular synthesis of DNA and RNA. After uncoating of the virions, the RNA polymerase is activated, and the early mRNA is transcribed. The molecular processes of virus synthesis resemble those of poxviruses. The synthesis of virions in the cytoplasm is dependent on the nucleus, the site of DNA and RNA synthesis (Goorha et al. 1977, 1978).

LYMPHOCYSTIS VIRUS IN FISH

Lymphocystis virus causes the proliferation of connective-tissue cells. The infected cells enlarge to a diameter ten times that of the original. The cytoplasm is filled with virions. The same phenomenon can be obtained in in vitro cultured cells. The growth is benign and does not develop into a tumor.

BIBLIOGRAPHY

Adlinger, H.K.; Stone, S.S.; Hess, W.R.; and Bachrach, H.L. Extraction of infectious deoxyribonucleic acid from African swine fever virus. *Virology 30*:750–752, 1966.

Enjuanes, L.; Carrascosa, A.L.; and Vinuela, E. Isolation and properties of the DNA of African swine fever virus. *J. Gen. Virol. 32*:479–492, 1976.

Goorha, R., and Granoff, A. Macromolecular synthesis in cells infected by frog virus 3. 1. Virus-specific protein synthesis and its regulation. *Virology 60*:237–250, 1974.

Goorha, R.; Murti, G.; Granoff, A.; and Tirey, R. Macromolecular synthesis in cells infected by frog virus 3. VIII. The nucleus is a site of frog virus 3 DNA and RNA synthesis. *Virology 84*:32–50, 1978.

Goorha, R.; Willis, D.B.; and Granoff, A. Macromolecular synthesis in cells infected by frog virus 3. VI. Frog virus 3 replication is dependent on the cell nucleus. *J. Virol. 21*:802–805, 1977.

Kang, H.S., and McAuslan, B.R. Virus-associated nucleases: location and properties of deoxyribonucleases and ribonucleases in purified frog virus 3. *J. Virol. 10*:202–210, 1972.

Ortin, J., and Vinuela, E. Requirement of cell nucleus for African swine fever virus replication in Vero cells. *J. Virol. 21*:902–905, 1977.

Sanchez-Botija, C.; McAuslan, B.R.; Tabares, E.; Wilkinson, P.; Ordas, A.; Friedmann, A.; Solana, A.; Fereira, D.; Ruiz-Gonzalvo, F.; Dalsgaard, C.; Marcotegui, M.A.; Becker, Y.; and

Shlomai, J. Studies on African swine fever virus: purification and analysis of virions. Commission of the European Communities EUR5626e, 1977.

Tabares, E.; Marcotegui, M.A.; Fernandez, M.; and Sanchez-Botija, C. Proteins specified by African swine fever virus. I. Analysis of the structural proteins and antigenic properties. *Arch. Virol. 66*:107–117, 1980*a*.

Tabares, E.; Martinez, J.; Ruiz-Gonzalvo, F.; and Sanchez-Botija, C. Proteins specified by African swine fever virus. II. Analysis of proteins in infected cells and antigenic properties. *Arch. Virol. 66*:119–132, 1980*b*.

Tabares, E., and Sanchez-Botija, C. Synthesis of DNA in cells infected with African swine fever virus. *Arch. Virol. 61*:49–59, 1979.

Tan, K.B., and McAuslan, B.R. Proteins of polyhedral cytoplasmic deoxyvirus. I. The structural polypeptides of FV_3. *Virology 45*:200–207, 1971.

Vilagines, R., and McAuslan, B.R. Proteins of polyhedral cytoplasmic deoxyvirus. II. Nucleotide phosphohydrolase activity associated with frog virus 3. *J. Virol.* 7:619–624, 1971.

7. HERPESVIRUSES

FAMILY HERPESVIRIDAE

The herpesvirus family (herpes, in Greek, means crawling or creeping, like a snake) named Herpesviridae, consists of viruses that have a similar morphology, but with a host range varying from lower vertebrates (reptiles, fish, and frogs) to humans. These viruses differ from each other antigenically and in the G + C content and organization of the viral DNA.

The virions are made up of 162 hollow capsomeres arranged in an icosahedral capsid covered by a lipid envelope containing more than 20 polypeptides. The capsid contains one DNA molecule of 100×10^6 daltons.

Subgroup herpesvirus occurs in numerous hosts:

In humans:	Herpes simplex virus types 1 and 2 (HSV-1 and 2)
	Varicella-zoster virus (human herpesvirus 3)
	Epstein-Barr virus (EBV) (infectious mononucleosis virus; human herpesvirus 4)
	Cytomegalovirus
In animals:	Cercopethiced herpesvirus types 1 (B virus) and three other monkey herpesviruses
	Pseudorabies virus (PRV) of pigs; also equine herpesvirus, herpesvirus of cattle, sheep, dogs, and cats
Avian herpesviruses:	Marek's disease virus

	Herpesvirus of turkeys (HVT)
	Avian infectious tracheobronchitis
Other herpesviruses:	Frog herpesviruses, including Lucké frog carcinoma; also herpesviruses of fish and snakes; herpesviruses of oysters and a fungus

The 1980 proposal for classification of the Herpesviridae family is as follows:

Alphaherpesvirinae	Human HSV-1 and 2
	Human varicella-zoster virus
	Herpes B virus of monkeys
	Alpha herpesvirus of cattle, pigs, horses, and cats
Betaherpesvirinae	Human cytomegalovirus
	Cytomegaloviruses of mice, pigs, and other mammalian species
Gammaherpesvirinae	EBV
	Gamma herpesviruses of monkeys, fowl, and rabbits

Icosahedral capsid of 162 capsomeres

Electron microscopy revealed the internal morphology of the virus (figure 10) and the icosahedral structure of the capsid which contains 162 capsomeres. The diameter of the capsid is 120–150 nm, and each capsomere has a diameter of 10 nm. Unenveloped capsids can be isolated from the nuclei of infected cells as well as empty capsids lacking the viral DNA; the nucleocapsid encapsidates one viral DNA molecule. Each capsomere has a molecular weight of about 10^6 (possibly made of 10 helical polypeptides, each of 100,000 daltons). The viral DNA molecule is about 50 nm in length and is arranged around a central protein core (designated torus, assuming that the DNA molecule is looped around the protein core). The coiled viral DNA molecule is inserted into the empty viral capsid through a hole that is later sealed with capsomeres to form a complete nucleocapsid. The nucleocapsid is finally enveloped by a lipid membrane when the capsid interacts with the inner side of the nuclear membrane.

The viral DNA is double stranded (ds)

The viral DNA can be released from the nucleocapsid by treatment with detergents (e.g., sodium dodecyl sulfate [SDS]) or proteolytic enzymes (e.g., pronase). Electron microscopy of the virion DNA revealed that the molecule is linear and double-stranded and has a molecular weight of 100×10^6 (Becker et al. 1968). Analysis on benzoylated-naphthoylated DEAE (BND) cellulose columns that separate dsDNA molecules (that elute with a 1M salt solution) to form single-stranded DNA molecules (that elute with formamide or caffein) suggested that at least 10% of the virion DNA molecules have single-stranded DNA sequences. It was also found that alkali treatment of HSV double-

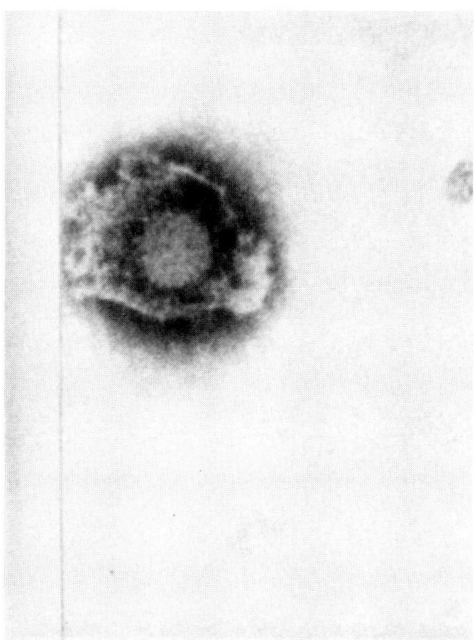

Figure 10. An enveloped herpes simplex virion.
(Department of Molecular Virology, The Hebrew University-Hadassah Medical School, Jerusalem, Israel.)

stranded DNA leads to denaturation of the DNA molecules into one complete single-stranded DNA strand and fragments of a second single-stranded DNA strand. It is still not clear whether the fragmentation of one single DNA strand is due to existing nicks in the DNA or to the presence of ribonucleotides or apurinic sites that enhance nicking of the DNA chain under alkaline conditions.

DNAs of different members of the Herpesviridae differ in their density

Analysis of the viral DNA extracted from members of the Herpesviridae by centrifugation in CsCl density gradients revealed marked differences in density. Canine and cottontail rabbit herpesvirus DNA were found to have a density of 1.692 g/cm^3, equivalent to a G + C content of 33%. At the other end of the scale, monkey herpesvirus DNA has a density of 1.733 g/cm^3, equivalent to a G + C of 75%. The Herpesviridae also show marked differences in the antigenicity of the proteins present in the virions, although the morphology of the virions is the same in all the members of the group.

Repetitive sequences of viral DNA

Alkali-denatured single-stranded DNA molecules of HSV were found to reanneal after removal of the alkali and incubation under appropriate salt and temperature conditions. Electron microscopy of self-annealed DNA molecules

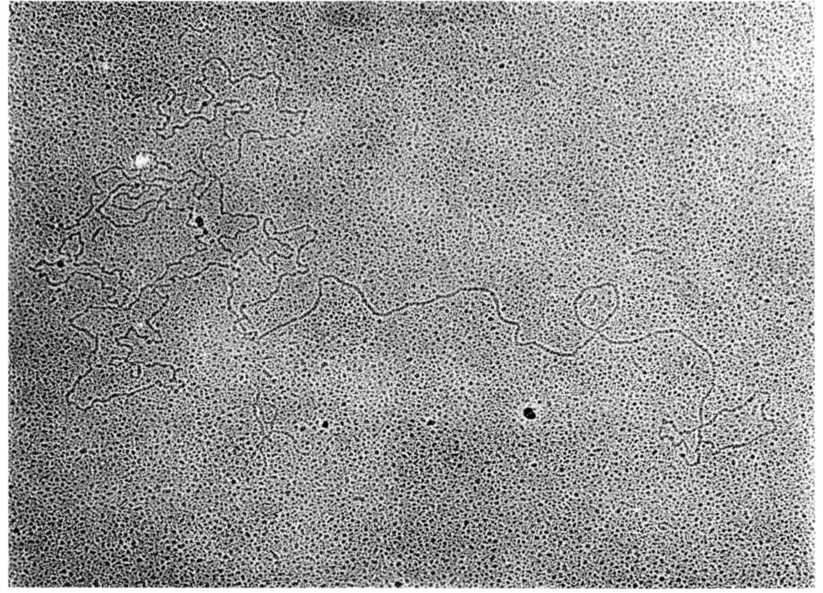

Figure 11. Herpes simplex virus-1 (HSV-1) single-stranded DNA in a self-annealing form; the small (s) and the large (L) single-stranded regions are the unique L and S components. The double-stranded stalk contains the repeat sequences TR_S and TR_L.
(By courtesy of Dr. A. Friedmann, Department of Genetics, Life Sciences Institute, The Hebrew University, Jerusalem, Israel.)

(Sheldrick and Berthelot 1974) revealed the formation of figure-8-shaped molecules—mostly single-stranded DNA present as small and large loops and some as a double-stranded DNA stalk in between the loops (figure 11). The small-loop S and the large-loop L are unique sequences of the viral DNA, while the repetitive sequences that are flanked on both sides of the S and L unique sequences are termed *terminal* and *internal* repeats of the L and S components (class 3 DNA) (figure 12). Analyses of the DNA of several herpesviruses revealed two additional forms of the viral DNA (Honess and Watson 1977) (figure 12): Class 1 DNA molecules have the unique L sequence flanked by two repeat sequences, while class 2 DNA molecules contain an S unique sequence flanked by repeat sequences and a unique L sequence with one internal short repeat sequence. Class 3 DNA has both L and S unique sequences, each flanked on both sides with repeat sequences. The G + C content of the repeat sequences is higher than the G + C content of the unique sequences.

Self-annealing of alkali denatured single-stranded DNA made it possible to isolate the double-stranded repeat sequences after treatment of the viral DNA with S_1 endonuclease, which digests single-stranded DNA only. Centrifugation of the repeated sequences in CsCl density gradients revealed that their

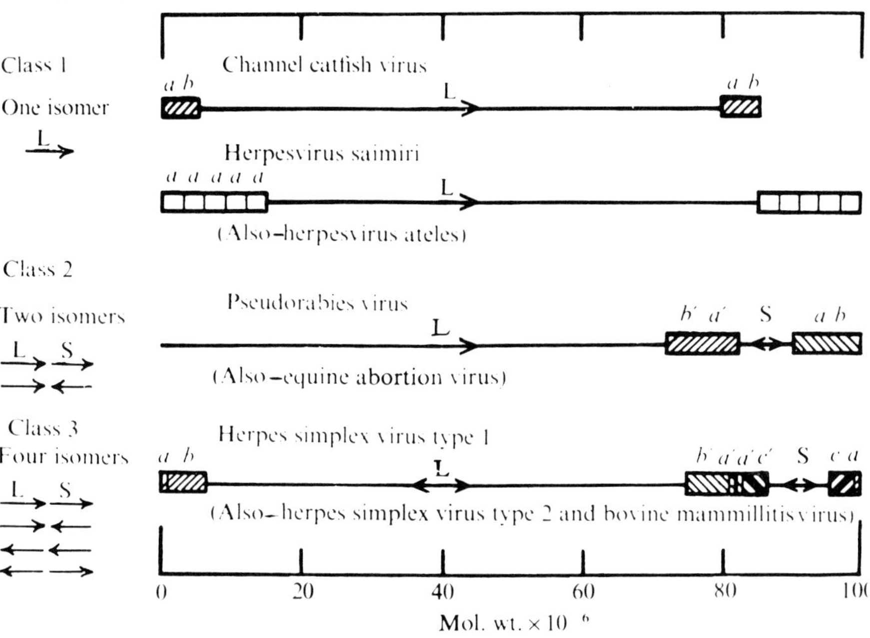

Figure 12. A tentative classification of several herpesviruses on the basis of differences in the content and arrangement of unique and redundant nucleotide sequences in populations of mature virus DNA. For each virus, L and S indicate long and short unique sequences, and *ab*, and so on indicate redundant sequences with *a′b′* being their complement. Arrows indicate the number of relative orientations of unique sequences represented in the population of mature DNA molecules.

(Honess and Watson 1977. Reprinted by permission from *J. Gen. Virol.* 37, Fig. 1, p. 17. Cambridge University Press, Cambridge.)

density was higher by 0.08 g/ml than the intact virion DNA, whereas the L and S sequences had a density slightly lower than the virion DNA.

Subpopulations of viral DNA in virions

Cleavage of virion DNA (class 3) with restriction endonucleases followed by electrophoresis of the double-stranded DNA fragments in agarose gels revealed that fragments are present in large amounts, while others are present in half- or quarter-molar amounts (reviewed by Roizman 1979). Thus herpesviruses of class 3 contain four types of viral DNA molecules that differ in the relative arrangement of the L and S unique sequences (left-hand side in figure 12). Analysis of class 2 herpesvirus DNA that lacks repeat sequences in the terminus of the L unique sequence but has two repeat sequences flanking the S component showed that only two isomers can be detected in which the L component remains in one orientation, while the S component has two possible orientations relative to L. The absence of the S unique sequence in class 1 DNA allows for only one isomer of the viral DNA (figure 12).

Viral DNA is infectious for permissive cells by transfection

Naked viral DNA extracted from virions can infect cells under conditions that allow entry of the DNA into the cell. This can be done by precipitation of the viral DNA with calcium phosphate and its sedimentation onto the cell membrane. The viral DNA is transported to the nucleus and replicates exactly as viral DNA from uncoated virions, although the efficiency of infection with the naked viral DNA is much lower. The virions do not carry a virus-coded transcription enzyme like the poxviruses. It is also possible to introduce fragments of the viral DNA into cells (e.g., the fragment carrying the viral thymidine kinase [TK] gene) by infecting TK^- cells with fragmented viral DNA. The viral TK gene cloned in a bacterial plasmid is incorporated into a TK^- cell that is then transformed into a TK^+ cell (Pellicer et al. 1978). Similarly, it is possible to transfect cells with fragmented viral DNA and select for cells that are transformed morphologically by the viral DNA.

Partial homology between the DNA of different Herpesviridae

The genetic relatedness between herpesviruses can be determined by hybridization of DNA preparations from different viruses. The homology of the two viruses ranges from 100% (when the viruses are identical) to zero (when the two viruses are unrelated). HSV types 1 and 2 (which are human pathogens) were found to have only 50% homology. Bovine mammillitis virus shares 8–10% homology with the DNA of pseudorabies virus of pigs and 5% homology with the DNA of equine herpesvirus. (EBV) of humans has 35% homology with a herpesvirus isolated from leukocytes of chimpanzees (H. papio). DNA of Marek's disease virus (MDV) of chickens showed 6–10% homology with the DNA of HVT which is used for immunizing chickens against MDV. It seems that homology between the two viruses is due to a viral antigen present in both viruses which makes the chickens immune to MDV.

Cloning of HSV DNA restriction fragments in bacterial plasmids for analysis of viral genes

Developments in the field of genetic engineering and the availability of a safe cloning plasmid pBR322 have made it possible to clone fragments of HSV DNA (and all other known viral genomes). Cloning of a viral DNA fragment involves the viral genome's being cleaved with a restriction enzyme and then inserted into a suitable bacterial plasmid cleaved in a similar way. HSV DNA fragments (e.g., cleaved with the *Bam*HI restriction enzyme) can be cloned in the *Bam*HI site of the pBR322 plasmid. Ligation of the cleaved pBR322 DNA with the *Bam*HI fragments of HSV DNA, by means of bacteriophage T4 ligase, leads to the formation of pBR322 plasmids that carry a fragment of HSV DNA as an insert. This plasmid now has two *Bam*HI cleavage sites, as compared to one in the original. The resulting mixture of ligated plasmid DNA molecules is used to transform bacteria which are seeded to form colonies. Since the *Bam*HI site in pBR322 DNA is within the gene for tetracycline resistance (tet^r), insertion of the HSV DNA *Bam*HI fragment into the plasmid leads to inactivation of the tet^r gene. Thus selection for ampicillin-resistant (amp^r: the other gene in pBR322 DNA) and tetracycline-sensitive (tet^s) phenotypes allows for the selection of bacterial colonies that carry the plasmid with the HSV DNA fragment inserted in it. Successful cloning of HSV DNA fragments allows large quantities of plasmids with the cloned fragment to be produced. By means of other plasmids that allow expression of the animal virus genes in bacteria, individual HSV genes can be studied in detail.

Virion proteins are antigenic

To determine the molecular weight of the viral structural proteins, mature enveloped virions, nucleocapsids, and empty capsids were purified from the cell homogenates, dissolved with detergents (usually SDS), and separated by electrophoresis in polyacrylamide gels. A typical protein analysis of herpes simplex virions (figure 13) shows more than 20 viral structural peptides present. HSV types 1 and 2 differ from each other in the molecular weights of several structural peptides that range from over 25,000 to about 200,000 daltons. Each capsomere has a molecular weight of 1 million and is made up of 10 identical polypeptides. Thus since the capsid consists of 162 capsomeres, 1,620 peptides, each with a molecular weight of about 100,000, are required to form 1 capsid.

The virion proteins are antigenic, and antibodies can be produced against each antigen when the proteins are injected into rabbits. It is possible, therefore, to study the relatedness between different HSVs by determining the cross reactivity between viral antigens. Table 1 shows the antigenic relatedness between HSV types 1 and 2, PRV, bovine mammillitis virus (BMV), and equine abortion virus (EAV), by comparing the number of precipitin bands found in agar gel immunodiffusion reactions between antisera to the herpesvirus and extracts of cells infected with each virus.

HSV-1 has 12 antigens: 6 are shared with HSV-2 and BMV, 3 with EAV, and 1 with PRV. HSV-2 has 7 antigens; 4 are shared with HSV-1. The other viruses have a much more limited antigenic similarity.

Herpesvirus replication in cells is controlled by the cell and by the virus.

Most herpesviruses are able to replicate in permissive cells and have a lytic cycle. However, infection of cells with herpesviruses can be abortive if the cells do not provide the virus with the systems necessary for replication. This can happen when the virus infects nonpermissive cells or permissive cells under nonpermissive conditions.

Adsorption and penetration require a receptor in the cell membrane and a corresponding attachment protein in the virion envelope. An intact virion envelope is also required, since in its absence the DNA-containing capsid cannot adsorb to the cell. The initial interaction between the virus and the cell is electrostatic in nature. The subsequent stage is fusion between the virion en-

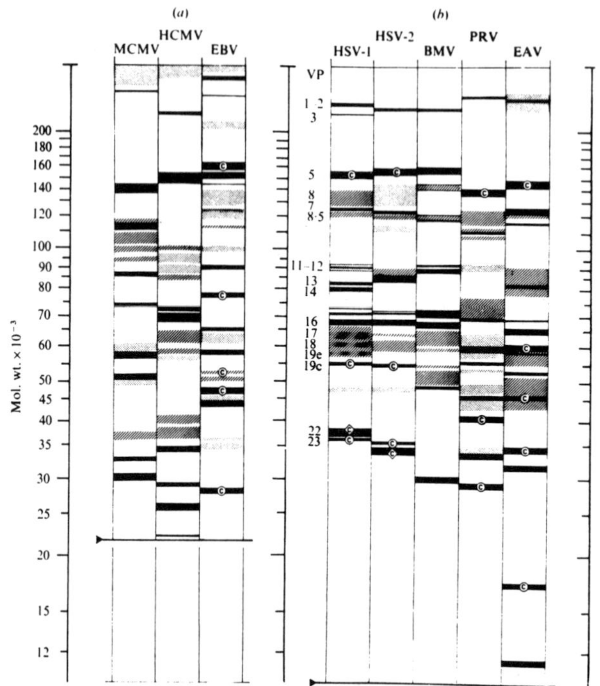

Figure 13. Diagrammatic comparison of the number, apparent molecular weight, and relative contributions of major polypeptides from purified enveloped virions of eight herpesviruses: murine cytomegalovirus (MCMV); human cytomegalovirus (HCMV); Epstein-Barr virus (EBV); herpes simplex virus types 1 and 2 (HSV-1, HSV-2); bovine mammalitis virus (BMV); pseudorabies virus (PRV); equine abortion virus (EAV).

(Honess and Watson 1977. Reprinted by permission from *J. Gen. Virol.* 37, Fig. 3, p. 24. Cambridge University Press, Cambridge).

Table 1. Number of precipitin bands observed against extracts of cells infected with HSV types 1 and 2, bovine mammalitis virus, equine abortion virus (EAV), and pseudorabies virus.

Antisera prepared against	HSV-1	HSV-2	BMV	EAV	PRV
HSV-1	≥ 12	≥ 6	6	3	1
HSV-2	≥ 5	≥ 7	4	2	1
PRV	2	2	1	3	≥ 3

(After: Honess and Watson 1977.)

velope and the cell membrane. Virion attachment is dependent on a gene in chromosome 3 (in human cells) that controls the synthesis of a cellular membrane protein which serves as a receptor for the virion.

THE VIRAL DNA IS UNCOATED BY CELLULAR ENZYMES. Entrance of the viral nucleocapsids into the cytoplasm results in an interaction with the lysosomes, and the lysosomal proteolytic enzymes dissolve the protein capsid and release the viral DNA. The lysosomal vacuole releases the viral DNA into the nucleus.

Early transcription of viral DNA by cellular RNA polymerase II

In the cell, the DNA-dependent RNA polymerase II is responsible for the transcription of the cellular genome and the production of precursors for the cellular mRNA species. This enzyme was found to be responsible for the transcription of the parental viral DNA. Treatment of HSV-infected cells with the toxin α-amanitin (produced by the mushroom *Amanita phalloides*) results in inhibition of the cellular RNA polymerase II and the prevention of virus replication. A cell mutant that has an RNA polymerase II enzyme resistant to α-amanitin is sensitive to HSV-1 replication in the presence of the inhibitor. Studies have shown that the cellular RNA polymerase II is, therefore, responsible for the synthesis of early viral mRNA.

Viral mRNA from a particular group of genes in the viral DNA.

It is possible to map and characterize the viral genes that code for the synthesis of early mRNA and early viral proteins. The viral mRNA synthesized in infected cells prior to viral DNA synthesis can be labeled with ^{32}P and extracted from the cells. These labeled RNA species are hybridized to viral DNA fragments obtained after cleavage with restriction enzymes, separation by electrophoresis on agarose gels, and blotting onto nitrocellulose paper, using the Southern technique (Southern 1975). Since the position of the DNA fragments in the viral genome is previously determined, it is possible to localize the viral genes that code for the early mRNA species. Labeled viral RNA extracted from infected cells revealed five molecular species of mRNA by electrophoresis in acrylamide gels.

To characterize the nature of the viral genes coding for early mRNA, the isolated mRNA species were added to in vitro protein-synthesizing systems (lysates from embryonic wheat or from rabbit reticulocytes). The proteins synthesized in vitro according to the information in the viral mRNA can be identified by the use of specific antibodies and by electrophoresis in polyacrylamide gels, using known proteins as markers.

The viral mRNA is symmetrically transcribed from both strands of the viral DNA, but only one RNA transcript is the functional mRNA. The reason for symmetrical RNA synthesis is not known.

The viral mRNA is synthesized in the form of a precursor which is then processed into mRNA. The 3′ end of the mRNA contains a poly(A) sequence, and a methylated cap is attached at the 5′ end. The viral mRNA is transported from the nucleus to the cytoplasm by the same mechanism which transports the cellular mRNA species.

VIRUS INFECTION LEADS TO DISAPPEARANCE OF NUCLEOLI. After viral DNA replication takes place, the nucleoli gradually disappear, most probably due to disaggregation. The cellular DNA-dependent RNA polymerase I activity diminishes, and the infected cell loses the ability to synthesize ribosomal RNA and ribosomal subunits.

SYNTHESIS OF VIRAL PROTEINS AS A PROCESS REGULATED BY VIRUS-CODED PROTEINS. The early viral proteins are synthesized in the infected cells under the direction of the early viral mRNA. These proteins are responsible mainly for the biosynthesis of viral DNA. The α proteins are responsible for the synthesis of the β proteins, which are responsible for the third and final group, the γ proteins. The analysis made by Roizman et al. (1978) (figure 14) shows that all the α and β proteins are the products of genes present in the unique L sequence, except for one β protein which is coded for by a gene in the internal repeat sequence. The genes for the γ proteins are localized in the terminal repeat sequence of S and in the internal repeat sequence of the L component. The exact mechanism of regulation is not known. About 30 viral proteins were found to be synthesized during the lytic cycle of the virus in infected cells, and more than 20 structural proteins are included in the mature virions. Current analyses of the viral proteins on two-dimensional gels revealed that more than 200 proteins are synthesized in HSV-1 infected cells. The properties of these proteins are under study.

DNA-BINDING PROTEINS IN THE SYNTHESIS OF VIRAL DNA. Nineteen DNA binding proteins were isolated from HSV-1 infected cells by chromatography on DNA-cellulose columns. In arginine-deprived HSV-1 infected BSC-1 cells in which viral DNA synthesis is unaffected but virions are not formed, only nine DNA binding proteins were isolated. Among these proteins, it was possible to identify the viral DNA polymerase (150,000 daltons and another unknown protein with DNA polymerase activity of 70,000 daltons), two DNases (40,000 and 27,000 daltons), and a thymidine kinase (40,000 daltons).

The interaction of the viral DNA polymerase with viral DNA is shown in figure 15.

CELLULAR GENE FUNCTION IN THE INITIATION OF VIRAL DNA BIOSYNTHESIS. Temperature sensitive (ts) cell cycle mutants have been produced which have a mutation in the gene controlling the ability of the cell to initiate the synthesis of chromosomal DNA in the S phase. Infection of such cells with HSV-1 at the nonpermissive temperature results in inhibition of viral DNA replication. The cellular protein controlling initiation of viral DNA synthesis is not yet known (Yanagi et al. 1978).

SEMICONSERVATIVE SYNTHESIS OF HSV DNA AT THREE POSSIBLE SITES. Electron microscopy of HSV-1 DNA molecules with a replicative loop led to the conclusion that in some DNA molecules the initiation site for DNA biosyn-

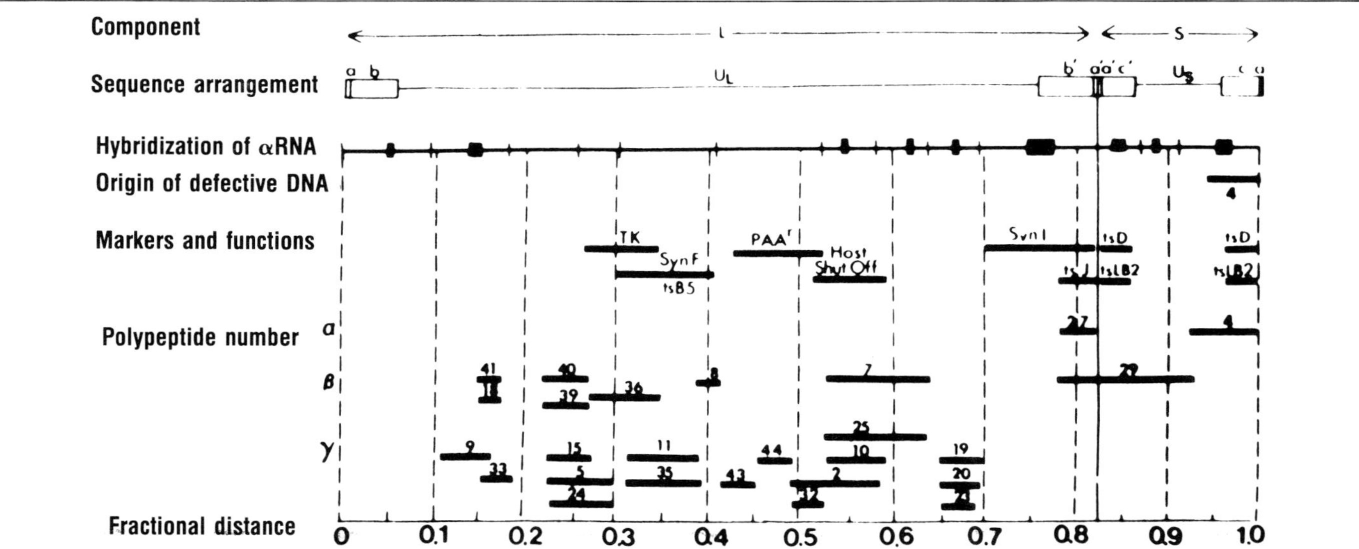

Figure 14. HSV map positions of (a) templates for α, β, and γ polypeptides, (b) selected ts mutants in α functions, (c) selected functions such as thymidine kinase (TK), phosphonoacetic acid resistance (PAA$^+$) cell fusion (syn I and syn F), and (d) cytoplasmic RNA from cells treated with cycloheximide from time of infection (α RNA).
(Roizman 1978. Reprinted by permission from *International Virology* 4, Fig. 1, p. 67. Centre for Agricultural Publishing and Documentation, Wageningen, The Netherlands.)

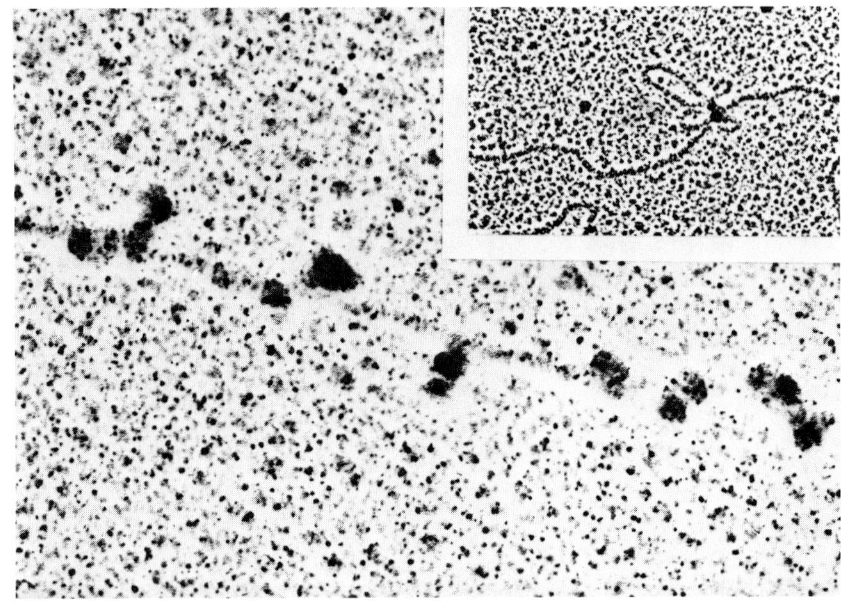

Figure 15. Electron micrograph of the binding of DNA polymerase and unwinding protein to HSV DNA. Purified HSV DNA was incubated with the viral DNA polymerase or with the purified T4 phage gene 32 unwinding protein *(inset)*. The preparations were fixed with glutaraldehyde and viewed in the electron microscope.
(Department of Molecular Virology, The Hebrew University-Hadassah Medical School, Jerusalem, Israel.)

thesis is situated 10 nm and in other molecules 20 nm from one of the molecular ends. One site was found to be in the center of the L unique sequence, the second site in the repeat sequence between the L and S sequences, and the third site seems to be at one of the molecular ends of the viral DNA (figure 16).

REPLICATION OF ONE HSV DNA MOLECULE IN 20 MIN. Studies on the synthesis of HSV DNA in nuclei of infected cells and in isolated nuclei under in vitro conditions revealed that 20 min are required for the semiconservative synthesis of one viral DNA molecule. Thus the enzymatic complex that replicates the viral DNA polymerizes 15,000 nucleotides/min on each HSV DNA strand and duplicates 5×10^6 daltons of DNA per min.

TEMPERATURE-SENSITIVE MUTANTS OF HSV DEFECTIVE IN DNA SYNTHESIS. Treatment of HSV-infected cells with mutagens (e.g., 5-bromodeoxyuridine) leads to mutations in the viral DNA. The mutagenized virus is plaque-purified and ts mutants are selected by testing for the ability of the virus isolates (plaques) to replicate at 31°C and 38°C (reviewed by Schaffer 1975; Subak-Sharpe and Timbury 1977). Those virus isolates that are restricted in their ability to grow at the higher temperature are ts mutants. From these ts mutants it is possible to identify the mutants that are blocked in their ability to synthesize viral DNA.

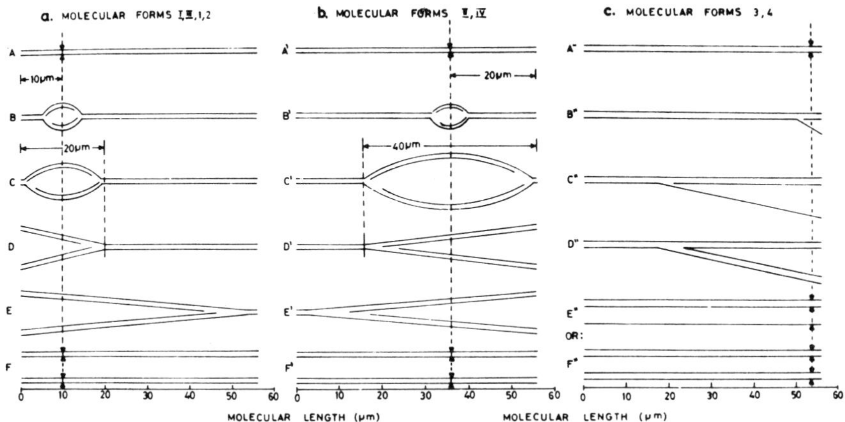

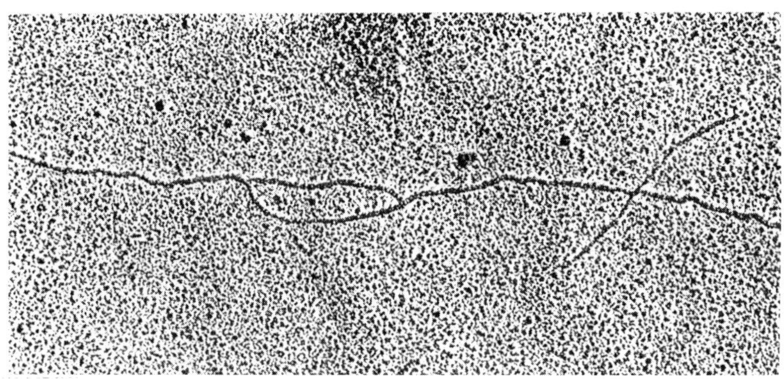

Figure 16. Top: Models of HSV DNA replication (a) initiation site (▲) is located at 10 μm from one of the molecular ends; (b) initiation site at 20 μm from one of the ends, and (c) initiation site at one of the molecular ends. A–F are consecutive steps in the replication process. Bottom: Morphology of the replicative loop of HSV DNA.

(Friedmann, Shlomai, and Becker 1977. Reprinted by permission from *J. Gen. Virol.* 34; A — Fig. 7, p. 520; B — Fig. 4(C), p. 515. Cambridge University Press, Cambridge.)

PHOSPHONOACETIC ACID (PAA)-RESISTANT MUTANTS PRODUCE A PAA-RESISTANT DNA POLYMERASE. PAA is a selective inhibitor that binds to the virus-coded DNA polymerase at the same site as the terminal phosphate of a deoxynucleoside triphosphate (see chapter 26). PAA-resistant mutants can be selected that produce an enzyme that does not bind PAA.

PLAQUE-MORPHOLOGY MUTANTS OF HSV. HSV strains are known that are capable of causing fusion of the outer membranes of infected cells, resulting in a large syncytium that contains many nuclei, a property designated Syn^+. A mutation that leads to inability to form syncytia (Syn^-) was described. In addition, strains of small-plaque morphology and large-plaque morphology were described.

Viruses that are recombinants of HSV-1 and HSV-2

Infection of the same cell with HSV types 1 and 2 leads to the appearance of recombinant viruses with a DNA molecule made of DNA fragments from both parent viruses. Since the proteins of HSV-1 and HSV-2, which have a similar function in the virus (e.g., structural proteins) have different molecular weights and migrate differently in polyacrylamide gels, it is possible to characterize the virus recombinants by analysis of their proteins (Stow and Wilkie 1978). Furthermore, insertion of a piece of DNA from one virus into the DNA of another can be detected by restriction-enzyme cleavage and electrophoresis in agarose gels since the cleavage sites change in the recombinant DNA molecules (figure 17). Such recombinants were used by Roizman et al. (1978) for the construction of the DNA map of HSV (figure 14).

Defective HSV due to an error in viral DNA biosynthesis

Infection of monkey cells in culture (BSC-1 or CV-1) with undiluted virus over a number of consecutive passages leads to the production of defective virus. Analysis of the infectious virus progeny produced at each passage level by plaque assay revealed that the virus titer gradually declined from about 10^7 pfu/ml at passage 1 to about 10^3 or 10^4 pfu/ml at passage 5 or 6, but the cells produced large quantities of noninfectious virus. The same phenomenon was observed in other cell lines, except that the noninfectious defective virus appeared at much higher (more than 14) passage levels. DNA molecules from noninfectious virions have a molecular weight of 100×10^6 daltons—the same as the DNA in infectious virions.

Restriction enzyme analysis revealed that defective viral DNA is a fragment of the infectious DNA, about 5×10^6 daltons, which is reiterated 20 times (Frenkel et al. 1976). The repeated fragments are arranged tail to head in the DNA molecules. There are three types of defective DNA molecules containing: (1) the terminal or internal repeat sequence of the S component (TR_s or IR_s); (2) reiterations of a DNA sequence arising from the center of the L component; and (3) sequences arising from two or more sites in the viral DNA.

The defective DNA arising from the TR_s (or IR_s) sequence has a higher density than that of infectious viral DNA. This is due to the high G + C content in the repeat sequence, and as a result, defective DNA can be separated from the wild type DNA by centrifugation in CsCl density gradients. Isolation of defective viral DNA of this type from the nuclei of infected cells, followed by electron microscopy, revealed DNA molecules resembling rolling circles. Such a rolling circle is demonstrated in figure 18, which shows linear DNA attached to circular DNA. It was proposed that the circular DNA represents a DNA fragment that escaped from the viral DNA and that was reiterated by the

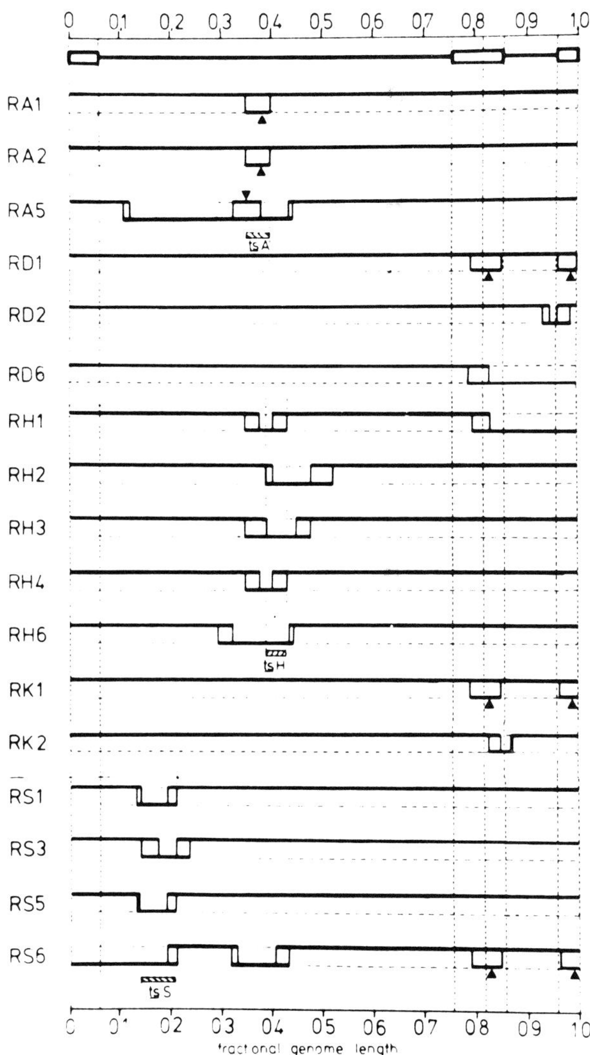

Figure 17. Summary of the genome structures of 17 temperature-sensitive (ts) recombinants generated using DNA of tsA, D, H, K, and S ts mutants of HSV-1. The sequences in a recombinant derived from the HSV-1 or HSV-2 parent are shown by thick continuous horizontal lines superimposed on the upper and lower (respectively) of the two horizontal lines representing the viral genome. The vertical dotted lines delimit the TR_L, U_L, IR_L, IR_S, U_S, and TR_S regions of the genome. Regions in which crossovers have occurred are shown by two vertical lines between the thick horizontal lines. ▲ indicates the location of the single-restriction endonuclease sites that identify certain small insertions of DNA.

(Stow and Wilkie 1978. Reprinted by permission from *Virology* 90, Fig. 6, p. 8.)

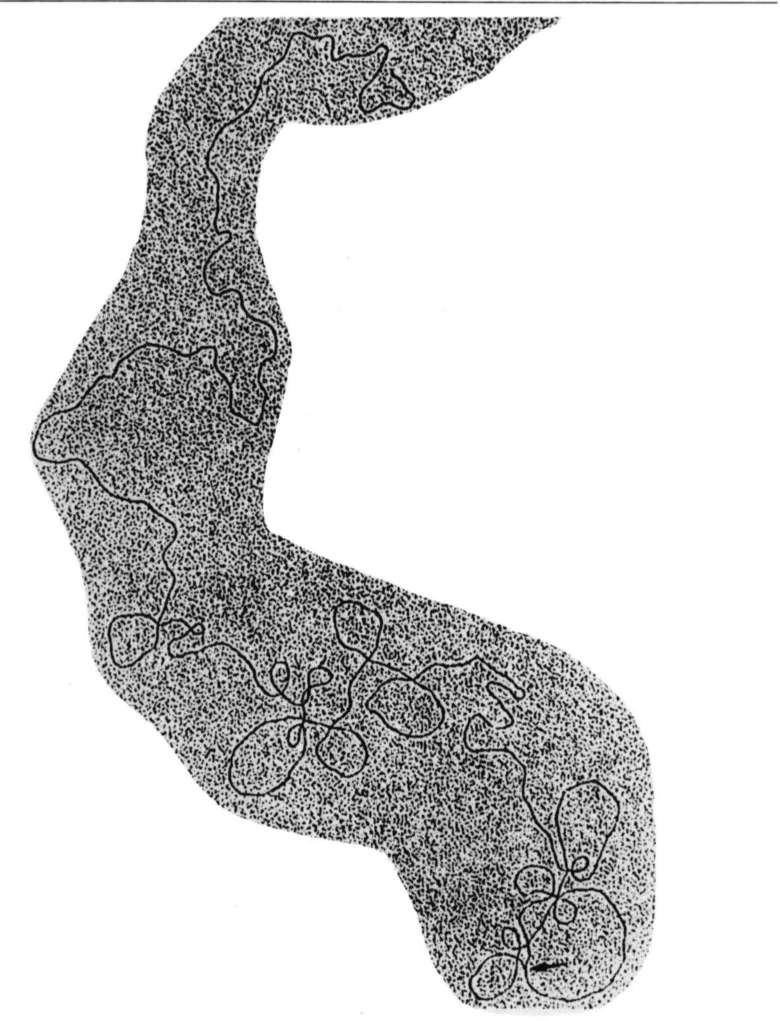

Figure 18. Electron micrograph of a circular-linear defective HSV DNA molecule. The molecule resembles a rolling-circle replicative intermediate. The arrow points to the origin of the linear molecule.
(Department of Molecular Virology, The Hebrew University-Hadassah Medical School, Jerusalem, Israel.)

rolling-circle mechanism described for bacteriophage ØX174 (Gilbert and Dressler 1968). In the case of defective HSV DNA, it was suggested that the displaced strand is copied by another DNA polymerase molecule and defective viral double-stranded DNA is synthesized. The defective DNA molecules with a molecular weight of 100×10^6 are encapsidated into virions at the same time as the wild-type DNA molecules since both are synthesized in cells producing defective virus. The same viral DNA polymerase is responsible for the synthesis of both wild-type and defective DNA.

HSV and latent infections in humans and animals

HSV-1 infections start with a primary infection in the mouth or eye that heals spontaneously but may recur at the same site years after the primary event. The recurrences may occur after a febrile infection, exposure to the sun, hormonal changes, etc., and result in the formation of lesions that heal spontaneously. It was found that the viral DNA resides in the trigeminal ganglion in humans (Baringer 1975). HSV-1 DNA was found in the trigeminal ganglia of people who died of various causes, including lymphomas and multiple sclerosis (Warren et al. 1977). Infection of test animals also revealed that the viral DNA is present in ganglia. The mechanisms that lead to the activation of the latent viral DNA in the ganglia and to the appearance of lesions in the skin are not yet known (reviewed by Stevens 1975).

Inactivated HSV transforms cells in vitro

It was demonstrated (Duff and Rapp 1973; Rapp and Reed 1976) that ultraviolet (uv) inactivation of HSV inhibits virus replication in infected hamster cells but leads to their transformation. Such cells, injected into the hamster, lead to the development of tumors. Uv irradiation leads to the inactivation of the viral gene coding for the HSV DNA polymerase; therefore, virus replication is prevented. The viral DNA recombining with the cellular DNA leads to cell transformation. The DNA of such transformed cells contains fragments of HSV DNA. These viral DNA fragments were found to arise from the middle of the L unique sequence of the viral DNA.

Partial inactivation of HSV DNA by dyes, such as neutral red or acridine orange, and visible light due to intercalation of the dye with base pairs in the DNA (Melnick and Wallis 1975) can also lead to cell transformation in vitro. Increased incubation temperature, ts mutants, as well as suboptimal temperatures can also cause transformation of mammalian cells infected with HSV types 1 and 2.

HSV-coded thymidine kinase gene biochemically transforms TK^- animal cells

HSV was found to have a gene that codes for the enzyme TK, which phosphorylates thymidine in infected cells. The virus-coded enzyme differs from the host-cell TK. Using mutant L cells which lack the TK enzymatic activity (designated L[TK^-]), it was found that the HSV-coded TK transforms the mutant cells to the TK^+ phenotype (Munyon et al. 1971). Additional studies revealed that transfection of the L(TK^-) cells with fragmented HSV DNA or with a DNA fragment cloned in the bacterial plasmid pBR322 transforms the L(TK^-) cells into the L(TK^+) phenotype (Pellicer et al. 1978). Transformation of the cells is due to insertion of the viral TK gene into the cellular genome. The descendants of the transformed L(TK^-) cell contain one copy of the viral TK gene in each cellular genome. This type of transformation is designated *biochemical transformation*.

The successful cloning of a 3.6 kilobase pair *Bam*HI fragment from HSV-1 DNA into the *Bam*HI site in plasmid pBR322 made it possible to study the HSV TK gene. It was recently suggested that the nucleotide sequence of the TK gene contains a promotor sequence at the start and a terminator sequence at the end of the gene (McKnight 1980).

Kit et al. (1981) have succeeded in producing a hybrid plasmid containing the HSV-1 TK gene in the form of a 2-kilobase pair fragment. This fragment was inserted into the plasmid pBR322 and was used to transform TK^- *E. coli*. The specificity of the TK enzyme molecules produced by *E. coli* was determined by inhibition with specific viral antibodies (Kit et al. 1981). Cloning of the HSV-1 TK gene to the lac Z gene of *E. coli*, allowed its expression in the bacterial host (Garapin et al. 1981).

HERPESVIRUSES AFFECT HUMANS AND ANIMALS

HSV-1 virus infects the mucous membranes in the mouth (viral stomatitis), the lips (herpes labialis), the eyes (herpeto-keratoconjunctivitis), the brain (herpetic meningoencephalitis); HSV-2 infects the genitals (herpesvirus genitalis). If untreated, herpetic encephalitis is lethal (Nahmias and Roizman 1973).

The virus infects the mucous membranes and skin by direct contact with an active herpetic lesion. Most children are infected at an early age with HSV-1 and develop immunity to the virus. However, even individuals with high antibody titers can develop recurrent virus infections. Local immunity is most probably the mechanism that restricts virus proliferation, since individuals with a defect in the immune system (e.g., cancer patients under treatment with immunosuppressive drugs) develop very large skin lesions due to unlimited virus spread.

The virus spreads in the bloodstream and along nerve axons

A localized virus infection can develop into a systemic infection with viremia (virus in the bloodstream) that spreads throughout the body and infects the internal organs. In an immunologically deficient patient with the genetic disorder ataxia telangiectasia, an infection of the mucous membranes in the mouth developed into a lethal systemic infection.

Alternatively, the virus spreads via the nerve axons in the body. It is assumed that the virus in the primary skin lesion enters a nerve in the affected area and migrates through the nerve axon to the trigeminal ganglion, and from there ascends into the brain where it causes meningoencephalitis. Once the virus reaches the brain, it spreads rapidly from cell to cell and causes extensive damage which, if not treated in time, can be fatal or leave the individual in a so-called vegetative state.

Connection between HSV-2 and cervical carcinoma?

Patients with diagnosed carcinoma of the cervix were found to have high antibody titers to HSV-2 in the serum as compared to healthy individuals. This led to the postulation that there is a relationship between HSV-2 and cervical carcinoma (Nahmias et al. 1971; Aurelian and Strand 1976). In-situ hybridization between HSV-2 DNA and biopsies taken from cervical carcinoma patients revealed the presence of viral DNA and RNA in the tumor cells only, and not in normal cells from the biopsy tissue.

Chemotherapy of HSV infections (see chapter 26)

Treatment of skin lesions

Both zinc acetate and phosphonoacetic acid were shown to selectively inhibit HSV-1 DNA polymerase in infected cells and in isolated viral DNA polymerase systems in vitro. Treatment of skin lesions with zinc acetate was reported to relieve the pain and promote faster healing of the lesion. Phosphonoacetate and phosphonoformate were found to be effective in the treatment of HSV-1 skin lesions in experimental animals. Unfortunately, it was found that PAA is concentrated in the bones of test animals, probably because phosphonocompounds have a role in the metabolism of bone. The Food and Drug Administration (FDA) in the United States requires more information on the effect of PAA on bone metabolism prior to permitting its use in human chemotherapy of HSV lesions. In the past, skin lesions were treated with acridine orange and visible light that led to inactivation of the virus (Melnick and Wallis 1977). With the demonstration that with such treatment the transforming ability of the virus is expressed, the treatment was abandoned.

Treatment of herpetic keratitis

The first effective antiviral drug was the halogenated derivative of thymidine, IUdR (iodouridine deoxyriboside) synthesized by Prussof (1960) and shown by Kaufman (1962) to cure herpetic keratitis in humans. Study of HSV-1 strains ten years after the introduction of IUdR showed increased resistance of the virus isolates to the drug and, in parallel, a decreased efficacy in treatment (Kaufman 1976). The analog adenine arabinoside, as well as interferon, are currently in use (chapter 26).

Treatment of patients afflicted with herpetic encephalitis

IUdR was used for the treatment of herpetic encephalitis with the hope that it would cure the brain infection. Confirmation of a herpes infection in the brain is required prior to treatment, but even in confirmed herpetic patients, IUdR is useless since it fails to penetrate into the brain cells where the virus multiplies. Therefore, the use of IUdR in herpetic encephalitis was discontinued.

Currently, adenine arabinoside is used for intravenous administration to patients with confirmed herpetic encephalitis. Analysis of the results of patient response to treatment revealed that patients treated very early in infection resulted in survival of many of the patients with minimal brain damage. Treatment of patients at later stages of the disease does not change the course of this lethal infection. Thus early laboratory diagnosis of herpetic encephalitis is a prerequisite for effective treatment.

Treatment of patients with generalized herpes with acyclovir (acycloguanosine)

The drug acycloguanosine (see chapter 26) is effective in treating patients with herpesvirus infections and has been used in children affected by generalized herpes simplex virus infections which, if left untreated, lead to death. Evidence is accumulating that such children fully recover after acyclovir treatment.

Other human herpesviruses

Varicella-zoster virus

Varicella virus causes chickenpox in children, while in elderly people the zoster virus can cause a recurrent infection (called *shingles*) along the dorsal root ganglia. The zoster virus, which is identical antigenically to varicella, must therefore inhabit the ganglia along this nerve as a latent infection. Varicella in children is usually a mild disease, but at the viremic stage, the virus could develop a systemic lethal infection. Recurrent zoster infections appear in cancer patients.

A live attenuated virus vaccine is used to immunize children against varicella, thus possibly also preventing zoster at a later age.

Cytomegalovirus (CMV): chronic infections in children

CMV causes a respiratory infection, but in most infected children there is no apparent disease, although in some jaundice or pneumonia may be manifested. The virus infects T lymphocytes and is present in them in a latent form. The virus is activated when immunosuppressive drugs are given to cancer or kidney patients. The virus infection is detected by megalic (giant) cells appearing in patients' urine which contain large quantities of virus. In pregnant women infected with CMV, in-utero infection of the fetus can cause either death or mental retardation. Pregnant women excreting CMV in their urine can also infect the newborn infant with CMV, which can cause acute pneumonia and death.

CMV CAUSES INFECTIOUS MONONUCLEOSIS (IM). Studies of IM patients negative to EBV revealed that they were infected with CMV. The virus infects T lymphocytes and leads to their proliferation.

EPSTEIN-BARR VIRUS (EBV)

Discovery

Burkitt (1963) described a lymphoma syndrome with a high incidence (mainly in children aged four–eight years) which was found in equatorial Africa, south

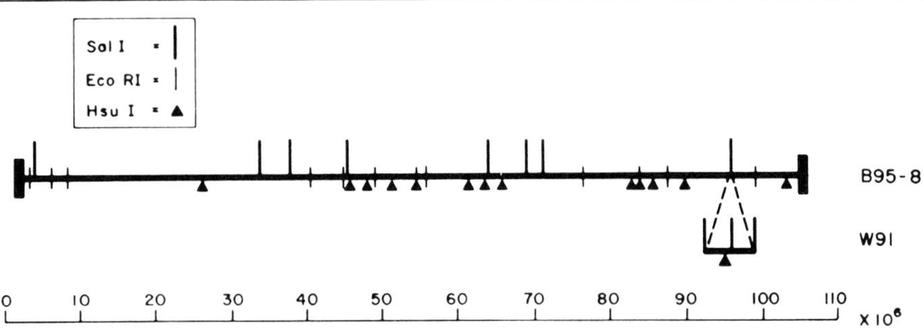

Figure 19. Linear drawing of EBV (B95-8) DNA showing the position of EcoRI, HsuI, and SalI restriction endonuclease cleavage sites. The location of the additional DNA present in the DNA of the W91 strain of EBV is shown as an insert.
(Kieff et al. 1979. Reprinted by permission from *Biochim. Biophys. Acta* 560, Fig. 4, p. 362.)

of the Sahara, in places such as Uganda. He stressed the bizarre anatomical distribution of the tumor, which also caused distortion of the children's faces. Since the lymphoma patients lived in an area infested with virus-carrying mosquitoes, Burkitt suggested that the cause might be a virus. Electron microscopy of lymphoblasts from biopsies of African lymphoma patients, grown under in vitro conditions (Epstein et al. 1964) revealed the presence of herpesvirus-like particles in some of the cells. This virus, which was named EBV was not only associated with Burkitt's lymphoma, but was also found to be a ubiquitous human virus.

The virus

EBV virions do not react with any of the antibodies to known herpesviruses. Lymphoblast cell lines that were adapted to growth under in vitro conditions synthesize very small amounts of EB virions. Incubation of certain cell lines in an arginine-deficient medium was found to stimulate the induction of virus particles (Henle and Henle 1968) since the mechanisms that prevent virus replication are inhibited. The virions that were isolated by the same techniques developed for the isolation of HSV were found to contain a DNA genome of 100×10^6 daltons and seven major viral proteins (Weinberg and Becker 1969). The viral DNA (figure 19) was reported by Kieff et al. (1979) to belong to class 1 DNA (figure 12).

Infectivity of EBV

Many lymphoblast cell lines were established in vitro from biopsies of patients with pathologically diagnosed Burkitt's lymphoma in Africa and New Guinea. In most laboratories, EBV is obtained from the P3HRl lymphoma cell line, since cells permissive to EBV that allow virus multiplication are not available. Another cell line producing EBV (B95-8) consists of in vitro *transformed* marmoset lymphocytes (Miller et al. 1972). It was established that Burkitt's lymphoma cells are monoclonal, which means that the cells are descendants of one cell that was transformed in the patient (Klein 1972). Most cell lines produce EBV intermittently, but certain cell lines (like Raji) cannot be induced to produce virions at all. Using radioactive virion DNA and a DNA-DNA reassociation technique, it was possible to demonstrate that Burkitt's lymphoblasts contain in the range of 20–50 genome equivalents of EBV DNA per cell (Zur Hausen 1979). However, three cell lines from confirmed Burkitt's lymphoma patients lack any detectable EBV DNA.

EBV can infect EBV-negative lymphoma cells and improve their in vitro growth properties.

EBV receptors on human B lymphocytes

EBV binds to B lymphocytes by attachment to a specific receptor on the membrane that is near the receptor for complement.

The adsorbed virus is incorporated into the cell cytoplasm, and the viral DNA induces the appearance of EB nuclear antigen (EBNA) (Klein 1974). The EBV infected lymphocytes are capable of unlimited proliferation under in vitro conditions, a property termed *immortalization*. The EBNA is a DNA binding protein with a molecular weight of 90,000 daltons that appears in a dimeric form. The role of EBNA in the virus infection is not known.

State of viral DNA in cancer cells

Treatment of lymphoblasts, each containing about 40 viral DNA genomes with cycloheximide, an inhibitor of protein synthesis, resulted in the disappearance of most of the viral DNA molecules. Such cells contain only one EBV DNA genome associated with the chromosomal DNA yet retain their transformed properties. It was recently demonstrated that the residual EBV DNA genomes are associated with DNA of the cellular chromosome 14 (Yamamoto et al. 1978). All other viral DNA molecules are episomal DNA which have a circular conformation (Adams and Lindahl 1975), are not associated with the host cell DNA, and are probably not essential for the transformation event.

The circular EBV DNA genomes present in the nuclei of Burkitt's lymphoblasts are replicated by the host cell enzymes during the S phase of the cell cycle (Hampar et al. 1973). Activation of the cells to produce virions has to do with the appearance of linear EBV DNA and transcription of the genomes followed by the synthesis of viral enzymes. The synthesis of the linear viral DNA is carried out by the viral DNA polymerase, a process sensitive to PAA. The synthesis of late viral proteins allows synthesis of capsids and virions. The activated cell in which EBV is made finally dies, and the virus progeny is released.

Infectious cycle of EBV in humans

EBV is a ubiquitous virus

It was found that EBV is prevalent in all parts of the world. It infects humans who develop antibodies to the viral capsid antigen (VCA). It was also reported that EBV can be isolated from the pharynx of individuals with infectious mononucleosis, and it was suggested by Pagano and Shaw (1979) that epithelial cells in the pharynx replicate the virus. Such a virus, if incubated with normal B lymphocytes, can infect the cell, induce EBNA, and the cell will immortalize—that is, it will be able to grow indefinitely under in vitro conditions.

EBV latently infects B lymphocytes in the peripheral blood. Incubation of lymphocytes under in vitro conditions leads to the induction of EBV in these latently infected cells, and the virus immortalizes normal B lymphocytes. This is the technique used for spontaneous isolation of transformed B lymphocytes.

EBV as the cause of infectious mononucleosis (IM) in young adults

Studies by G. and W. Henle on antibodies to EBV in the human population revealed that most children develop antibodies to EBV VCA at an early age, except for a small group who are not exposed to the virus and remain seronegative. As young adults, they can be infected by oral contact with a person carrying EBV, and they develop infectious mononucleosis (Henle et al. 1968, 1974). EBV infects the epithelial cells and B lymphocytes in the throat and surrounding lymph nodes, leading to elevated temperature and swelling of the lymph nodes, most probably due to the proliferation of B cells for which EBV is a mitogen. The T lymphocytes in IM patients also respond to the proliferating B lymphocytes and appear in the peripheral blood in large numbers as monocytes. IM can be fatal. The proliferation of the T cells and their large numbers in the blood are the reasons for the long-term weakness of the patients. T cell proliferation subsides after a few months, and the blood picture returns to normal. Since EBV is transmitted from mouth to mouth, IM is also named the *kissing disease*.

The role of EBV in cancer

Two types of cancer that are linked to EBV occur in humans: (1) Burkitt's lymphoma in Africa and (2) nasopharyngeal carcinoma (NPC) in Southern China and North Africa.

Burkitt's lymphoma

Since EBV DNA or fragments of it are intimately involved with the chromosomal DNA of Burkitt's lymphoblasts, the question arises of whether transformation of perhaps one lymphocyte into a cancer cell may be due to the presence of EBV in the cell. Three lines of evidence contradict the claim that EBV is the cause of such cell transformation:

1. In a few patients with pathologically diagnosed Burkitt's lymphoma, the B lymphocytes did not contain any detectable EBV DNA (the resolution of the DNA:DNA hybridization was 0.1 genome equivalent).
2. Both EBV DNA-negative and DNA-positive B lymphoblast cell lines

from patients with Burkitt's lymphoma were found to contain an inducible retrovirus.

3. A study was done in Uganda in a region with a high incidence of Burkitt's lymphoma to determine the connection between EBV infection and the incidence of Burkitt's lymphoma. Blood samples were taken from 42,000 children during a five-year period, and out of these, only 32 children developed Burkitt's lymphoma (de-Thé 1979). This low incidence further negates the question of whether EBV is the cause of Burkitt's lymphoma.

Nasopharyngeal carcinoma (NPC)

This tumor involves epithelial cells. The tumor can be diagnosed serologically long before its detection by other means. Appearance in the blood of antibodies to EBV antigens is currently used to diagnose NPC. Irradiation at the site where the tumor usually appears is used as the method of treatment, and this prevents the development of NPC. This tumor is prevalent among Southern Chinese, regardless of where they may have migrated to. A similar tumor was described in North Africans. Thus in certain ethnic groups, EBV can cause transformation of epithelial cells and the formation of tumors.

Figure 20 summarizes the possible involvement of EBV in human cancers and its role in disease and all immortalization.

Herpesvirus papio (HVP): a monkey virus related to EBV

This virus was isolated from baboons (*Papio hamadryas*) that developed lymphomas (Falk et al. 1976). Hybridization of EBV DNA with baboon and chimpanzee herpesvirus revealed 35–45% homology in the viral DNAs. Baboon lymphoblastoid cell lines carrying HVP do not produce EBNA in the nuclei, although superinfection of the cells with EBV showed that the baboon lymphoid cell line is competent to synthesize EBNA. The HVP can immortalize monkey B cells in vitro. Injections of such lymphocytes into adult monkeys led to a disease similar to IM.

Other monkey herpesviruses

Herpesvirus saimiri and herpesvirus ateles were isolated from circulating lymphocytes in New World monkeys by L.V. Melendez and his colleagues (reviewed by Fleckenstein 1979). These viruses contain DNA of 100×10^6 daltons, but the virion population is made up of two subgroups: (1) Molecules with the M genome composed of about 30% heavy (H) DNA (G + C = 71%; density 1.729 g/ml) with repetitive sequences, and 70% light (L) DNA (G + C = 36%; density 1.695 g/ml) with unique sequences; (2) the H genome contains heavy sequences only.

The M-DNA is infectious in cell culture and capable of transforming primate cells. Circular episomal viral DNA of 131.5×10^6 daltons was found in the transformed lymphoid cell lines. The circular viral DNA consists of two L regions (54×10^6 and 31.5×10^6 daltons) and two H regions (25.6×10^6 and

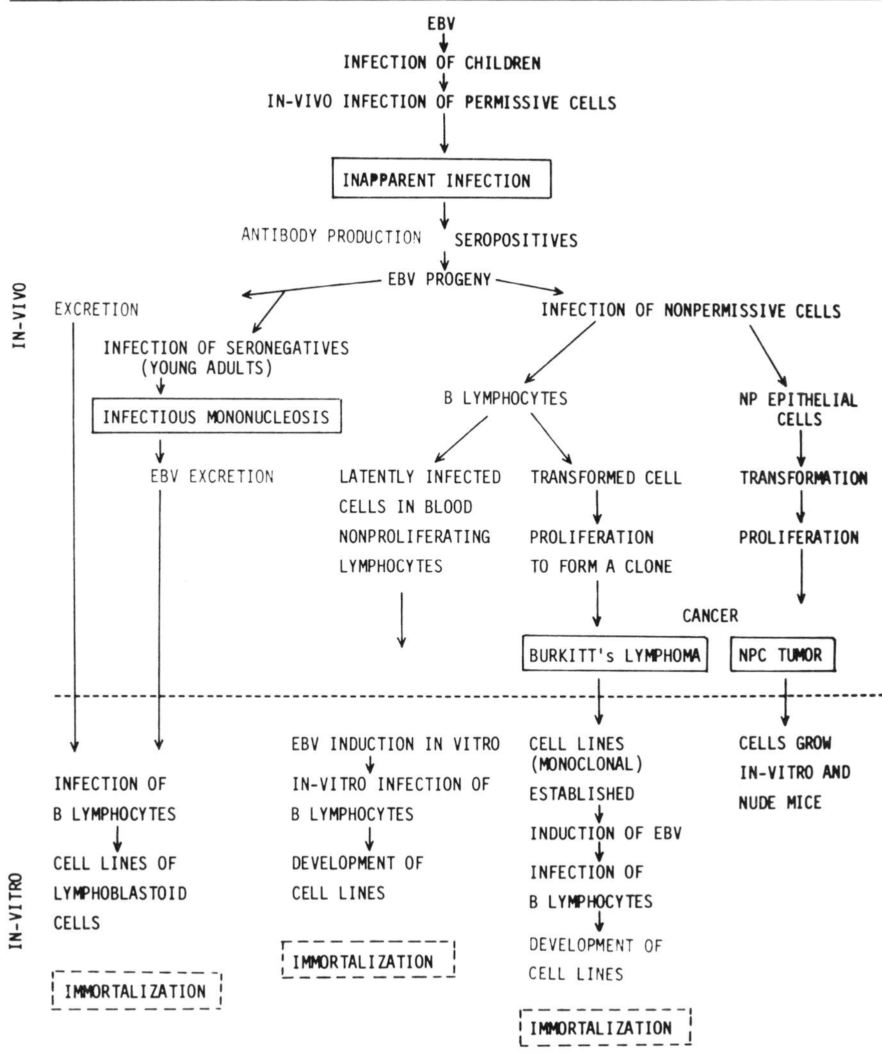

Figure 20. Scheme for EBV infection of nonpermissive cells in vivo and in vitro. (NP = nasopharyngeal; NPC = nasopharyngeal carcinoma).

20×10^6 daltons). The two L sequences have the same orientation (Werner et al. 1977).

H. saimiri and H. ateles, which cause malignant lymphoma in various New World primate species, are related: There is 35% homology between the L genomes of the two viruses, while there is only 10% between the H genomes. The natural hosts of H. saimiri and H. ateles (squirrel monkeys and spider monkeys, respectively) are infected at an early age and develop antibodies to the virus. The T lymphocytes in vivo remain latently infected. Infection of either cells or 10 pfu of virus into other species like owl monkeys causes the development of lymphoproliferative disease.

RHESUS HERPESVIRUS. Leukocytes of rhesus monkeys were found to contain a latent virus, unrelated to other known herpesviruses.

HERPES B VIRUS. This virus causes latent infections in monkeys and is also present in the saliva. When such a monkey bites someone, the B virus migrates to the brain of the victim and causes encephalitis, which can be fatal. Individuals working with monkeys may be exposed to this virus.

Herpesvirus of pigs

PSEUDORABIES VIRUS CAUSES AUJESKY DISEASE IN PIGS. This virus has a DNA genome that belongs to class 2 DNA (figure 12). The virus is latent in pigs but can develop into a prominent disease when the pigs are transported or kept in large numbers on farms.

Herpesviruses in cattle, horses, and dogs

Herpesviruses can cause abortion in animals. The viruses differ antigenically from one another.

Fish herpesvirus

A herpesvirus isolated from the channel catfish was found to have a DNA genome of 80×10^6 daltons. The viral DNA which is classified as class 1 DNA (figure 12) contains only the L component of the herpesvirus DNA.

Herpesvirus in frogs causes Lucké renal carcinoma

A tumor of the adenocarcinoma type, containing a herpesvirus, was discovered in the leopard frog (*Rana pipiens*) by R. Lucké in 1934. Injection of purified virus into frog embryos was shown by Mizell et al. (1969) to cause tumors in the kidney. The tumor appeared in the frogs in the summer season, but herpesvirus could not be found until it was shown that transfer to a low temperature (4–9°C) caused induction of the virus.

Marek's disease in chickens

This disease in chickens, described by Marek in Hungary, was represented as a lymphoma of the internal organs that causes damage to the nervous tissues (reviewed by Nazerian 1979). It was found that certain strains of chickens are resistant to Marek's disease, while others are highly sensitive. Marek's disease virus replicates in the cells of the feather follicles and is released with the feathers. Respiration of the virus in the debris contaminating the chicken house leads to a respiratory infection of sensitive chickens, development of lymphomas, and death with brain damage. It was also shown that the development of the lymphoma depends on the presence or absence of the bursa fabricius. Removal of the bursa from resistant chickens renders them sensitive to MDV, and removal of the bursa from sensitive chickens leads to the development of resistance to infection. The immunological mechanisms involved in the sensitivity and resistance to the disease are not understood.

Herpesvirus of turkeys was found to immunize and protect MDV-sensitive chickens when injected into young chickens. HVT is only slightly related to MDV but nevertheless protects against infection with MDV. The use of HVT live vaccine for immunization of poultry has markedly reduced the incidence of Marek's disease on chicken farms.

BIBLIOGRAPHY

Adams, A., and Lindahl, T. Epstein-Barr virus genomes with properties of circular DNA molecules in carrier cells. *Proc. Natl. Acad. Sci. USA 72*:1477–1481 (1975).

Aurelian, L., and Strand, B. Herpesvirus type 2 related antigens and their relevance to humoral and cell mediated immunity in patients with cervical cancer. *Cancer Res. 36*:810–820, 1976.

Baringer, J.R. Herpes simplex virus infection of nervous tissue in animals and man. *Prog. Med. Virol. 20*:1–26, 1975.

Becker, Y.; Dym, H.; and Sarov, I. Herpes simplex virus DNA. *Virology 36*:185–192, 1968.

Burkitt, D. A lymphoma syndrome in tropical Africa. *Int. Rev. Exp. Pathol. 2*:67–138, 1963.

Duff, R., and Rapp, F. The induction of oncogenic potential by herpes simplex virus. *Perspect. Virol. 8*:199–210, 1973.

Epstein, M.A.; Achong, B.G.; and Barr, Y.M. Virus particles in cultured lymphoblasts from Burkitt's lymphoma. *Lancet i*:702–703, 1964.

Falk, L.; Deinhardt, F.; Nonoyama, M.; Wolfe, L.G.; Bergholz, C.; Lapin, B.; Yakovleva, L.; Agrba, V.; Henle, G.; and Henle, W. Properties of a baboon lymphotropic herpesvirus related to EBV. *Int. J. Cancer 18*:798–807, 1976.

Fleckenstein, B. Oncogenic herpesviruses of non-human primates. *Biochim. Biophys. Acta 560*:301–342 (1979).

Frenkel, N.; Locker, H.; Batterson, W.; Hayward, G.S.; and Roizman, B. Anatomy of herpes simplex virus DNA. VI. Defective DNA originates from the S component. *J. Virol. 20*:527–531, 1976.

Friedmann, A.; Broit, M.; and Becker, Y. Annealing of alkali-resistant HSV DNA strands and isolation of S and L components. *In: Oncogenesis and Herpesviruses III* (eds: G. de-The, W. Henle, and F. Rapp), IARC Scientific Publications No. 24, Lyon, 1978, pp. 137–148.

Friedmann, A.; Shlomai, J.; and Becker, Y. Electron microscopy of herpes simplex virus DNA molecules isolated from infected cells by centrifugation in CsCl density gradients. *J. Gen. Virol. 34*:507–522, 1977.

Garapin, A.C.; Colbere-Garapin, F.; Cohen-Solal, M.; Horodniceanu, F.; and Kourilsky, P. Expression of herpes simplex virus type 1 thymidine kinase gene in *Escherichia coli*. *Proc. Natl. Acad. Sci. USA 78*:815–819, 1981.

Gilbert, S., and Dressler, D. DNA replication: the rolling circle model. *Cold Spring Harbor Symp. Quant. Biol. 33*:473–484, 1968.

Hampar, B.; Derge, J.G.; Martos, L.M.; Tagamets, M.A.; Chang, S.Y.; and Chakrabarty, M. Identification of a critical period during S phase for activation of the Epstein-Barr virus by 5-iododeoxyuridine. *Nature New Biol. 244*:214–217, 1973.

Henle, G.; Henle, W.; and Diehl, V. Relation of Burkitt's tumor-associated herpes-type virus to infectious mononucleosis. *Proc. Natl. Acad. Sci. USA 59*:94–101, 1968.

Henle, W., and Henle, G. Effect of arginine deficient media on the herpes-type virus associated with cultured Burkitt tumor cells. *J. Virol. 2*:182–191, 1968.

Henle, W.; Henle, G.; and Horwitz, C.A. Epstein-Barr virus-specific diagnostic tests in infectious mononucleosis. *Hum. Pathol. 5*:551–565, 1974.

Honess, R.W., and Watson, D.H. Unity and diversity in the herpesviruses. *J. Gen. Virol. 37*:15–37, 1977.

Kaufman, H.E. Clinical cure of herpes simplex keratitis by 5-iodo-2′-deoxyuridine. *Proc. Soc. Exp. Biol. Med. 109*:251–252, 1962.

Kaufman, H.E. Ocular antiviral therapy in perspective. *J. Infect. Dis. Suppl. 133*:A96–A100, 1976.
Kieff, E.; Given, D.; Powell, A.; King, W.; Dambough, T.; and Raab-Traub, N. Epstein-Barr virus: structure of the viral DNA and analysis of viral RNA in infected cells. *Biochim. Biophys. Acta 560*:355–373, 1979.
Kit, S.; Otsuka, H.; Qavi, H.; and Hazen, M. Herpes simplex virus thymidine kinase activity of thymidine kinase-deficient *Escherichia coli* K-12 mutant transformed by hybrid plasmids. *Proc. Natl. Acad. Sci. USA 78*:582–586, 1981.
Klein, G. Herpesviruses and oncogenesis. *Proc. Natl. Acad. Sci. USA 69*:1056–1064, 1972.
Klein, G. Studies on the Epstein-Barr virus genome and the EBV-determined nuclear antigen in human malignant disease. *Cold Spring Harbor Symp. Quant. Biol. 39*:783–790, 1974.
McKnight, S.L. The nucleotide sequence and transcript maps of the herpes simplex virus thymidine kinase gene. *Nucleic Acids Res. 8*:5949–5964, 1980.
Melnick, J.L., and Wallis, C. Photodynamic inactivation of herpesvirus. *Perspect. Virol. 9*:297–314, 1975.
Melnick, J.L., and Wallis, C. Photodynamic inactivation of herpes simplex virus: a status report. *Ann. NY Acad. Sci. 284*:171–181, 1977.
Miller, G.; Shope, T.; Lisco, H.; Still, D.; and Lipman, M. Epstein-Barr virus: transformation, cytopathic changes and viral antigens in squirrel monkey and marmoset leukocytes. *Proc. Natl. Acad. Sci. USA 69*:383–387, 1972.
Mizell, M.; Joplin, I.; and Isaacs, J.J. Tumor induction in developing frog kidneys by a zonal centrifuge purified fraction of the frog herpes-type virus. *Science 165*:1134–1137, 1969.
Munyon, W.; Kraiselburd, E.; Davis, D.; and Mann, J. Transfer of thymidine kinase to thymidine kinaseless L cells by infection with ultraviolet-irradiated herpes simplex virus. *J. Virol.* 7:813–820, 1971.
Nahmias, A.J.; Naib, Z.M.; and Josey, W.E. Herpesvirus hominis type 2 infection associated with cervical cancer and prenatal disease. *Perspect. Virol. 8*:73–88, 1971.
Nahmias, A.J., and Roizman, B. Infection with herpes simplex viruses 1 and 2. (Review in three parts). *N. Engl. J. Med. 289*:667, 719, 781, 1973.
Nazerian, K. Marek's disease lymphoma of chickens and its causative herpesvirus. *Biochim. Biophys. Acta 560*:375–395 (1979).
Pagano, J.S., and Shaw, J.E. Molecular probes and genome homology. *In: The Epstein-Barr Virus* (M.A. Epstein and B.G. Achong, eds.), Springer-Verlag, Berlin, Heidelberg, New York, 1959, pp. 109–146.
Pellicer, A.; Wigler, M.; Axel, R.; and Silverstein, S. The transfer and stable integration of the HSV thymidine kinase gene into mouse cells. *Cell 14*:133–141, 1981.
Prusoff, W.H. Studies on the mechanism of action of 5-iododeoxyuridine, an analog of thymidine. *Cancer Res. 20*:92–95, 1960.
Rapp, F., and Reed, C. Experimental evidence for the oncogenic potential of herpes simplex virus. *Cancer Res. 36*:800–806, 1976.
Roizman, B. The structure and isomerization of herpes simplex virus genomes. *Cell 16*:481–494, 1979.
Roizman, B.; Knipe, D.; Morse, L.S.; Ruyechan, W.; and Kovler, M.B. The molecular organization and expression of the herpes simplex virus (HSV) 1 and 2 DNAs. *In: International Virology IV*. Centre for Agricultural Publishing and Documentation, Wageningen, The Netherlands, 1978, pp. 67.
Schaffer, P.A. Temperature-sensitive mutants of herpesviruses. *Curr. Top. Microbiol. Immunol. 70*:51–100, 1975.
Sheldrick, P., and Berthelot, N. Inverted repetitions in the chromosome of herpes simplex virus. *Cold Spring Harbor Symp. Quant. Biol. 39*:667–678, 1974.
Southern, E.M. Detection of specific sequences among DNA fragments separated by gel electrophoresis. *J. Mol. Biol. 98*:503–518, 1975.
Stevens, J.G. Latent herpes simplex virus and the nervous system. *Curr. Top. Microbiol. Immunol. 70*:31–50, 1975.
Stow, N.D., and Wilkie, N.M. Physical mapping of temperature-sensitive mutations of herpes simplex virus type 1 by intertypic marker rescue. *Virology 90*:1–11, 1978.
Subak-Sharpe, J.H., and Timbury, M.C. Genetics of herpesviruses. *In: Comprehensive Virology* (H. Fraenkel-Conrat and R.R. Wagner, eds.), Plenum Press, New York, 1977, Vol. 9, pp. 89–132.

de-The, G. Demographic studies implicating the virus in the causation of Burkitt's lymphoma; prospects for nasopharyngeal carcinoma. *In: The Epstein-Barr Virus* (eds: M.A. Epstein and B.G. Achong), Springer-Verlag, Berlin, Heidelberg, New York, 1979, pp. 417–437.

Warren, K.G.; Gilden, D.H.; Brown, S.; Devlin, M.; Wroblewska, Z.; Subak-Sharpe, J.; and Koprowski, H. Isolation of herpes simplex virus from human trigeminal ganglia including ganglia from one patient with multiple sclerosis. *Lancet* ii:637–639, 1977.

Weinberg, A., and Becker, Y. Studies on EB virus of Burkitt lymphoblasts. *Virology 39*:312–321, 1969.

Werner, E.-J.; Bornkamm, G.W.; and Fleckenstein, B. Episomal viral DNA in a herpes-virus saimiri-transformed lymphoid cell line. *J. Virol. 22*:794–803, 1977.

Yamamoto, K.; Mizuno, F.; Matsuo, T.; Tanaka, A.; Nonoyama, M.; and Osaka, T. Epstein-Barr virus and human chromosomes: close association of the resident viral genome and the expression of the virus-determined nuclear antigen (EBNA) with the presence of chromosome 14 in human-mouse hybrid cells. *Proc. Natl. Acad. Sci. USA 75*:5155–5159, 1978.

Yanagi, K.; Talavera, A.; Nishimoto, T.; and Rush, M.G. Inhibition of herpes simplex virus type 1 replication in temperature-sensitive cell cycle mutants. *J. Virol. 25*:42–50, 1978.

Zur Hausen, H. Biochemical detection of the virus genome. *In: The Epstein-Barr Virus* (eds. M.A. Epstein and B.G. Achong), Springer-Verlag, Berlin, Heidelberg, New York, 1979, pp. 147–153.

RECOMMENDED READING

Becker, Y. (ed.) *Herpesvirus DNA*. Martinus-Nijhoff Publishers BV, The Hague, 1981.

Epstein, M.A., and Achong, B.G. *The Epstein-Barr Virus*. Springer-Verlag, Berlin, Heidelberg, New York, 1979.

Epstein, M.A., and Achong, B.G. Recent progress in Epstein-Barr virus research. *Annu. Rev. Microbiol. 31*:421–446, 1977.

Gallo, R.G. (ed.) *Recent Advances in Cancer Research, Cell Biology, Molecular Biology and Tumor Virology*. CRC Press Inc., Florida, U.S.A., 1980.

Kaplan, A.S. (ed.) *The Herpesviruses*. Academic Press, New York, 1973.

Lonsdale, D.M.; Brown, S.M.; Lang, J.; Subak-Sharpe, J.H.; Koprowski, H.; and Warren, K.G. Variations in herpes simplex virus isolated from human ganglia and a study of clonal variation in HSV-1. *Ann. NY Acad. Sci. 354*:291–308, 1980.

Nahmias, A.J.; Dowdle, W.R.; and Schinazi, R.F. (eds.) *The Human Herpesviruses*. Elsevier, New York, 1981.

O'Callaghan, D.J., and Randall, C.C. Molecular anatomy of herpesviruses: recent studies. *Prog. Med. Virol. 22*:123–151, 1976.

Rapp, F. *Oncogenic Herpesviruses*. CRC Press Inc., Florida, U.S.A., 1980.

Rawls, W.E.; Bacchetti, S.; and Graham, F.L. Relation of herpes simplex viruses to human malignancies. *Curr. Top. Microbiol. Immunol. 77*:71–95, 1977.

Roizman, B., and Furlong, D. The replication of herpesviruses. *In: Comprehensive Virology* (H. Fraenkel-Conrat and R.R. Wagner, eds.) Plenum Press, New York, 1977, Vol. 9, pp. 89–132.

8. ADENOVIRUSES

Adenoviruses were discovered in adenoid tissue removed from a healthy individual and grown in tissue culture. In spontaneously dying cells, virions with a typical morphology (figure 21) were seen by electron microscopy.

Family adenoviridae

The virion (m.w. 170×10^6) is a nonenveloped isometric particle, 70–90 nm in diameter, composed of 252 capsomeres arranged in icosahedral symmetry. Each capsomere is 7–9 nm in diameter. Inside the capsid is a single, linear double-stranded DNA molecule containing about 35,000 base pairs. Some adenoviruses are oncogenic in animals and in animal cells in vitro. Mammalian adenoviruses (mastadenoviruses) have a common antigen that differs from avian adenoviruses (aviadenoviruses).

Genus Mastadenovirus

Human adenovirus: types 1–33.
Mammalian viruses: monkey viruses, 24 serotypes; adenoviruses of cattle, dogs, horse, mouse, opossum, sheep, goats, and pigs

Genus Aviadenovirus

ADENOVIRUS OF FOWL, TURKEY, GOOSE, PHEASANT, AND DUCK. More than 80 adenovirus serotypes (of which 35 serotypes are human adenoviruses)

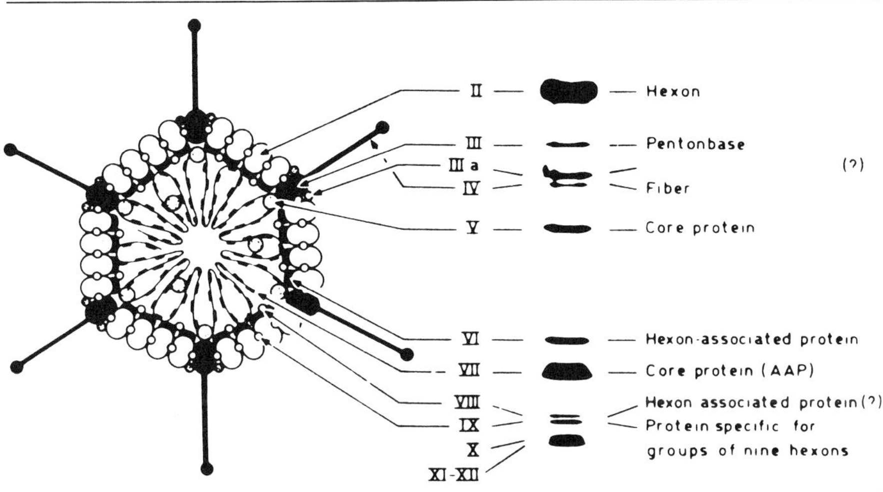

Figure 21. A refined topographical model of adenovirus type 2.
(Everitt et al. 1975. Reprinted by permission from Virology, Vol. 67, Fig. 7, p. 206.)

were isolated from different hosts. Human adenoviruses can agglutinate red blood cells from rats or monkeys.

Structure of the virus

The virions contain 252 capsomeres, of which 240 are nonvertex capsomeres (hexons) and 12 are vertex capsomeres (penton base). Each penton base contains a filamentous protein fiber around which five hexon capsomeres are arranged. Treatment of the capsids with SDS results in their dissolution, and electrophoresis of these proteins in polyacrylamide gels makes possible their identification (figure 21).

1. The hexon is made of three polypeptides, each with a molecular weight of 120,000 (polypeptide II).
2. Polypeptide IX is associated with the capsomeres that assemble in the nucleus into groups of nine hexons.
3. Polypeptides VI and VIII are internally located in the capsid.
4. Polypeptides III, IIIa, and IV are present in the vertices of the icosahedron; polypeptide III is the penton base, and polypeptide IV is the fiber.
5. Polypeptides V and VII are associated with the DNA molecule in the capsid. Polypeptides IIIa, VI, VIII, and IX are the four connecting proteins of the virion.

Infection of the cell

The virions attach to the cell membrane and are incorporated into the cytoplasm where they are uncoated in a phagocytic vacuole by lysosomal enzymes.

The viral DNA is released into the nuclei, the site of virus replication. The uncoating of the viral DNA occurs in the infected cells during the second hour after infection and does not require cellular protein synthesis.

The viral DNA is infectious and can be incorporated into the cytoplasm, provided the DNA is precipitated with calcium phosphate so as to bind to the cell membrane. The precipitate is engulfed by the invagination of the cell membrane and is transported to the nucleus.

Properties of adenovirus DNA

The linear double-stranded DNA of the mastadenoviruses has a molecular weight of $20 - 25 \times 10^6$. Those viruses were divided into subgroups according to their G + C contents, which range from 48–59%: subgroup A – G + C = 48–49%; subgroup B – G + C = 49–52%; and subgroup C – G + C = 57–59%. Adenoviruses from monkeys have a G + C content of 60%. The aviadenovirus DNA has a molecular weight of 30×10^6 and a G + C content of 54–55%.

The linear DNA contains inverted terminal repetitions of around 100 nucleotides; thus the molecular ends of the molecule are identical. It is possible to separate the two viral DNA strands by hybridization to the synthetic polymer poly(U,G). The DNA can be separated into light (l) and heavy (h) strands in cesium chloride density gradients (Tibbetts and Pettersson 1974).

Adenovirus DNA can attain a circular conformation due to a protein of 55,000 daltons that is covalently linked to the 5′ ends of both DNA strands.

The structure of adenovirus DNA was revealed by cleavage with restriction enzymes. The six fragments produced by EcoRl digestion of adenovirus 2 were mapped on the viral DNA genome (figure 22, fragments A–F). Fragment A represents 58.5% and is at the left-hand side of the genome. Cleavage of DNA is useful in the characterization of the function of the viral genes. The restriction enzyme cleavage sites are mapped on the genome, using coordinates that may be referred to as either 0 to 1.0 (figure 22) or 0 to 100 (figure 23).

Arrangement of viral genes

The site of the viral genes in the DNA genome can be determined by hybridization of mRNA molecules synthesized early and late in the virus replicative cycle to isolated DNA fragments. The nature of the gene product is revealed using a reticulocyte homogenate or wheat-germ extract for translation of the viral mRNA in vitro. The gene products in relation to the physical map of adenovirus DNA are shown in figure 22.

Transcription of viral DNA

The adenovirus type 2 genome does not code for an enzyme for the synthesis of viral mRNA. The cellular DNA-dependent RNA polymerase II is responsible for the transcription of the viral genomes. The cellular RNA polymerase III

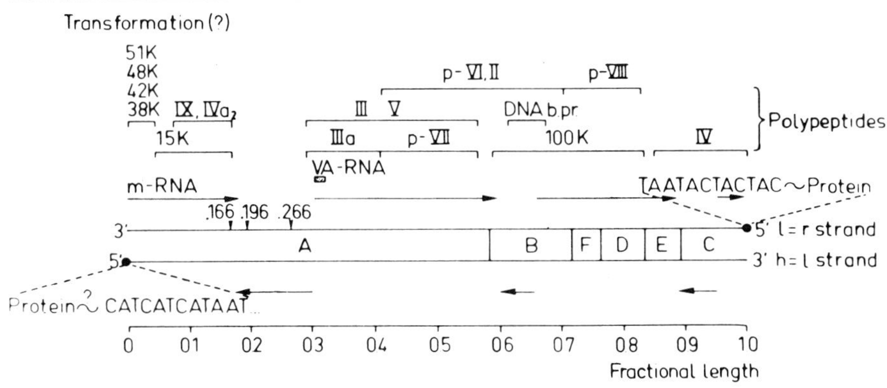

Figure 22. Schematic representation of adenovirus 2 genome. The l(r) and h(l) strands are indicated with the terminal (inversely repetitive) sequences in the inserts. The horizontal arrows indicate the direction of transcription of mRNA in different segments of the genome. A–F designate the EcoRI restriction fragments of Ad2 DNA. Roman numerals refer to the map location of the structural proteins of the virion. The location of the virus-associated (VA)-RNA and of the leader sequences (0.166, 0.196, and 0.266) are also included.
(Doerfler 1978. Reprinted by permission from *International Virology* 4, Fig. 2, p. 66. Centre for Agricultural Publishing and Documentation, Wageningen, The Netherlands.)

that transcribes the tRNA genes in the host cell genome, transcribes the viral genes for the 5.2 S and 5.5 S RNA molecules. These two virus-associated (VA) RNA species (about 156 nucleotides long) are transcribed from viral genes situated just in front of the late genes, close to coordinates 29–30 on the adenovirus 2 map (figure 23B). Both species of VA RNA appear early in infection; the 5.2 S species levels off, but the 5.5 S species increases during the late stage. Their function is not known (Söderlund et al. 1976).

The viral mRNAs are produced by the removal of internal sequences from precursor nuclear RNAs by a cut and splice mechanism (Kitchingman and Westphal 1980; Blanchard et al. 1978).

Transcription of adenovirus DNA proceeds in either a rightward (r) direction on the light (l) strand or in a leftward (l) direction on the heavy (h) strand (figures 22 and 23), and genes on both DNA strands are transcribed for mRNA.

In recent years, extensive studies on the transcription of early and late adenovirus mRNA have revealed a complex sequence of molecular events. These include initiation of transcription at separate promoters, capping of mRNA at the 5′ termini, splicing and polyadenylation of the RNA transcripts at the 3′ termini, regulatory processes leading to a switch of the transcription from the early to the late viral genes, as well as regulatory processes that affect the transcription of the host cell genome and selective transport of viral mRNA. The transcription of adenovirus DNA was reviewed by Ziff (1980) as described in the following discussion and in figure 23.

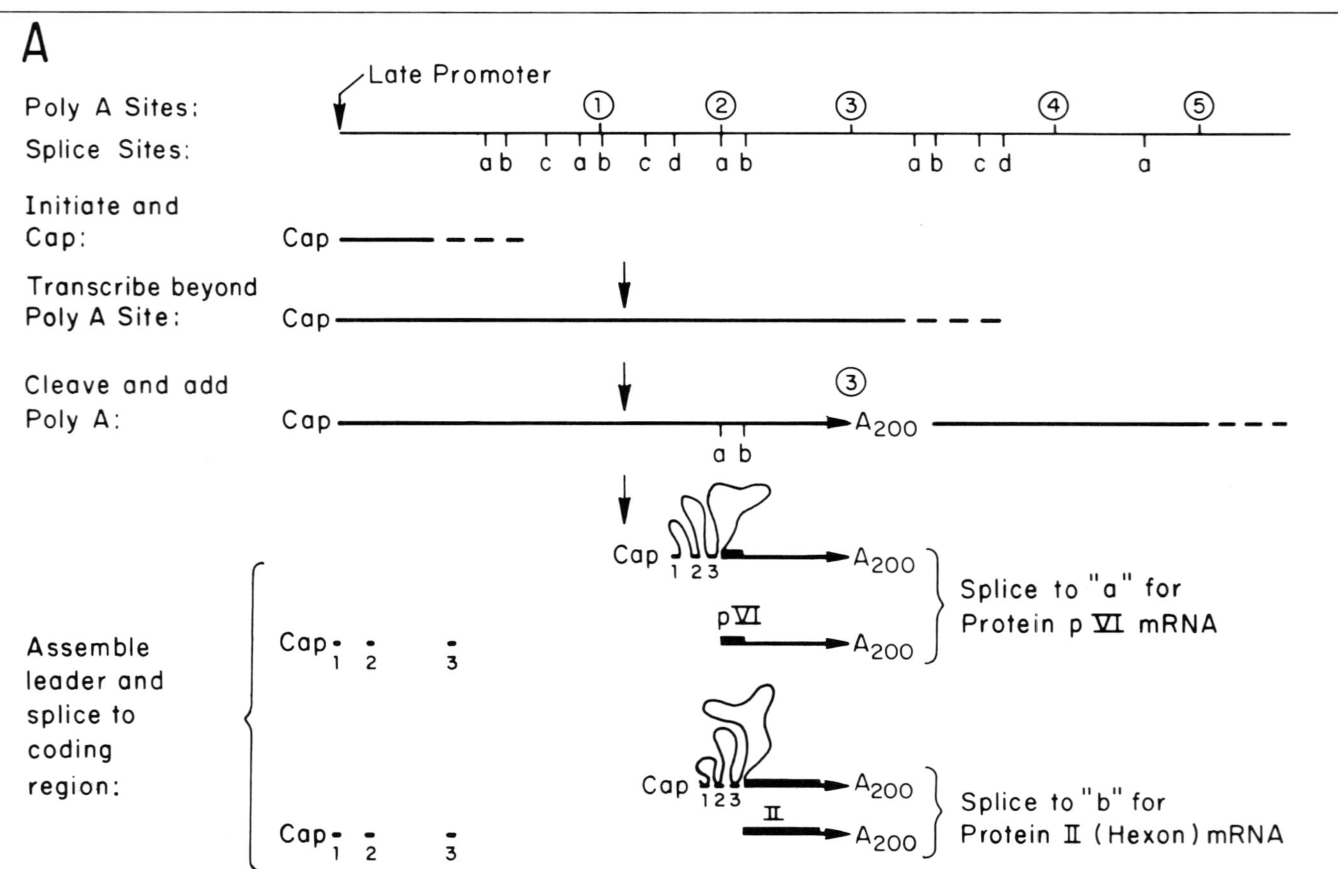

A
Late Promoter
Poly A Sites:
Splice Sites:
Initiate and Cap:
Cap
Transcribe beyond Poly A Site:
Cap
Cleave and add Poly A:
Cap
A_{200}
Assemble leader and splice to coding region:
Cap
Splice to "a" for Protein p VI mRNA
Splice to "b" for Protein II (Hexon) mRNA

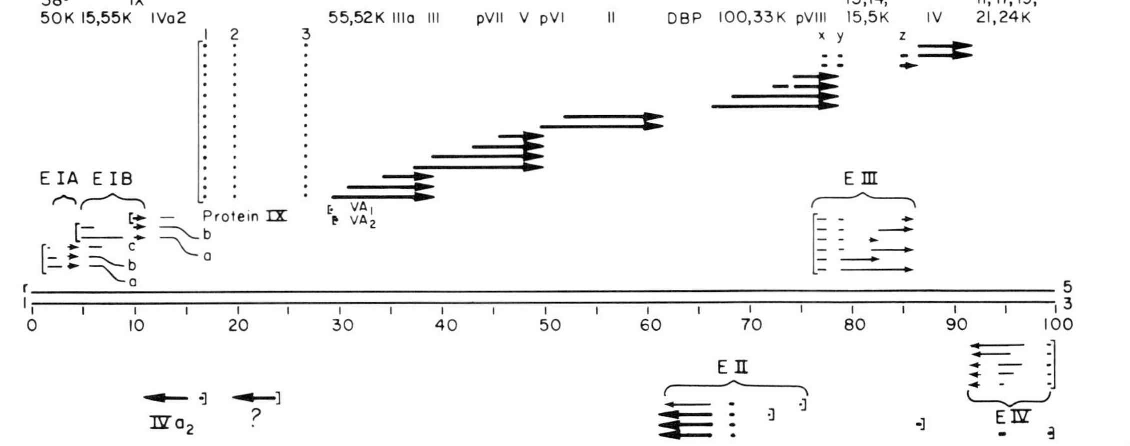

Figure 23. Synthesis of messenger RNA by adenovirus-2. The physical map of Ad-2 is divided into 100 units, each consisting of ~350 nucleotides. RNAs synthesized from left to right utilize the DNA r-strand as template and are 'rightward' transcripts. Those transcribed in the opposite direction are 'leftward'. *A,* Transcription and processing of late messenger RNA. Precursors to late messenger RNA are rightward molecules initiated at the major late promoter and are capped while nascent RNA. Five sites, ① to ⑤, exist for the polyadenylation of these transcripts. However, only one poly(A) site is used for any given transcript, although on average the five sites are used with approximately equal frequency. RNA polymerase transcribes beyond the functional poly(A) site, ③ in this example, and continues until it reaches the far end of the genome at coordinate 100. Soon after the polymerase passes ③, while the RNA is still a nascent molecule, the transcript is cleaved to expose the poly(A) acceptor. RNA sequences encoded at coordinates 16, 19 and 27 are spliced together to form a non-coding tripartite 5′ leader. This leader is subsequently spliced adjacent to a coding sequence. Several alternatives are possible for this final splicing step when any particular poly(A) site has been utilized, for example joining the leader to *a* or to *b*. In the example shown, splicing to *a* juxtaposes the leader to the coding sequences for the pVI protein, while splicing to *b* splices the leader adjacent to the hexon coding sequence. The mRNA produced by splicing to *a* contains the coding sequences for both pVI and hexon. However, eukaryotic ribosomes do not re-initiate translation at internal positions, and translation of this multicistronic mRNA is confined to the first cistron, pVI. Hexon is translated from mRNA with leader spliced to *b*. Two 'choices', selection of the poly(A) site and the leader splicing site, therefore govern the mRNA species which is produced from the primary transcript. This processing pathway allows 14 or more different proteins to be encoded within a single transcription unit, and places their expression under the control of one promoter. *B,* Adenovirus-2 messenger RNA maps. Early messenger RNAs fall within five separate transcription units, EIA-EIV. Each early transcription unit encodes multiple species which differ in the positions of their internal splices. The mRNAs encoding the proteins IX and IVa2 are expressed at an intermediate stage of infection. Species from the major late transcription unit fall into five 3′ co-terminal families with all members of a given family sharing a common poly(A) site. The mRNAs from this transcription unit also share a common 5′-capped tripartite leader. For mRNAs within a 3′-co-terminal family, this leader is spliced adjacent to different coding sequences. Other leader sequences, *x, y* and *z,* appear in some fibre (protein IV) mRNA molecules. The protein translation products of various mRNAs are aligned with their genes. The VA RNAs are RNA polymerase III products.
(Ziff 1980. Reprinted by permission from *Nature* 287, Fig. 1, p. 492. Copyright © 1980 Macmillan Journals Limited.)

Synthesis of early mRNA

At least five separate transcription units for early mRNA species have been described. Four of these regions are as follows (see figure 23):

EI region: EIA and EIB transcription units at the left end of the adenovirus genome encode multiple mRNA species with common 5′ and 3′ termini but with different internal arrangements of their splices. These two transcription units encode functions capable of cell transformation. The EIA region codes for two 13 S mRNA species (EIA-a and EIA-b) (figure 23B). These mRNA species are rapidly transported to the cytoplasm and have a short half-life. The EIB region codes for two species: 22 S mRNA EIB-a and 13 S mRNA EIB-b. This region may encode the Ad-2-specific tumor antigen. The EIA and EIB regions are also transcribed during late stages of infection: EIA-c, a new 9 S mRNA species, is transcribed from EIA. The amount of EI-b increases dramatically in the cytoplasm by 50 to 100-fold.

The *EII region* is transcribed, starting in coordinate 7.5 or 75 (the origin of the leader sequence) into mRNA species that is translated to yield the 72,000 dalton peptide. This peptide is the virus-coded DNA binding protein that participates in virus replication. During the late stage of adenovirus-2 replication, the EII region is transcribed with leaders originating from coordinates 0.72 to 0.86 or 72 to 86.

The *EIII region* yields mRNAs with complex splicing patterns. Region EIII is located wholly within the major late transcription unit and encodes a glycoprotein that appears on the cell surface.

The *EIV region* yields mRNAs with complex splicing patterns that are initiated at the right terminus of the adenovirus-2 DNA molecule (close to coordinate 1.00 or 100), and transcription is on the 1 strand. The synthesis of mRNA from this region is regulated by the 72,000 dalton DNA binding protein coded for by the EII region.

Intermediate mRNA

The mRNA of Ad-2 protein IX is synthesized later than the early mRNAs but before late mRNA synthesis starts. These mRNA molecules from the fifth transcription unit of adenovirus DNA are unspliced and have a 3′ terminus that coincides with that of EIB mRNAs but with its 5′ end encoded 1800 nucleotides downstream from the EIB cap site at coordinate 9.7 (or 0.097).

Late mRNA

The synthesis of late mRNAs is closely coupled to the onset of DNA replication. The late mRNAs contain a common tripartite 5′ leader encoded at coordinates 0.16, 0.19, and 0.27 on the viral DNA (figures 22 and 23B). At this stage, transport of cellular mRNA to the cytoplasm stops, although polyadenylated cellular RNA continues to be synthesized in the nucleus. The 5′ terminus of the primary late transcript was mapped at coordinates 0.163–0.165, but the promoter is thought to be a TATAAA sequence (the Hogness box). The percursors to mRNAs from the major late transcription unit are large rightward-transcribed nuclear molecules up to 25 kilobases long (Goldbert et al. 1977). The 3′ ends of the mRNAs are encoded at five poly(A) sites that have been mapped at coordinates 0.38, 0.50, 0.62, 0.78, and 0.92 (figure 23). The poly(A) is added to the mRNA molecule prior to splicing. The primary transcripts initiated at coordinate 0.16 extend beyond the five poly(A) sites and proceed to the end of the genome (coordinate 1.00); the 3′ ends are generated by cleavage, and products of cleavage from both the 5′ and 3′ ends have been identified. The adenovirus-2 late 3′ acceptors contain the AAUAAA region of homology found near the poly(A) of cellular mRNAs.

The late mRNAs are organized into five banks of overlapping mRNA molecules (figure 23B). Two posttranscriptional possibilities (figure 23A), probably govern which segment of viral RNA will be transported to the cytoplasm as mRNA: selection of the poly(A) site from one of the five positions followed by selection of the site to which the leader is spliced. These events take place in the nucleus. The splicing polarity is from the 5′ end to the 3′ end. The mechanism of splicing has not been fully established.

Replication of viral DNA

Viral DNA synthesis that marks the beginning of the late phase of virus replication starts about 8 hr after infection and reaches a maximal rate at 15 hr postinfection. With adenoviruses 2 and 5, 90% of the newly synthesized DNA in the nucleus is viral 13 hr after infection. DNA replication starts at or near either end of the viral genome. After initiation, a daughter strand is synthesized in the 5′-to-3′ direction with concomitant displacement of the parental strand of the same polarity (figure 24). Synthesis of the complementary daughter strand is initiated at or near the 3′ end of the displaced parental strand and also proceeds in the 5′-to-3′ direction. Thus the right (r) daughter strand is synthesized from right to left and the left (l) daughter strand is synthesized from left to right (reviewed by Winnacker 1978).

The DNA polymerase responsible for the synthesis of the viral DNA is not known but it is thought that both cellular polymerases α and γ might be involved. In addition, a viral gene product, a polypeptide of 75,000 daltons, was characterized as a DNA binding protein that may be involved in the replication of the viral DNA. This protein was found to be present in the DNA replication complex extracted from nuclei of infected cells (Arens et al. 1977).

A protein with a molecular weight of 80,000 was found to be covalently linked to the 5′ end of the nascent daughter strand of replicating adenovirus DNA (Challberg et al. 1980). This protein may represent a precursor to the 55,000 dalton protein that links the 5′ ends of mature adenovirus DNA. Both

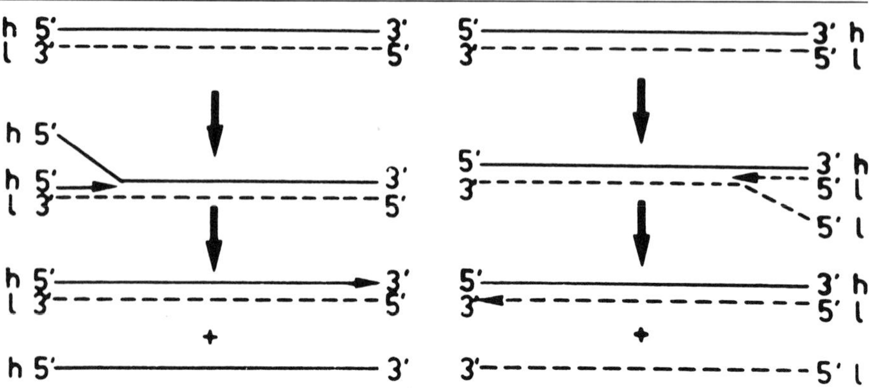

Figure 24. Model of adenovirus DNA replication. Labels "h" and "l" refer to heavy (leftward) and light (rightward) strands of adenovirus DNA, respectively.
(Doerfler 1978. Reprinted by permission from *International Virology* 4, Fig. 1, p. 65. Centre for Agricultural Publishing and Documentation, Wageningen, The Netherlands.)

the 80,000 and 55,000 dalton proteins form similar phosphodiester bonds between the β-OH of a serine residue in the protein and the 5′-OH of the terminal deoxycytidine residue of the viral DNA.

The model of adenovirus DNA replication (Rekosh et al. 1977; Challberg et al. 1980) suggests that the primary initiation event is the formation of an ester linkage between the α-phosphoryl group of dCTP and the β-OH of a serine residue in the 80,000-dalton protein. The 3′-OH of the dCMP residue is then positioned at the end of the genome to serve as a primer for chain elongation.

Assembly of virions

The hexon, which is the major capsid unit, is made up of three 120,000 dalton polypeptides (polypeptide II) and probably assembles in the cytoplasm. The hexons are transported to the nucleus and are assembled into groups of nine (ninemeres). The ninemeres together form the triangular surfaces of the icosahedron. Polypeptide IX may associate with the ninemeres as a cementing substance. Polypeptides VI and VIII are found mainly inside the capsid. The vertices of the icosahedron contain three polypeptides—III, IIIa, and IV—of which III forms the penton base and IV forms the fiber. The hexon ninemeres aggregate into a capsid skeleton. The sequence of assembly is given in figure 25 (Philipson 1979). The viral DNA molecule, associated with two core proteins (V and VII) is inserted into the empty capsid with the help of scaffolding proteins of 32,000 and 40,000 daltons. The newly formed young virions that are not infectious lack polypeptide IVa_2 and the 32K and 40K scaffolding proteins. Finally, after further cleavage of precursor proteins, the mature virion is formed in the cytoplasm.

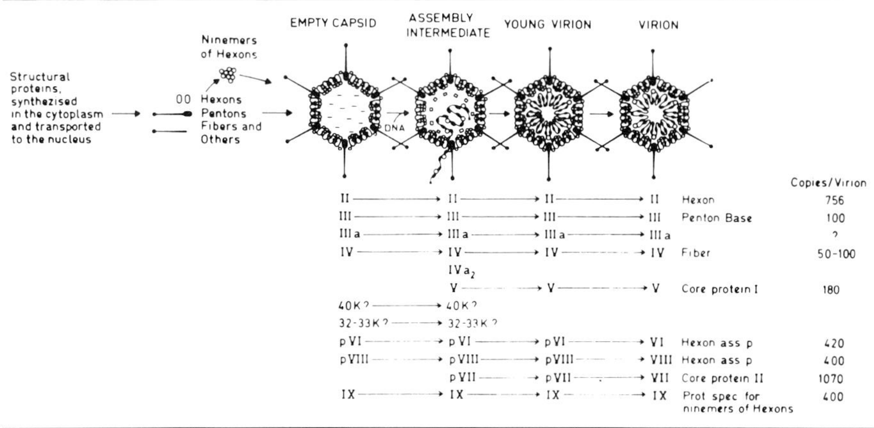

Figure 25. Tentative assembly pathway for adenoviruses. The four stages in virus assembly have been indicated at the top, and the polypeptides present in each structure are indicated below. The number of copies of the polypeptides per virion is indicated to the right.
(Philipson, 1979. Reprinted by permission from *Adv. Virus Res.* 25, Fig. 11, p. 394. Copyright 1979 by Academic Press Inc., New York.)

Prevention of virion formation

Removal of arginine from the medium of infected cells results in an abortive virus infection. Incubation of infected cells at 42°C inhibits viral replication.

Infection of hamster cells in vitro with adeno-12 virus results in an abortive infection (Doerfler 1975). The viral DNA is fragmented by cellular nucleases. Infection of the BHK cell line derived from the kidney of a baby hamster with adenovirus-12 leads to pulverization of the chromosomes. In monkey cells, the virus is capable of replicating the viral DNA, but part of the late viral mRNA is not synthesized. Coinfection of adenovirus infected monkey cells with SV40 helps to overcome the block in adenovirus replication (Lucas and Ginsberg 1972).

Cell transformation

Adenoviruses are also divided into four subgroups on the basis of their oncogenicity for newborn hamsters. Subgroup A is the most, and subgroup B the least, oncogenic. The nononcogenic adenoviruses that can transform cultured rodent cells have been divided into subgroups C and D, based on differences in antigenicity of the tumor (T) antigen. The classification into subgroups A and B is compatible with the subdivision of adenovirus serotypes according to the GC content of their genome, but this does not apply to subgroups C and D (Wadell et al. 1980). Early experiments revealed that rat fibroblasts are nonpermissive for adenovirus. Later it was found that hamster cells are transformed

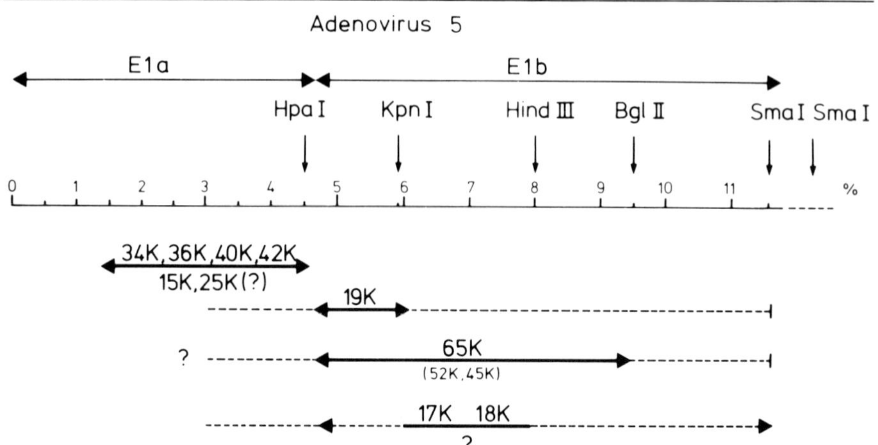

Figure 26. Approximate coding regions of proteins encoded by the transforming segment m of adenovirus 5 DNA (El, extending from 1.5% to 11.5%). The positions for the 17K, 18K, and 25K proteins are still uncertain.
(Van der Eb et al. 1979. Reprinted by permission from *Cold Spring Harbor Symp. Quant. Biol.* 34, Fig. 13, p. 396. Copyright 1980 by The Cold Spring Harbor Laboratory).

by the virus or by fragments of the viral DNA, leading to the conclusion that most probably only part of the viral DNA is required for cell transformation.

Transformed cells produce the viral T antigen. When the in vitro transformed hamster cells are injected into hamsters, a tumor develops and antibodies against the viral T antigen are made. These antibodies in the serum do not affect tumor growth. Virions were never isolated from adenovirus-transformed cells.

Transformation of cells by viral DNA

The genes responsible for cell transformation are localized in the sequences between 0 and 0.10 map units (figures 22, 23, and 26) and comprise 7–10% of the entire adenovirus genome. In adenovirus 5, 7.5% of the genome (0 to 0.075 map units) constitutes the transforming DNA. It is possible to transform rat-kidney cells by transfection of these cells with the 0 to 0.075 map units viral DNA. In the transformed cells, the viral T antigen is synthesized. Further studies revealed that the sequences between 0 and 0.045 map units are capable of transforming rat cells, and the cells produce the viral T antigen. However, these cells are less transformed than the cells transformed by the 7.5% left-hand-side fragment of the viral genome.

The mRNA species transcribed from the transforming viral DNA fragment were isolated, and translation of the mRNA in vitro revealed that five polypeptides are coded by the viral DNA fragment with molecular weights of 14,000; 33,000; 35,000; 38,000; and 40,000. An additional polypeptide of

19,000 daltons is coded for by a gene between 0.075 to 0.11 map units. Four of these large polypeptides (given as 34K, 36K, 40K, and 42K in figure 26) resemble each other and might be products of the same gene (Van der Eb et al. 1979).

Analysis of cells transformed by adenovirus-2 revealed cell lines that contained the complete EcoRI-B fragment of the viral DNA (which codes for the 72,000 dalton DNA-binding protein) integrated in the cellular DNA. In one cell line, this viral gene was expressed, while in other cell lines the viral DNA was methylated and not expressed (Vardimon and Doerfler 1981).

No evidence for specific integration sites for adenovirus DNA

Analysis of adenovirus-2 and adenovirus-12-transformed hamster cell lines and adenovirus-12-induced hamster and rat tumors and tumor-derived cell lines revealed that the viral DNA is integrated randomly in the transformed cell DNA. Deuring and associates (1981) analyzed the nucleotide sequence at the site of the junction between the integrated viral DNA and the cellular DNA by cloning in a λgtWES·λB′ vector. They used DNA from an adenovirus-12-induced hamster tumor that carries five integrated copies of adenovirus genomes. At the site of junction, the adenovirus-12 DNA sequence was found to be preserved, starting with basepair (bp) 46 of the authentic left end of the viral DNA. The first 82 bp of the recloned fragment are cellular sequences that contain patch-like octa- to undecanucleotide pair homologies to distant sequences in the viral DNA.

THE CHROMOSOMAL SITE OF THE INTEGRATED VIRAL DNA. Chromosome 1 contains the integration site for adenovirus DNA (see chapter 4).

GENETICS OF ADENOVIRUSES. Mutagenization of adenoviruses makes possible the isolation of five classes of mutants: (a) ts mutants; (b) plaque morphology mutants; (c) chemical resistant mutants; (d) mutants in the structural viral proteins, and (e) deletion mutants.

Adeno-SV40 hybrid viruses

Ad7-SV40 hybrid virions (designated $E46^{+}$ or PARA) require coinfection with nonhybrid adenovirus for replication. Human adenovirus-7 underwent 22 subsequent passages in monkey kidney cells. When SV40 virus (chapter 9) was discovered to be present in the monkey kidney cells, two additional passages were performed in the presence of antibodies to SV40 which were added to the culture medium. At passage level 28, the virus was injected into newborn hamsters, and tumors were induced that contained SV40 tumor (T) antigen. Thus it was demonstrated that adenovirus DNA contained the genetic information of SV40. The adeno and SV40 DNAs in the hybrid virions are covalently linked. There are defective Ad2-SV40 hybrids that produce SV40 virions as well as hybrids that do not.

Table 2. Location of adenovirus DNA in adeno-SV40 hybrids

Adeno-SV40 hybrids	Coordinates on the adenovirus DNA genome (0 → 1.0)	
	Beginning	End
Ad2+ND3	0.11	0.17
Ad2+ND1	0.11	0.29
Ad2+ND5	0.11	0.39
Ad2+ND2	0.11	0.43
Ad2+ND4	0.11	0.59

(After Lewis et al. 1974.)

Nondefective adeno-SV40 hybrids are capable of independent replication without helper virions and produce tumors in hamsters. By a complicated procedure of plaque isolation in African green monkey kidney cells—growing up the progeny and testing them for SV40 antigen induction, and then selecting the appropriate SV40 antigen-inducing virions for subsequent plaque isolation—the nondefective hybrid strain designated Ad2 + ND1 was obtained. Four other adeno-SV40 hybrids were isolated (table 2) that differ from each other in adenovirus content. These recombinants contain from 7 to 23.9% of SV40 DNA inserted into adenovirus DNA in three different locations, while 4.5 to 40% of the adenovirus DNA is deleted (Lewis et al. 1974; Doerfler 1975; Lewis 1977).

Diseases in humans caused by adenoviruses

Spread of the virus occurs only from man to man via the respiratory and alimentary routes. Types 1, 2, 5, and 6 are endemic and persist in the tonsils and adenoids after infecting most children by the age of three.

Acute respiratory infections are caused by adenovirus strains 3, 4, and 7, and to a lesser extent, types 14 and 21. The pharyngeal infection is accompanied by headache, cough, elevated temperature, and lymphoadenopathy. These strains can also cause conjunctivitis in children. Strains 2, 11, and 21 are associated with hemorrhagic cystitis and adeno-7 is associated with meningoencephalitis.

ADENOVIRUS TYPE 7 CAN CAUSE EPIDEMIC OUTBREAKS OF RESPIRATORY DISEASE IN DIFFERENT PARTS OF THE WORLD. A live vaccine against adenovirus-7 has been developed and is being used among military recruits in the United States. One hundred sixty-four isolates of adenovirus-7 were analyzed by restriction endonuclease analysis of the DNA to evaluate their distribution and pathogenicity. Four distinct genome types were demonstrated: the Ad7 prototype, the Ad7a vaccine strain, Ad7b, and Ad7c. Analysis of the relative occurrence of the four genome types showed that Ad7c was isolated in the Netherlands from 1958–1969, while Ad7b has been recovered from 1970.

Ad7c was isolated in Sweden up to 1972, and afterward only Ad7b strains were detected. The Ad7 prototype and Ad7a genome type account for less than 5% of the analyzed isolates and are probably less virulent than the Ad7b and Ad7c genome types. (Wadell, G.; de Jong, J.C.; and Wolontis, S., Abstracts of the Fifth International Congress of Virology, 1981, p. 306.)

BIBLIOGRAPHY

Arens, M.; Yamashita, R.; Padmanabhan, R.; Tsuruo, T.; and Green, M. Adenovirus deoxyribonucleic acid replication. Characterization of the enzyme activities of a soluble replication system. *J. Biol. Chem.* *252*:7947–7954, 1977.

Blanchard, J.-M.; Weber, J.; Jelinek, W.; and Darnell, J.E. In vitro RNA-RNA splicing in adenovirus 2 mRNA formation. *Proc. Natl. Acad. Sci. USA* *75*:5344–5348, 1978.

Challberg, M.D.; Desiderio, S.V.; and Kelly, T.J. Adenovirus DNA replication in vitro: Characterization of a protein covalently linked to nascent DNA strands. *Proc. Natl. Acad. Sci. USA* *77*:5105–5109, 1980.

Deuring, R.; Winterhoff, U.; Tamanoi, F.; Stahel, S.; and Doerfler, W. Site of linkage between adenovirus type 12 and cell DNAs in hamster tumour line CLAC3. *Nature* *293*:81–85, 1981.

Doerfler, W. Integration of viral DNA into the host genome. *Curr. Top. Microbiol. Immunol.* *71*:1–78, 1975.

Everitt, E.; Lutter, L.; and Philipson, L. Structural proteins of adenoviruses. XII. Location and neighbor-relationship among proteins of adenovirion type 2 as revealed by enzymatic iodination, immunoprecipitation and chemical cross-linking. *Virology* *67*:197–208, 1975.

Goldberg, S.; Weber, J.; and Darnell, J.E. The definition of a large viral transcription unit late in Ad2 infection of HeLa cells: mapping by effects of ultraviolet irradiation. *Cell* *10*:617–621, 1977.

Kitchingman, G.R., and Westphal, H. The structure of adenovirus 2 early nuclear and cytoplasmic RNAs. *J. Mol. Biol.* *137*:23–48, 1980.

Lewis, A.M. Defective and nondefective Ad2-SV40 hybrids. *Prog. Med. Virol.* *23*:96–139, 1977.

Lewis, A.M.; Breeden, J.H.; Wewerka, Y.L.; Schnipper, L.E.; and Levine, H.S. Studies of hamster cells transformed by adenovirus 2 and the nondefective Ad2-SV40 hybrids. *Cold Spring Harbor Symp. Quant. Biol.* *39*:651–656, 1974.

Lucas, J.L., and Ginsberg, H.S. Transcription and transport of virus-specific ribonucleic acids in African green monkey kidney cells abortively infected with type 2 adenovirus. *J. Virol.* *10*:1109–1117, 1972.

Philipson, L. Adenovirus proteins and their messenger RNAs. *Adv. Virus Res.* *25*:357–405 (1979).

Rekosh, D.M.K.; Russell, W.C.; Bellet, A.J.D.; and Robinson, A.J. Identification of a protein linked to the ends of adenovirus DNA. *Cell* *11*:283–295, 1977.

Söderlund, H.; Pettersson, U.; Vennström; and Philipson, L. A new species of virus-coded low molecular weight RNA from cells infected with adenovirus type 2. *Cell* 7:585–593, 1976.

Tibbetts, C., and Pettersson, U. Complementary strand-specific sequences from unique fragments of adenovirus type 2 DNA for hybridization-mapping experiments. *J. Mol. Biol.* *88*:767–784, 1974.

Van der Eb, A.J.; van Ormandt, H.; Schrier, P.I.; Lupker, J.H.; Jochemsen, H.; van den Elsen, P.J.; Deheys, R.J.; Maat, J.; van Beveren, C.P.; Dijkema, R.; and de Waard, A. Structure and function of the transforming genes of human adenoviruses and SV40. *Cold Spring Harbor Symp. Quant. Biol.* *44*:383–399, 1979.

Vardiman, L., and Doerfler, W. Patterns of integration of viral DNA in adenovirus type 2-transformed hamster cells. *J. Mol. Biol.* *147*:227–246, 1981.

Wadell, G.; Hammarskjöld, M.-L.; Winberg, G.; Varsanyi, T.M.; and Sundell, G. Genetic variability of adenoviruses. *Ann. NY Acad. Sci.* *354*:16–42, 1980.

Winnacker, E.L. Adenovirus DNA: structure and function of a novel replicon. *Cell* *14*:761–773, 1978.

Ziff, E.B. Transcription and RNA processing by the DNA tumour viruses. *Nature* *287*:491–499, 1980.

RECOMMENDED READING

Berns, K.O. Molecular biology of the adeno-associated viruses. *Curr. Top. Microbiol. Immunol.* *65*:1–20, 1974.

Darnell, J. Transcription units for mRNA production in eukaryotic cells and their DNA viruses. *Prog. Nucleic Acid Res. Mol. Biol.* *22*:327–353, 1979.

Ginsberg, H.S., and Young, C.S.H. Genetics of adenoviruses. *In: Comprehensive Virology* (H. Fraenkel-Conrat and R.R. Wagner, eds.), Plenum Press, New York, 1977, Vol. 9, pp. 27–88.

Philipson, L., and Lindberg, U. Reproduction of adenoviruses. *In: Comprehensive Virology* (H. Fraenkel-Conrat and R.R. Wagner, eds.), Plenum Press, New York, 1974, Vol. 3, pp. 143–228.

Rogers, E. In pursuit of the transforming genes of adenoviruses. *Nature* *251*:668–669, 1974.

Ross, S.R.; Levine, A.J.; Galos, R.S.; Williams, J.; and Shenk, T. Early viral proteins in HeLa cells infected with adenovirus type 5 host range mutants. *Virology* *103*:475–492, 1980.

Sharp, P.A., and Flint, S.J. Adenovirus transcription. *Curr. Top. Microbiol. Immunol.* *74*:137–166, 1976.

Westphal, H. In-vitro translation of adenovirus messenger RNA. *Curr. Top. Microbiol. Immunol.* *73*:125–139, 1976.

Young, J.F., and Mayor, H.D. Adeno-associated virus—an extreme state of viral defectiveness. *Prog. Med. Virol.* *25*:113–132, 1979.

9. PAPOVAVIRUSES

Family papovaviridae

The virions (figure 27) are nonenveloped with a diameter of 45 nm (polyoma virus) or 55 nm (papilloma virus) and have an icosahedral capsid of 72 capsomeres in a skew arrangement; elongated capsids were also observed. The virion contains one circular, double-stranded DNA molecule of $3–5 \times 10^6$ daltons. Most of the viruses are oncogenic; some can agglutinate erythrocytes by binding to neuraminidase-sensitive receptors.

Genus Papillomavirus

Papillomaviruses infect man, cows, deer, foxes, dogs, goats, sheep, and hamsters
Rabbit (Shope) papillomavirus

Genus Polyomavirus

Polyoma virus (mouse)
BK and JC viruses isolated from human brain
SV40 (simian virus 40) isolated from African green monkey kidney cell cultures
RIKV (rabbits)
K (mice)
SA12 (baboons)

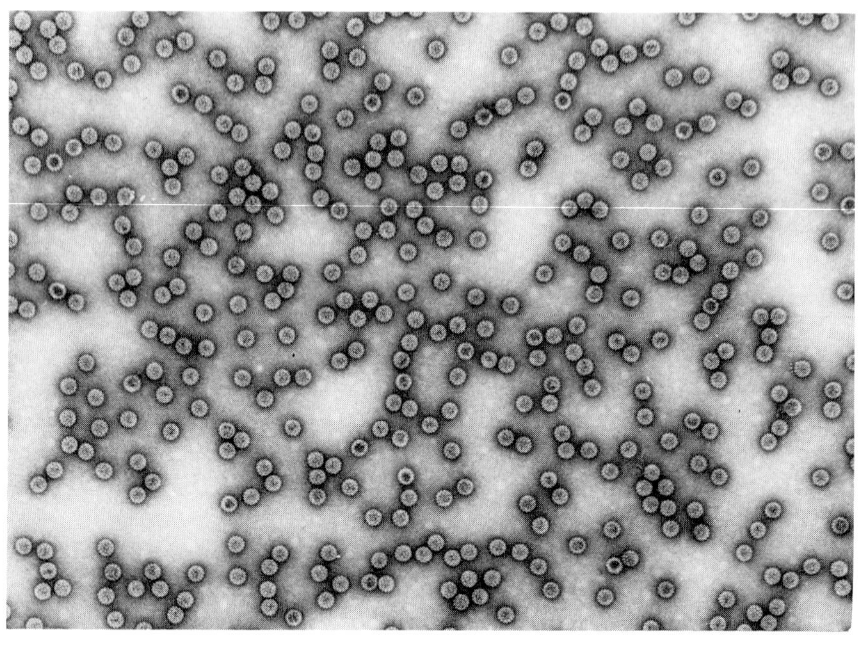

Figure 27. An electron micrograph of virions of human wart virus ($\times$ 60,000) (By courtesy of Dr. Daniel Dekegel, Institut Pasteur du Brabant, Brussels, Belgium.)

Papova stands for: *Pa*pilloma, *Po*lyoma, *Va*cuolating agent (early name for SV40).

The host cell determines the type of infection

Polyomaviruses have a lytic replicative cycle in mouse embryo cells, mouse kidney cells, and 3T3 cell cultures of mouse origin.

SV40 virus lytically replicates in African green monkey kidney cultures or lines derived from monkey kidney cells like BSC-1, CV-1, or Vero cells. Other cells are not permissive for these viruses.

The organization of the viral DNA

The viral genome is a circular DNA molecule of 5×10^6 daltons in papilloma viruses and 3×10^6 daltons in polyomaviruses. The naked viral DNA genome is infectious and transforming for cells.

Simian Virus 40

Molecular conformation of viral DNA

Extraction of viral DNA from SV40 virions and centrifugation in neutral sucrose gradients revealed that the intact double-stranded DNA molecules are superhelical with a sedimentation coefficient of 21 S. Each Form I DNA molecule has 19 helical turns. Form II viral DNA consists of nicked relaxed circular DNA molecules with a sedimentation coefficient of 16 S. Linear double-stranded DNA molecules can also be extracted from virions. These are fragments of cellular DNA that were introduced into the viral capsids (pseudovirions) and have a sedimentation coefficient of 11–15 S. The circular viral DNA molecule has a molecular weight of 3.2–3.6 $\times$ 10^6 and contains 4,800–5,500 nucleotide pairs that can code for proteins with a total molecular weight of 160,000–180,000. The G + C content in the viral DNA is 41%, close to that of the cellular DNA.

It is possible to separate Forms I and II of SV40 DNA by treatment with ethidium bromide, which is able to intercalate with the two DNA forms in different amounts. Since the supercoil in Form I DNA binds more ethidium bromide, the density is increased and the two types of molecules can be separated in density gradients (Levine et al. 1976).

Strand separation and 5′ → 3′ orientation in the DNA strands

It is possible to separate the two strands of the circular viral DNA by hybridization of the denatured DNA strands to mRNA transcribed by *E. coli* RNA polymerase that transcribes RNA from one DNA strand only. The DNA strand that hybridizes to mRNA is called the *minus* strand or the E (early) strand. The viral DNA strand that does not hybridize to this viral RNA is the *plus* strand or the L (late) strand. The E strand codes for the early viral mRNA, while the L strand codes for the late viral mRNA. The RNA:DNA hybrid molecules can be separated from the L strand by chromatography on hydroxylapatite columns that separate between double-stranded and single-stranded molecules. Alkali treatment of the RNA-DNA hybrids allows for isolation of the E strand of the DNA after the degradation of the RNA.

To determine the orientation of the nucleotides in each of the two viral strands, the circular double-stranded DNA molecules are cleaved with EcoRl to yield linear molecules. This DNA is incubated with the reverse transcriptase enzyme that attaches a radioactive (^{32}P-labeled) nucleotide to the 3′ end of the DNA molecule. The DNA is then cleaved with the restriction enzyme HpaI into four fragments that are separated by electrophoresis in agarose gels. The fragments are denatured and hybridized to *E. coli* polymerase synthesized viral RNA. It was found that the radioactive nucleotide was introduced at the ends of the minus strand, which codes for early mRNA. It was concluded that the nucleotides in the minus (E) strands are arranged clockwise from 5′ to 3′, and the plus (L) strand is oriented from 3′ to 5′.

Mapping of SV40 DNA with bacterial restriction enzymes

Kelly and Nathans (1977) utilized bacterial restriction enzymes to map SV40 DNA. The single EcoRl site was used as the zero point of the physical map of

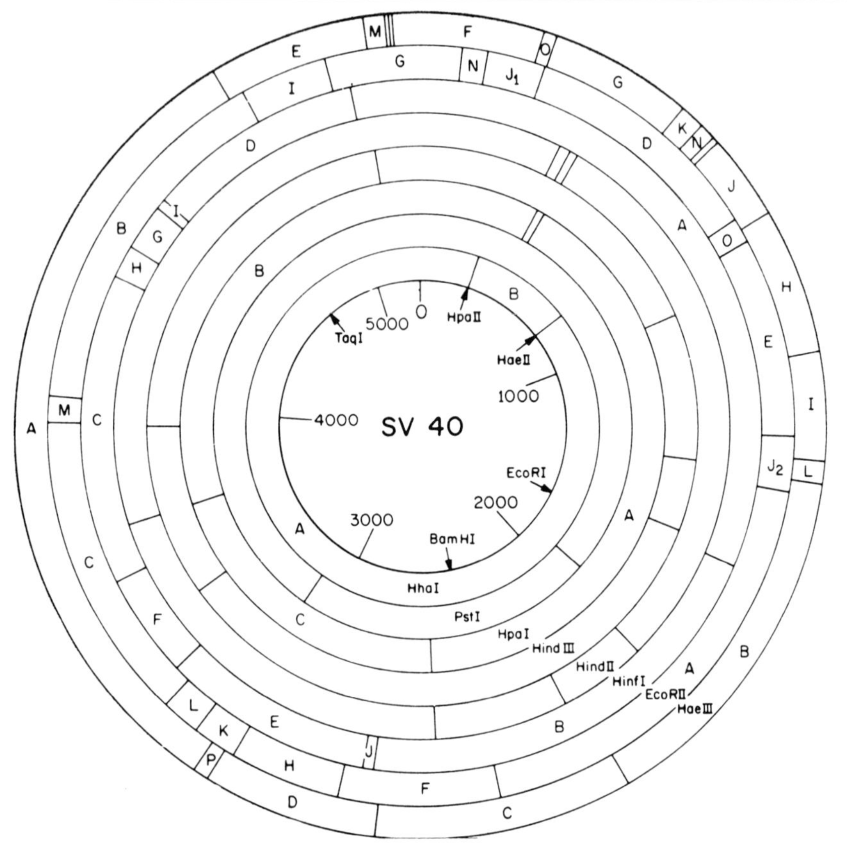

Figure 28. The cleavage sites of restriction enzymes on SV40 DNA.
(Reprinted by permission from *Focus* 1, No. 2, January/February 1978. Copyright Bethesda Research Laboratories, Inc., Rockville, Maryland.)

SV40 DNA. Figure 28 shows the positions of cleavage sites of a number of restriction enzymes on SV40 DNA.

The SV40 DNA genome contains 5,224 base pairs. The restriction enzymes *Hind*II and *Hind*III cleave the DNA into 13 fragments (A to M, figure 28). The cleavage site for the enzyme BglI is located at 0.663 map units: at this site there is a 27-base-pairs palindrome near which the initiation site for DNA synthesis is located. DNA synthesis starts at 0.67 map units, continues bidirectionally, and terminates at 0.17 map units. The viral DNA is divided into the early region between 0.67 and 0.17 map units in a counterclockwise direction. This sequence codes for both the T and t antigens. The second half of the DNA which codes for the late viral functions is arranged in a clockwise orientation to produce structural proteins VP_1, VP_2, and VP_3, as shown in figure 29.

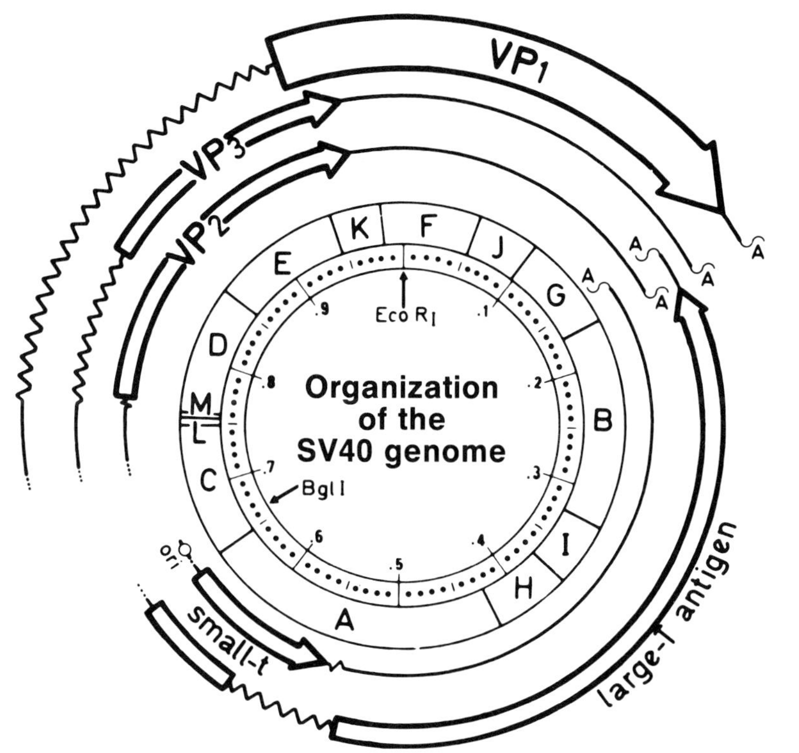

Figure 29. Standard physical map of SV40 DNA and localization of the main biological functions. The single cleavage site of the restriction enzyme EcoRI is used as a reference point for the physical map (inner circle). The next circle shows the position of the HindII and III restriction fragments A to M.

(Fiers et al. 1978. Reprinted by permission from *Nature* 273, Fig. 1, p. 114. Copyright © 1978 Macmillan Journals Limited.)

The nucleotide sequence in SV40 DNA

W. Fiers and collaborators sequenced the nucleotides in the entire SV40 DNA molecule (figure 30). Only 15.2% of the base sequences do not code for proteins; the rest of the genome codes for the T and t antigens and the structural proteins, the sequence of which was also defined.

Fiers's analysis of the SV40 DNA revealed that the t antigen is made by the nucleotide sequence at 0.67 to 0.55 map units on the complete early mRNA molecule, which spans from 0.67 to 0.15 map units. The T antigen is coded for by the early mRNA that has a deletion in the sequences 0.60 to 0.54 map units. It is assumed that the two nucleotide sequences (0.67 to 0.60 and 0.54 to 0.15 map units) of the mRNA are spliced to each other.

The late mRNA for VP_2/VP_3 starts at 0.77 and ends at 0.97 map units (in a

Early region

CGCCTCGGCCTCTGAGCTATTCCAGAAGTAGTGAGGAGGCTTTTTTGGAGGCCTAGGCTTT 60

Hind C ↓ A

TGCAAAAAGCTTTGCAAAG**ATG**GAT.AAA.GTT.TTA.AAC.AGA.GAG.GAA.TCT.TTG.CAG.CTA.AT 120

G.GAC.CTT.CTA.GGT.CTT.GAA.AGG.AGT.GCC.TGG.GGG.AAT.ATT.CCT.CTG.ATG.AGA.AAG.GCA.TA 180

T.TTA.AAA.AAA.TGC.AAG.GAG.TTT.CAT.CCT.GAT.AAA.GGA.GGA.GAT.GAA.GAA.AAA.ATG.AAG.AA 240

A.ATG.AAT.ACT.CTG.TAC.AAG.AAA.ATG.GAA.GAT.GGA.GTA.AAA.TAT.GCT.CAT.CAA.CCT.GAC.TT 300

T.GGA.GGC.TTC.TGG.GAT.GCA.ACT.GAG.GTA.TTT.GCT.TCT.TCC.TTA.AAT.CCT.GGT.GTT.GAT.GC 360

A.ATG.TAC.TGC.AAA.CAA.TGG.CCT.GAG.TGT.GCA.AAG.AAA.ATG.TCT.GCT.AAC.TGC.ATA.TGC.TT 420

G.CTG.TGC.TTA.CTG.AGG.ATG.AAG.CAT.GAA.AAT.AGA.AAA.TTA.TAC.AGG.AAA.GAT.CCA.CTT.GT 480

G.TGG.GTT.GAT.TGC.TAC.TGC.TTC.GAT.TGC.TTT.AGA.ATG.TGG.TTT.GGA.CTT.GAT.CTT.TGT.GA 540

A.GGA.ACC.TTA.CTT.CTG.TGG.TGT.GAC.ATA.ATT.GGA.CAA.ACT.ACC.TAC.AGA.GAT.TTA.AAG.CT 600

C**TAA**GGTAAATATAAAATTTTTAAGTGTATAATGTGTTAAACTACTGATTCTAATTGTTT 660

GTGTATTTTAG.ATT.CCA.ACC.TAT.GGA.ACT.GAT.GAA.TGG.GAG.CAG.TGG.TGG.AAT.GCC.TTT.A 720

AT.GAG.GAA.AAC.CTG.TTT.TGC.TCA.GAA.GAA.ATG.CCA.TCT.AGT.GAT.GAT.GAG.GCT.ACT.GCT.G 780

AC.TCT.CAA.CAT.TCT.ACT.CCT.CCA.AAA.AAG.AAG.AGA.AAG.GTA.GAA.GAC.CCC.AAG.GAC.TTT.C 840

CT.TCA.GAA.TTG.CTA.AGT.TTT.TTG.AGT.CAT.GCT.GTG.TTT.AGT.AAT.AGA.ACT.CTT.GCT.TGC.T 900

TT.GCT.ATT.TAC.ACC.ACA.AAG.GAA.AAA.GCT.GCA.CTG.CTA.TAC.AAG.AAA.ATT.ATG.GAA.AAA.T 960

AT.TCT.GTA.ACC.TTT.ATA.AGT.AGG.CAT.AAC.AGT.TAT.AAT.CAT.AAC.ATA.CTG.TTT.TTT.CTT.A 1020

CT.CCA.CAC.AGG.CAT.AGA.GTG.TCT.GCT.ATT.AAT.AAC.TAT.GCT.CAA.AAA.TTG.TGT.ACC.TTT.A 1080

GC.TTT.TTA.ATT.TGT.AAA.GGG.GTT.AAT.AAG.GAA.TAT.TTG.ATG.TAT.AGT.GCC.TTG.ACT.AGA.G 1140

AT.CCA.TTT.TCT.GTT.ATT.GAG.GAA.AGT.TTG.CCA.GGT.GGG.TTA.AAG.GAG.CAT.GAT.TTT.AAT.C 1200

Hind A ↓ H

CA.GAA.GAA.GCA.GAG.GAA.ACT.AAA.CAA.GTG.TCC.TGG.AAG.CTT.GTA.ACA.GAG.TAT.GCA.ATG.G 1260

AA.ACA.AAA.TGT.GAT.GAT.GTG.TTG.TTA.TTG.CTT.GGG.ATG.TAC.TTG.GAA.TTT.CAG.TAC.AGT.T 1320

TT.GAA.ATG.TGT.TTA.AAA.TGT.ATT.AAA.AAA.GAA.CAG.CCC.AGC.CAC.TAT.AAG.TAC.CAT.GAA.A 1380

AG.CAT.TAT.GCA.AAT.GCT.GCT.ATA.TTT.GCT.GAC.AGC.AAA.ACC.AAA.AAA.CCA.TAT.GCC.AAC 1440

AG.GCT.GTT.GAT.ACT.GTT.TTA.GCT.AAA.AAG.CGG.GTT.GAT.AGC.CTA.CAA.TTA.ACT.AGA.GAA.C 1500

Hind H ↓ I

AA.ATG.TTA.ACA.AAC.AGA.TTT.AAT.GAT.CTT.TTG.GAT.AGG.ATG.GAT.ATA.ATG.TTT.GGT.TCT.A 1560

CA.GGC.TCT.GCT.GAC.ATA.GAA.GAA.TGG.ATG.GCT.GGA.GTT.GCT.TGG.CTA.CAC.TGT.TTG.TTG.C 1620

CC.AAA.ATG.GAT.TCA.GTG.GTG.TAT.GAC.TTT.TTA.AAA.TGC.ATG.GTG.TAC.AAC.ATT.CCT.AAA.A 1680

AA.AGA.TAC.TGG.CTG.TTT.AAA.GGA.CCA.ATT.GAT.AGT.GGT.AAA.ACT.ACA.TTA.GCA.GCT.GCT.T 1740

Hind I ↓ B

TG.CTT.GAA.TTA.TGT.GGG.GGG.AAA.GCT.TTA.AAT.GTT.AAT.TTG.CCC.TTG.GAC.AGG.CTG.AAC.T 1800

TT.GAG.CTA.GGA.GTA.GCT.ATT.GAC.CAG.TTT.TTA.GTA.GTT.TTT.GAG.GAT.GTA.AAG.GGC.ACT.G 1860

GA.GGG.GAG.TCC.AGA.GAT.TTG.CCT.TCA.GGT.CAG.GGA.ATT.AAT.AAC.CTG.GAC.AAT.TTA.AGG.G 1920

AT.TAT.TTG.GAT.GGC.AGT.GTT.AAG.GTA.AAC.TTA.GAA.AAG.AAA.CAC.CTA.AAT.AAA.AGA.ACT.C 1980

AA.ATA.TTT.CCC.CCT.GGA.ATA.GTC.ACC.ATG.AAT.GAG.TAC.AGT.GTG.CCT.AAA.ACA.CTG.CAG.G 2040

```
C C.A G A.T T T.G T A.A A A.C A A.A T A.G A T.T T T.A G G.C C C.A A A.G A T.T A T.T T A.A A G.C A T.T G C.C T G.G A A.C
          2050            2060            2070            2080            2090            2100
G C.A G T.G A G.T T T.T T G.T T A.G A A.A A G.A G A.A T A.A T T.C A A.A G T.G G C.A T T.G C T.T T G.C T T.C T T.A T G.T
          2110            2120            2130            2140            2150            2160
T A.A T T.T G G.T A C.A G A.C C T.G T G.G C T.G A G.T T T.G C T.C A A.A G T.A T T.C A G.A G C.A G A.A T T.G T G.G A G.T
          2170            2180            2190            2200            2210            2220
G G.A A A.G A G.A G A.T T G.G A C.A A A.G A G.T T T.A G T.T T G.T C A.G T G.T A T.C A A.A A A.A T G.A A G.T T T.A A T.G
          2230            2240            2250            2260            2270            2280
T G.G C T.A T G.G G A.A T T.G G A.G T T.T T A.G A T.T G G.C T A.A G A.A A C.A G T.G A T.G A T.G A T.G A T.G A A.G A C.A
          2290            2300            2310            2320            2330            2340
G C.C A G.G A A.A A T.G C T.G A T.A A A.A A T.G A A.G A T.G G T.G G G.G A G.A A G.A A C.A T G.G A A.G A C.T C A.G G G.C
          2350            2360            2370            2380            2390            2400
A T.G A A.A C A.G G C.A T T.G A T.T C A.C A G.T C C.C A A.G G C.T C A.T T T.C A G.G C C.C C T.C A G.T C C.T C A.C A G.T
          2410            2420            2430            2440            2450            2460
C T.G T T.C A T.G A T.C A T.A A T.C A G.C C A.T A C.C A C.A T T.T G T.A G A.G G T.T T T.A C T.T G C.T T T.A A A.A A A.C
          2470            2480            2490            2500            2510            2520
                                                                                  Hind B ↓
C T.C C C.A C A.C C T.C C C.C C T.G A A.C C T.G A A.A C A[T A A]A A T G A A T G C A A T T G T T G T T G T T ...3'
          2530            2540            2550            2560            2570
```

Figure 30. Nucleotide sequence of the early region. Only the strand with the same polarity as the early mRNA is shown (in a 5′ to 3′ orientation). The sequence starts at the centre of the 27-base-pair-long palindrome at position 0.663 on the standard map (*Bgl*I site). The end is the junction between the *Hind* fragments B and G. The different *Hind* fragments are indicated. The initiation codon for t and T antigens is boxed, as are the respective termination codons. The reading frame for translation is indicated by dots.

(Fiers et al. 1978. Reprinted by permission from *Nature* 273, Fig. 2, p. 115. Copyright © 1978 Macmillan Journals Limited.)

clockwise direction), while the VP_1 mRNA starts at 0.94 and ends at 0.17 map units (also in a clockwise direction). To each mRNA molecule an RNA sequence transcribed from the sequence between 0.72 and 0.76 map units (clockwise) is added. This RNA sequence is the leader nucleotide sequence that is spliced to the 5′ end of the mRNA. The mRNA molecules that are translated to the VP_2/VP_3 structural proteins have a sedimentation coefficient of 19 S in sucrose gradients and the mRNA molecules for VP_1, 16 S.

Thus the SV40 DNA contains one early gene that codes for the t and T antigens and a late gene that codes for the three viral capsid proteins. A special RNA leader sequence is transcribed from a distinct sequence in the DNA and is spliced to the mRNA molecules.

SV40 DNA—a minichromosome

Extraction of the virion DNA under mild conditions allows for the isolation of a minichromosome; the viral DNA is isolated together with histones that are bound to it in the form of nucleosomes, like the nucleosomal structure of the cellular chromosomes. The histones bound to viral DNA were identified as the cellular histones $F2_{a1}$, $F2_{a2}$, $F2_b$, and F3.

When SV40 contains cellular DNA

Consecutive passage of SV40 at a high multiplicity of infection in African green monkey kidney cells (Vero cells) allows for the isolation of viruses containing DNA from which 13% of the viral DNA sequences were deleted and replaced by 7–12% of cellular DNA sequences. These viruses are defective and cannot replicate. Analysis of the DNA of these viruses revealed that the site for initiation of DNA synthesis was conserved (Brockman 1977).

Virion-associated proteins

Three structural proteins compose the viral capsid: VP_1 - 43–48 × 10^3 daltons, VP_2 – 30–38 × 10^3 daltons, and VP_3 – 20–23 × 10^3 daltons. The cellular histones bound to the viral DNA are $F2_{a1}$ with a molecular weight of 10–12 × 10^3; $F2_{a2}$ – 12–14 × 10^3; $F2_b$ – 12–14 × 10^3; and F_3 – 14–16 × 10^3.

All the structural capsid proteins are phosphorylated, and the phosphate is bound to the amino acid serine in the peptide chains.

Replicative cycle of SV40

The SV40 virions adsorb to the membrane of the host cell and are engulfed into the cytoplasm where the viral DNA is uncoated and transferred to the nucleus, the site of virus replication and assembly.

EARLY VIRUS FUNCTIONS: EARLY TRANSCRIPTION OF VIRAL DNA. Transcription of the early viral gene is initiated at 0.65 map units by the cellular RNA polymerase II. The transcription of the viral DNA is in a counterclockwise direction and 48% of the DNA is transcribed. At the 3′ end of the mRNA molecules a poly(A) sequence is synthesized. The sedimentation coefficient of the early mRNA in a sucrose gradient is 19 S.

Ts mutants were isolated in which the mutations appear in the fragments H (0.42 to 0.37 map units) and I (0.37 to 0.32 map units) and sometimes in the B fragment (figure 29). These ts mutants were classified as group A (Fiers et al. 1978).

The products of the A gene are the T and t antigens

The nucleotide sequence of the gene that codes for the early viral function was determined by Fiers and associates (1978) and its amino acid sequence was also determined (figure 31). The tumor-specific (T) antigen that elicits synthesis of antibodies in hamsters infected with SV40 was isolated and characterized. The T antigen is a protein with a molecular weight of 90,000–100,000 that is phosphorylated to a form that can activate the infected cell. This antigen is involved in the maintenance of the transformed state of the cell. It is a regulatory protein that controls the amount of synthesized protein by regulating the rate of early mRNA transcription.

The carboxy terminal end of the T antigen was mapped in the nucleotide 2549 in the early gene and the genetic information spans from 0.54 to 0.175 map units (counterclockwise direction). This nucleotide sequence is enough to code for a polypeptide with a molecular weight of 72,000, while the T antigen has a molecular weight of 90,000. Mapping of the T antigen as described earlier and not close to the origin of transcription is based on the isolation of a group of nondefective deletion mutants that lacked the nucleotide sequences between 0.54 to 0.59 map units. These deletion mutants produce a normal T antigen.

A group of early polyoma virus mutants described by Benjamin (1972) are host range mutants that replicate only in cells that carry the DNA fragment that is complementary to the defective DNA sequence in the mutants. The mutated gene is thus complemented functionally in the host cell. These virus mutants are not capable of transforming cells, and the mutations were mapped in the early gene of the viral DNA in the sequence mapping between 0.54–0.59 map units. Other virus mutants deleted in the same sequences in the viral DNA are also unable to transform cells.

The protein responsible for the ability of a SV40-transformed cell to attain independent growth was identified as a protein with a molecular weight of 15,000–20,000 and was designated small t antigen (figure 29). This protein is not made by the mutants with the deletion in 0.59–0.54 map units.

The large T and small t proteins have a common antigenicity. All the methionine-containing peptides arising from tryptic digest of t antigen appear in the T antigen when analyzed by two-dimensional electrophoresis. Both polypeptides have the same N terminus. Antigen t starts in nucleotide 80 of the early gene and is read until the UAA codon in position 602. This polypeptide has 174 amino acids and a molecular weight of 20,503. The N terminus is the amino acid methionine and the polypeptide contains 19 lysine residues, 8 arginine, 14 aspartic acid, and 11 residues of glutamic acid; therefore it is a basic protein. It also has sulfur-containing amino acids: 10 methionine residues and 11 cystein residues, mainly at the end of the molecule.

The T antigen starts in the same codon as the t antigen and is identical in sequence up to the middle of the sequence that codes for the t antigen. Afterward, as a result of a splice, the mRNA continues from 0.53 map units to the termination codon UAA in position 2550. In the mRNA coding for the T antigen, the 5′ side of the splice removes the termination codon that terminates the small t antigen mRNA. As a result, the translation of the large T mRNA can continue almost up to the 3′ end of the molecule. The mutant dl 1001 lacking the HindIII fragment H and I (figure 29) codes for a T antigen of 33,000 daltons. Since HindIII fragments H and I code for a polypeptide of 24,000 daltons, it was concluded by Fiers and associates (1978) that 300 amino acids are read from the HindIII A fragment, about 100 amino acids are present also in the t antigen, and 188 are coded for by sequences prior to the junction with HindIII H fragment (0.534–0.426 map units), suggesting that translation continues from nucleotide 672 onward.

T antigen is a DNA-binding protein but not a basic protein. A high proline concentration is present at the carboxy terminal end. The T antigen binds to the SV40 DNA replication origin. Since the 5′ end of the large T antigen is in the sequence to which it binds on the SV40 DNA, the T antigen regulates its own synthesis.

The mRNAs for the T and t antigens have closely mapping or identical 5′ and 3′ termini but differ in their splicing (figure 32). The smaller of the two early mRNAs encodes large T.

Late region

```
G G C C T C G G C C T C T G C A T A A A T A A A A A A A A T T A G T C A G C C A T G G G G C G G A G A A T G G G C G G A A
C T G G G C G G A G T T A G G G G C G G G A T G G G C G G A G T T A G G G G C G G G A C T A T G G T T G C T G A C T A A
T T G A G A T G C A T G C T T T G C A T A C T T C T G C C T G C T G G G G A G C C T G G T T G C T G A C T A A T T G A G
A T G C A T G C T T T G C A T A C T T C T G C C T G C T G G G G A G C C T G G G G A C T T T C C A C A C C C T A A C T G
A C A C A C A T T C C A C A G C T G G T T C T T T C C G C C T C A G A A G G T A C C T A A C C A A G T T C C T C T T T C
A G A G G T T A T T T C A G G C C A T G G C T G C G C C G G C T G T C A C G C C A G G C C T C C G T T A A G G T T C G T
A G G T C A T G G A C T G A A A G T A A A A A A A C A G C T C A A C G C C T T T T T G T G T T T G T T T T A G A G C T T
                                        Hind C ↓ L
T T G C T G C A A T T T T G T G A A G G G G A A G A T A C T G T T G A C G G G A A A C G C A A A A A A C C A G A A A G G
Hind L ↓ M                     Hind M ↓ D
T T A A C T G A A A A A C C A G A A A G T T A A C T G G T A A G T T T A G T C T T T T T G T C T T T T A T T T C A G G T
C C[A T G]G G T.G C T.G C T.T T A.A C A.C T G.T T G.G G G.G A C.C T A.A T T.G C T.A C T.G T G.T C T.G A A.G C T.G C T.G
C T.G C T.A C T.G G A.T T T.T C A.G T A.G C T.G A A.A T T.G C T.G C T.G G A.G A G.G C C.G C T.G C T.G C A.A T T.G A A.G
T G.C A A.C T T.G C A.T C T.G T T.G C T.A C T.G T T.G A A.G G C.C T A.A C A.A C C.T C T.G A G.G C A.A T T.G C T.G C T.A
T A.G G C.C T C.A C T.C C A.C A G.G C C.T A T.G C T.G T G.A T A.T C T.G G G.G C T.C C T.G C T.G C T.A T A.G C T.G G A.T
T T.G C A.G C T.T T A.C T G.C A A.A C T.G T G.A C T.G G T.G T G.A G C.G C T.G T T.G C T.C A A.G T G.G G G.T A T.A G A.T
T T.T T T.A G T.G A C.T G G.G A T.C A C.A A A.G T T.T C T.A C T.G T T.G G T.T T A.T A T.C A A.C A A.C C A.G G A[A T G]G
C T.G T A.G A T.T T G.T A T.A G G.C C A.G A T.G A T.T A C.T A T.G A T.A T T.T T A.T T T.C C T.G G A.G T A.C A A.A C C.T
T T.G T T.C A C.A G T.G T T.C A G.T A T.C T T.G A C.C C C.A G A.C A T.T G G.G G T.C C A.A C A.C T T.T T T.A A T.G C C.A
      Hind D ↓ E
T T.T C T.C A A.G C T.T T T.T G G.C G T.G T A.A T A.C A A.A A T.G A C.A T T.C C T.A G G.C T C.A C C.T C A.C A G.G A G.C
T T.G A A.A G A.A G A.A C C.C A A.A G A.T A T.T T A.A G G.G A C.A G T.T T G.G C A.A G G.T T T.T T A.G A G.G A A.A C T.A
C T.T G G.A C A.G T A.A T T.A A T.G C T.C C T.G T T.A A T.T G G.T A T.A A C.T C T.T T A.C A A.G A T.T A C.T A C.T C T.A
C T.T T G.T C T.C C C.A T T.A G G.C C T.A C A.A T G.G T G.A G A.C A A.G T A.G C C.A A C.A G G.G A A.G G G.T T G.C A A.A
T A.T C A.T T T.G G G.C A C.A C C.T A T.G A T.A A T.A T T.G A T.G A A.G C A.G A C.A G T.A T T.C A G.C A A.G T A.A C T.G
A G.A G G.T G G.A A G.C T C.A A A.G C C.A A A.G T C.C T A.A T G.T G C.A G T.C A G.G T G.A A T.T T A.T T G.A A A.A A T
T T.G A G.G C T.C C T.G G T.G G T.G C A.A A T.C A A.A G A.A C T.G C T.C C T.C A G.T G G.A T G.T T G.C C T.T T A.C T T.C
                                        Hind E ↓ K
T A.G G C.C T G.T A C.G G A.A G T.G T T.A C T.T C T.G C T.C T A.A A A.G C T.T[A T.G]A A.G.A T.G.G C.C.C C.A.A C.A.A A.A.
A G.A.A A.A.G G.A.A G.T.T G.T.C C.A.G G.G.G C.A.G C.T.C C.C.A A.A.A A.A.C C.A.A A.G.G A.A.C C.A.G T.G.C A.A.G T.G.C C.A.
A A.G.C T.C.G T.C.A T.A.A A.A.G G.A.G G.A.A T.A.G A.A.G T.T.C T.A.G G.A.G T[T.A A]A.A C T.G G A.G T A.G A C.A G C.T T C.
A C T.G A G.G T G.G A G.T G C.T T T.T T A.A A T.C C T.C A A.A T G.G G C.A A T.C C T.G A T.G A A.C A T.C A A.A A A.G G C.
        Hind K ↓ F
T T A.A G T.A A A.A G C.T T A.G C A.G C T.G A A.A A A.C A G.T T T.A C A.G A T.G A C.T C T.C C A.G A C.A A A.G A A.C A A.
                         Eco RI ↓
C T G.C C T.T G C.T A C.A G T.G T G.G C T.A G A.A T T.C C T.T T G.C C T.A A T.T T A.A A T.G A G.G A C.T T A.A C C.T G T.
G G A.A A T.A T T.T T G.A T G.T G G.G A A.G C T.G T T.A C T.G T T.A A A.A C T.G A G.G T T.A T T.G G G.G T A.A C T.G C T.
A T G.T T A.A A C.T T G.C A T.T C A.G G G.A C A.C A A.A A A.A C T.C A T.G A A.A A T.G G T.G C T.G G A.A A A.C C C.A T T.
C A A.G G G.T C A.A A T.T T T.C A T.T T T.T T T.G C T.G T T.G G T.G G G.A A C.C T T.T G G.A G C.T G C.A G G.T G T G.
                                                                                               Hind F ↓
T T A.G C A.A A C.T A C.A G G.A C C.A A A.T A T.C C T.G C T.C A A.A C T.G T A.A C C.C C A.A A A.A A T.G C T.A C A.G T T.
```

```
J
G A C.A G T.C A G.C A G.A T G.A A C.A C T.G A C.C A C.A A G.G C T.G T T.T T G.G A T.A A G.G A T.A A T.G C T.T A T.C C A.
G T G.G A G.T G C.T G G.G T T.C C T.G A T.C C A.A G T.A A A.A A T.G A A.A A C.A C T.A G A.T A T.T T T.G G A.A C C.T A C.
A C A.G G T.G G G.G A A.A A T.G T G.C C T.C C T.G T T.T T G.C A C.A T T.A C T.A A C.A C A.G C A.A C C.A C A.G T G.C T T.
                                                                                                         Hind J ↓
C T T.G A T.G A G.C A G.G G T.G T T.G G G.C C C.T T G.T G C.A A A.G C T.G A C.A G C.T T G.T A T.G T T.T C T.G C T.G T T.
G
G A C.A T T.T G T.G G G.C T G.T T T.A C C.A A C.A C T.T C T.G G A.A C A.C A G.C A G.T G G.A A G.G G A.C T T.C C C.A G A.
T A T.T T T.A A A.A T T.A C C.C T T.A G A.A A G.C G G.T C T.G T G.A A A.A A C.C C C.T A C.C C A.A T T.T C C.T T T.T T G.
T T A.A G T.G A C.C T A.A T T.A A C.A G G.A G G.A C A.C A G.A G G.G T G.G A T.G G G.C A G.C C T.A T G.A T T.G G A.A T G.
T C C.T C T.C A A.G T A.G A G.G A G.G T T.A G G.G T T.T A T.G A G.G A C.A C A.G A G.G A G.C T T.C C T.G G G.G A T.C C A.
G A C.A T G.A T A.A G A.T A C.A T T.G A T.G A G.T T T.G G A.C A A.A C C.A C A.A C T.A G A.A T G.C A G[T G A]A A A A A A
T G C T T T A T T T G T G A A A T T T G T G A T G C T A T T G C T T T A T T T G T A A C C A T T A T A A G C T G C A A T
       Hind G ↓
A A A C A A G T T ...3'
```

Figure 31. Nucleotide sequence of the late region. Only the strand with the same polarity as late mRNA is shown (in a 5′ to 3′ orientation). The start and endpoint are the same as in fig. 30. The cleavage sites of *Hind*II and III and *Eco*RI restriction enzymes are shown (see also figure 29). The presumed initiation codon for VP_2, VP_3, and VP_1 are boxed, as well as the termination codons for VP_2/VP_3 and for VP_1. The reading frame in the translated region is indicated by dots (note that the last part of the VP_2/VP_3 gene overlaps the beginning of the VP_1 gene).

(Fiers et al. 1978. Reprinted by permission from *Nature* 273, Fig. 3, p. 116. Copyright © 1978 Macmillan Journals Limited.)

Polyadenylation of the 3′ end of early SV40 mRNA

The polyadenylated 3′ ends of late and early mRNA converge at coordinate 0.17 and overlap (reviewed by Ziff 1980). The RNA polymerase transcribes a polynucleotide chain longer than the mRNA molecule which is cleaved and polyadenylated at the 3′ end of the processed mRNA molecule (figure 32).

Late transcription

The transcription of the late gene starts immediately after the initiation of viral DNA replication that requires an active protein product of the early A gene—namely, the large T antigen. When the synthesis of viral DNA is prevented, capsid proteins are not made.

The 5′ end of the late mRNA maps at the coordinate 0.72, but the exact position is not known. The 3′ end of the mRNA maps in the junction between HindIII G fragment and HindIII B fragment (figure 29).

Two major classes of late mRNA—18 S–19 S coding for VP2 and VP3—and 16 S coding for VP1, are made from viral RNA sequences extending from the cap sites to the poly(A) site synthesized in the infected nuclei. Two alternative forms of 19 S mRNA differ in the splicing of the 5′ leader. In VP1 mRNA, a larger intervening sequence is removed, and the leader is attached to the AUG codon which initiates VP1 translation. This AUG lies within the VP2 gene, but the VP1 and VP2 gene sequences that overlap are read in different reading frames. The VP3 mRNA is not fully characterized but overlaps VP2 mRNA and uses an internal AUG codon of VP2 for initiation of translation (reviewed by Ziff 1980; figure 32).

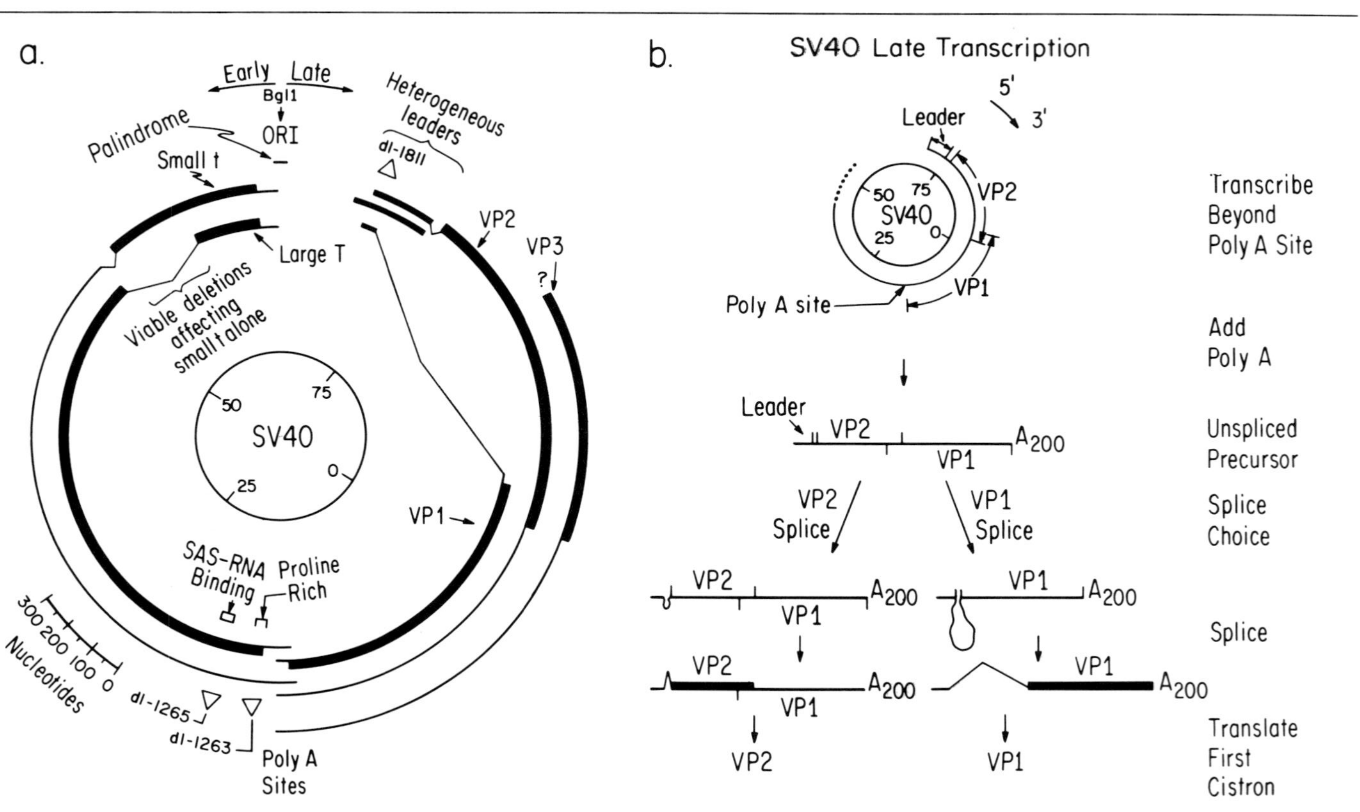
a.
Early
Late
Bgl1
ORI
Palindrome
Small t
Heterogeneous leaders
dl-1811
Large T
VP2
VP3
?
Viable deletions affecting small t alone
SV40
50
75
25
0
VP1
SAS-RNA Binding
Proline Rich
300 200 100 0
Nucleotides
dl-1265
dl-1263
Poly A Sites
b.
SV40 Late Transcription
5'
3'
Leader
VP2
VP1
Poly A site
Transcribe Beyond Poly A Site
Add Poly A
Leader
VP2
VP1
A200
Unspliced Precursor
VP2 Splice
VP1 Splice
Splice Choice
VP2
VP1
A200
VP1
A200
Splice
VP2
VP1
A200
VP1
A200
VP2
VP1
Translate First Cistron

Transcription of polyoma virus DNA

Transcription of the early region of polyoma virus DNA resembles that of SV40. The transcripts are spliced similarly to those of SV40 early mRNA to produce the large T and small t antigens. Polyoma encodes a third early protein with a molecular weight of 55,000 which has an amino terminal similar to that of large T and small t (mol. wt. 100,000 and 22,000, respectively), but a different carboxy terminus. The splice removes the terminating sequence of the small t antigen, using a reading frame different from that of the large T antigen. As a result, the peptide is terminated prior to the 3′ end and is therefore designated as intermediate (or middle) T antigen (figure 33). The middle and possibly small t antigens are important for transformation but not the large T antigen (Novak et al. 1980).

The late region of the viral DNA contains three genes for the three structural capsid proteins VP1, VP2, and VP3 which were mapped by using ts mutants. All the tryptic peptides of VP3 were found in VP2. Since deletion mutants at the beginning of VP2 do not affect the size of VP3, it was concluded that the VP3 initiates within the VP2 gene and is read in the same reading frame. The VP2 gene starts in nucleotide 543, VP3 in nucleotide 897, and they are read in a clockwise direction until the termination codon UAA is in position 1599. The molecular weight of VP2 is 38,533 and VP3 weighs 26,967. These two genes have a 122 nucleotide sequence identical with VP1 (see figure 33) (Ziff 1980). During the lytic infection the amount of late mRNA in the cytoplasm and nucleus is 10–20 times the amount of early mRNA. It is possible to separate the 16 S late mRNA that codes for VP2 and VP3 and the 19 S RNA that codes for VP3.

Figure 32. SV40 messenger RNAs. *A*, Messenger RNA map. The 5′ termini of SV40 mRNAs map near the origin of DNA replication (ORI). ORI contains an inverted repeat palindrome centred on the *Bgl*I restriction endonuclease cleavage site. Early mRNA species are transcribed anticlockwise on this map, and late clockwise. The polyadenylated 3′ ends of early and late mRNAs overlap. Two early mRNAs, which differ in the positions of their internal splices, encode, respectively, the SV40 tumour antigens small t and large T, which are amino co-terminal. The T mRNA lacks the translational terminator sequence for the t protein and T translation continues essentially to the 3′ end of the mRNA. Late mRNA species encode the structural proteins VP1, VP2 and VP3. VP2 and VP3 are carboxy co-terminal, and both differ from VP1. A single DNA sequence, however, encodes the amino terminus of VP1 and the carboxy terminus of VP2 and VP3, but two different reading frames are used. The leaders spliced to the coding regions of the late mRNAs are heterogeneous, with multiple 5′ termini. The VP3 splicing pattern is not well established. The effects of DNA deletions, the presence of a SV40 associated small RNA (SAS RNA) and possible functions of the proline-rich carboxy terminus of T are discussed in the text. *B*, Pathway of synthesis of SV40 late mRNAs. Newly formed late nuclear transcripts are molecules which extend from the cap site(s) beyond the poly(A) site, but are shorter than the full length of the genome. These primary transcripts are apparently cleaved and polyadenylated at the poly(A) site to yield unspliced molecules with the 5′ and 3′ ends of mature mRNA. As with Ad-2 late mRNAs (Fig. 23), multiple pathways exist for splicing this precursor to allow synthesis of a family of 3′-co-terminal mRNAs. In the example given here, the alternatives of splicing the leader adjacent to either the VP2 or the VP1 coding sequences are shown. The cistron closest to the cap is translated.

(Ziff 1980. Reprinted by permission from *Nature* 287, Fig. 2, p. 494. Copyright © 1980 Macmillan Journals Limited.)

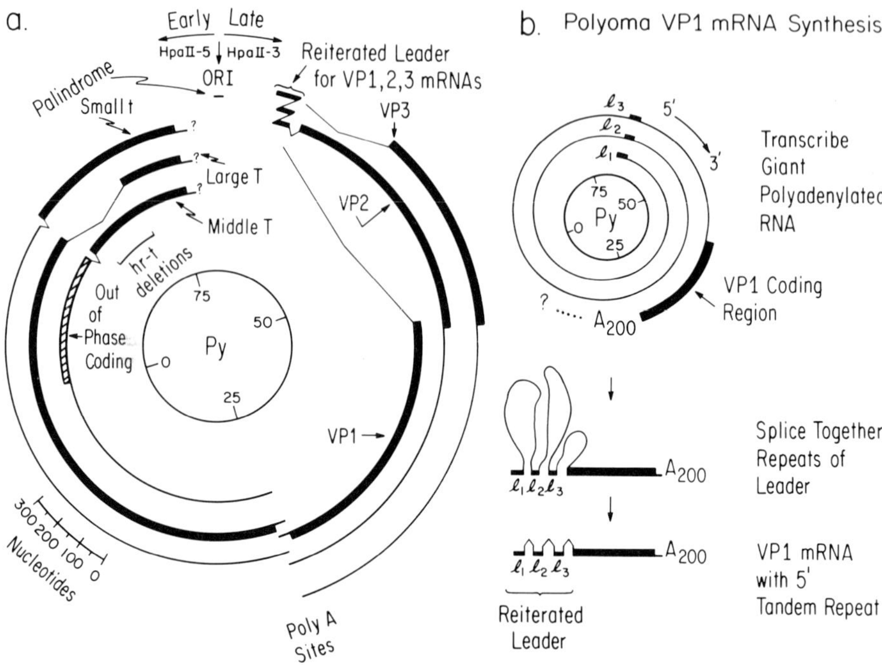

Figure 33. Polyoma messenger RNAs. *A,* Messenger RNA map. As with SV40 (figure 32), the 5′ termini of polyoma mRNAs map near the origin of DNA replication (ORI). The sequence of the ORI contains an inverted repeat palindrome which overlaps the junction of fragments 3 and 5 from *Hpa*II restriction endonuclease digests of polyoma DNA. Early mRNAs are from anti-clockwise transcripts, and late from clockwise transcripts. Polyoma synthesizes three early mRNA species which encode the tumour antigens small t, large T, and middle T. Although the exact splice points for the early mRNAs have not been determined, the t and T messengers have splicing patterns comparable to those of the equivalent SV40 mRNAs. With the mRNA for T, splicing allows translation to continue beyond the t gene. Translation enters the region encoding T, however, in a second out of phase reading frame generating a novel carboxy-terminal sequence for T. The arrangement of coding sequences in the polyoma late messengers is analogous to those of SV40 (see figure 32) and the polyoma mRNAs also have heterogeneous capped 5′ termini (not shown). The 5′ leader, however, contains a novel tandemly reiterated sequence. *B,* Polyoma VP1 mRNA synthesis. At the late stage of infection, polyoma polyadenylated nuclear RNAs are large molecules encoded by polymerases which make several transits of the circular genome. These polymerases can transcribe the late poly(A) acceptor sequence without addition of poly(A), on one transit, but use the poly(A) acceptor sequence on a subsequent pass (compare with late Ad-2 in Fig. 23). It is not established whether this 3′ end is formed by RNA cleavage. The mature mRNA contains a 5′ leader with tandem repeats (1_1, 1_2, 1_3) of a sequence that exists only once in the genome. The reiteration results from the splicing together of each occurrence of this sequence within the giant polyadenylated RNA and joining the repeated structure to the VP1 (shown here), VP2 or VP3 coding sequences.

(Ziff 1980. Reprinted by permission from *Nature* 287, Fig. 3, p. 496. Copyright © 1980 Macmillan Journals Limited.)

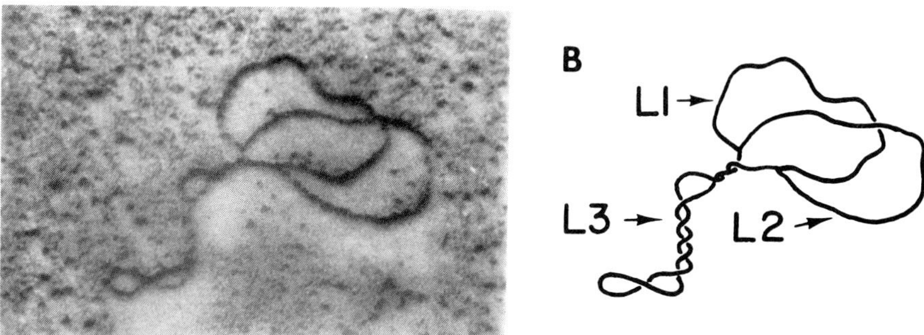

Figure 34. Replicative intermediate of SV40 DNA (A) electron micrograph, (B) drawing of the replicating DNA molecule.
(Sebring et al. 1971. Reprinted by permission from *J. Virol.* 8, Fig. 4, p. 483.)

Synthesis of viral DNA

The initiation sequence for DNA replication is mapped in 0.67 map units and contains 27 base pairs which form a palindrome. The synthesis of viral DNA is semiconservative and bidirectional. Two replication forks move along the DNA until they reach position 0.17 map unit. The initiation of DNA synthesis is done by the binding of a cellular protein, the synthesis of which is induced in the infected cell by SV40 infection or the T or t antigens. The synthesis of the viral DNA is carried out by a cellular α-DNA polymerase. To allow replication, the superhelical viral DNA must uncoil and attain afterward the original conformation. An enzyme is present in the infected cell that can nick the Form I DNA to Form II. This enzyme has a molecular weight of 70,000–90,000, and on incubation with SV40 DNA Form I, it is possible to obtain by electrophoresis in agarose 20 molecular forms, relative to the number of supercoils opened by the enzyme, leading to the conclusion that SV40 DNA has 20 coils (figure 34) (Sebring et al. 1971). The time required for SV40 DNA synthesis was calculated to range from 5 to 25 min.

Replication complexes of the viral DNA were isolated from the nuclei of infected cells and were found to be able to continue DNA synthesis under in vitro conditions (Su and DePamphilis 1976). Initiation of DNA synthesis does not occur under in vitro conditions.

Effect of SV40 infection on cellular DNA synthesis

After the synthesis of the early viral proteins in the infected cell, the enzymatic system involved in cellular DNA synthesis is stimulated, and the synthesis of host chromosomal DNA occurs at the same time as the biosynthesis of viral DNA.

Transformation of cells by SV40

THE TRANSFORMATION MECHANISM. The circular viral DNA was found to be capable of integrating into the chromosomal DNA of the host cell. As a result of this DNA recombination process, the infected cell undergoes transformation. In the transformed cell, the viral T and t antigens are expressed (Martin and Khoury 1976).

Chromosome 7 of human cells contains the site for SV40 DNA integration. The integrated SV40 DNA in chromosome 7 DNA codes for the viral T and t antigens. Chromosome 7 also contains the gene that codes for the tumor specific transplantation antigen (TSTA), a cellular antigen that appears on the surface membrane of the transformed cell. This antigen is an expression of a cellular gene that functions as the result of transformation of the cell by SV40. Chromosome 17 contains a second site for SV40 DNA integration (see chapter 4).

Release of SV40 from transformed cells by fusion with permissive cells

The virus was rescued from transformed cells that do not produce virus after fusion of the transformed cell with a permissive cell. In the resulting heterokaryons, the viral DNA is transferred from the nucleus of the transformed cell to the nucleus of the permissive cell where it replicates.

SV40 mutants

TS MUTANTS. Treatment of virus-infected cells with mutagens like nitric acid, hydroxylamine, or nitrosoguanidine leads to the isolation of many ts mutants: (1) mutation in the gene A (tsA); (2) mutants in late function.

Three groups of mutants (A to C) were identified that complement each other when the cells are infected with two mutants of the different groups. Another group of mutants (D) was described that cannot complement mutants from groups A to C. When the viral DNA is isolated from virions of different ts mutant groups and used for infection of cells, the virus replication is not sensitive to temperature. It is possible that one of the DNA-binding proteins is the product of the mutated gene (reviewed by Kelly and Nathans 1977).

Mapping of mutations with fragments of viral DNA

Lai and Nathans (1974) used restriction enzymes to generate viral DNA fragments for complementation of ts mutants. The fragment of the DNA that complements the lesion in the mutant DNA allows virus replication. This technique preceded the sequence analysis of SV40 DNA done by Fiers and collaborators.

PROPERTIES OF HUMAN PAPOVAVIRUSES

A human disease of the central nervous system (CNS) associated with polyomaviruses is progressive multifocal leukoencephalopathy (PML), which with focal demyelination associated with oligodendrocytes, is fatal. Patients die within a year after onset. In the nuclei of the oligodendrocytes, inclusion bodies were seen that contained virions with the morphology of papovaviruses. Two virus strains were isolated from two patients: one in human embryonic glial cells (designated JC) and the other in cultured VERO cells from the African green monkey (designated BK). The two isolates have a common T antigen similar to that of SV40 (Padgett and Walker 1976).

The viruses are spread in human populations, but only 7% of people above the age of 15 years have antibodies to these viruses. Kidney transplants in patients lead to the activation of the virus, probably due to the use of corticosteroids or immunosuppressive drugs. It is possible that the virus resides in the kidney.

JC and BK viruses are capable of transforming monkey cells in vitro. However, it is not known if these viruses are responsible for any human tumor or even if they are responsible for PML, although there is strong evidence that JC virus causes PML.

Papilloma virus: the human wart virus

Common warts (Verrucae vulgaris) in the skin of humans, as well as the genital warts (Condylomata acuminata) are caused by a human papovavirus (zur Hausen 1977). Eight distinct types of human papilloma viruses as designated by H. zur Hausen are as follows:

HPV-1 causes mainly plantars' warts.
HPV-2 causes mainly hand warts.
HPV-3 causes mainly juvenile flat warts (epidermodysplasia verruciformis).
HPV-4 causes mainly mosaic warts.
HPV-5 and HPV-8 cause warts with malignant conversion.
HPV-6 causes Condyloma acuminata, genital type.
HPV-7 causes mainly butcher warts.
Laryngeal papilloma is unidentified.

The viral DNA does not integrate into the cell DNA in the basal layer of the skin. The virus appears in keratinized cells on the surface of the wart. Inside the warts, the cells are proliferating while synthesizing the viral DNA without expressing the late viral genes. The structural genes are expressed only when the cells start to synthesize keratin, as a stage in their differentiation process.

The virus causing the genital warts is transmitted sexually. The warts tend to disappear spontaneously.

Laryngeal papillomas have been seen in children and adults. Sometimes they attain a large form or disappear spontaneously.

Organization of the DNA genome

The viral DNA is circular and double-stranded. Form I of the DNA is supercoiled and Form II DNA is circular with a nick in one strand. Form III DNA is linear. The molecular weight of the viral DNA is $4–9 \times 10^6$ daltons.

Viral proteins

The major capsid protein has a molecular weight of 63,000 to 53,000. The histones $F2_{a1}$, $F2_{a2}$, $F2_b$, and F3 are also present in the virions.

Animal warts

The wart viruses are spread among rabbits, hamsters, sheep, goats, cattle, horses, dogs, and monkeys.

Bovine papillomavirus (BPV)

This virus causes, in addition to skin warts, warts in the esophagus, male sex organ, and as a tumor of the bladder (enzootic hematuria).

BPV transforms bovine cells and mouse cells in culture. Study of benign equine tumors revealed the presence of 50 to 500 viral DNA equivalents in each diploid set of chromosomes.

The benign esophageal warts can become malignant in cattle, possibly due to consumption of the plant *Peridium aquilinume* that was found to contain a carcinogenic toxin.

BIBLIOGRAPHY

Benjamin, T.L. Physiological and genetic studies of polyoma virus. *Curr. Top. Microbiol. Immunol. 59*:107–134, 1972.

Brockman, W.W. Evolutionary variants of simian virus 40. *Prog. Med. Virol. 23*:69–95, 1977.

Fiers, W.; Contreras, G.; Haegeman, G.; Rogiers, R.; Van de Voorde, A.; Van Heuverswyn, H.; Van Herreweghe, J.; Volckaert, G.; and Ysebaert, M. Complete nucleotide sequence of SV40 DNA. *Nature 273*:113–120, 1978.

Kelly, T.Y., and Nathans, D. The genome of simian virus 40. *Adv. Virus Res. 21*:85–173, 1977.

Levine, A.J.; van der Vliet, P.C.; and Sussenbach, J.S. The replication of papovavirus and adenovirus DNA. *Curr. Top. Microbiol. Immunol. 73*:66–124, 1976.

Lai, C.-J., and Nathans, D. Mapping the genes of simian virus 40. *Cold Spring Harbor Symp. Quant. Biol. 39*:53–60, 1974.

Martin, M.A., and Khoury, G. Integration of DNA tumor virus genomes. *Curr. Top. Microbiol. Immunol. 73*:35–65, 1976.

Novak, U.; Dilworth, S.M.; and Griffin, B.E. Coding capacity of a 35% fragment of the polyoma virus genome is sufficient to initiate and maintain cellular transformation. *Proc. Natl. Acad. Sci. USA 77*:3278–3282, 1980.

Padgett, B.L., and Walker, D.L. New human papovaviruses. *Prog. Med. Virol. 22*:1–35, 1976.

Sebring, E.D.; Kelly, T.J. Jr.; Thoren, M.M.; and Salzman, N.P. Structure of replicating simian virus 40 deoxyribonucleic acid molecules. *J. Virol. 8*:478–490, 1971.

Su, R.T.; and DePamphilis, M.L. In-vitro replication of simian virus 40 DNA in a nucleoprotein complex. *Proc. Natl. Acad. Sci. USA 73*:3466–3470, 1976.

Ziff, E.B. Transcription and RNA processing by the DNA tumour viruses. *Nature 287*:491–499 (1980).

RECOMMENDED READING

Aloni, Y. Splicing of viral mRNAs. *Prog. Nucleic Acid Res. Mol. Biol. 25*:1–31, 1981.

Eckhart, W. Genetics of polyoma virus and simian virus 40. *In: Comprehensive Virology* (H. Fraenkel-Conrat and R.R. Wagner, eds.) Plenum Press, New York, 1977, Vol. 9, pp. 1–26.

Finch, J.T., and Crawford, L.V. Structure of small DNA containing viruses. *In: Comprehensive Virology* (H. Fraenkel-Conrat and R.R. Wagner, eds.) Plenum Press, New York, 1975, Vol. 5, pp. 119–154.

Rigby, P. The transforming genes of SV40 and polyoma. *Nature 282*:781–784, 1979.

Rigby, P.W.J. SV40 and polyoma viruses: their analysis by DNA recombination in-vitro and their use as vectors in eukaryotic systems. *In:* "Biochemistry of Genetic Engineering" (P.B. Garland and R. Williamson, eds). *Biochem. Soc. Symp. 44*:89–102, 1979.

Salzman, N.P., and Khoury, G. Reproduction of papovaviruses. *In: Comprehensive Virology* (H. Fraenkel-Conrat and R.R. Wagner, eds) Plenum Press, New York, 1974, Vol. 3, pp. 63–142.

Seif, I.; Khoury, G.; and Dhar, R. The genome of human papovavirus BKV. *Cell 18*:963–977, 1979.

Winocur, E.; Keshet, I.; Nedjar, G.; and Vogel, T. Origins of SV40 genetic variation. *Ann. NY Acad. Sci. 354*:43–52, 1980.

zur Hausen, H. Human papillomaviruses and their possible role in squamous cell carcinoma. *Curr. Top. Microbiol. Immunol. 78*:1–30, 1977.

10. HEPATITIS B VIRUS

AUSTRALIA ANTIGEN AND ITS CONNECTION WITH HEPATITIS B VIRUS

The study by B.S. Blumberg on the polymorphism of human antigens led to the discovery in 1964 of the viral cause of hepatitis, a disease of the liver in humans (Blumberg 1977). The experimental approach used by Blumberg was to search for antibodies to antigens in the peripheral blood of patients who had received a large number of blood transfusions. He looked for antibodies produced against unique antigens in the blood of the donors that would differ from the blood group antigens, which are always matched with that of the recipient. In this study, sera from individuals all over the world—from the Eskimos in the north of Alaska to the aborigines in Australia—were tested. A hemophiliac in New York who had received many blood transfusions was found to have antibodies in his serum that gave a precipitation line in the Ouchtherlony test with an antigen present in the serum of an Australian aborigine. For this reason, the antigen was designated Australia (Au) antigen. Studies of sera received from hospitalized children with Down's syndrome (mongoloidism) revealed that one of the children who initially lacked antibodies to Au antigen subsequently developed such antibodies after being ill with hepatitis. Thus a correlation between active hepatitis and antibodies to Au antigen was made. In 1967, the chief technician in Blumberg's laboratory, while working on the purification of the antigen, developed hepatitis. Antibodies in her serum gave a precipitation like with Au antigen, and for the first time hepatitis was diagnosed with the help of an antigen. Shortly thereafter, it was concluded that transfusion of blood containing Au antigen results in infection of the patient with hepatitis virus. Since 1969, every blood donation used in the USA for transfusion must be tested for Au antigen, and the positive blood samples are discarded. The Au antigen, also called the hepatitis-associated antigen (HAA) or SH-antigen, was subsequently found to be the surface antigen, HBsAg, of hepatitis B virus (HBV).

Two other hepatitis viruses are known: hepatitis A virus, which is an enterovirus, and the non A-non B hepatitis virus, which is unclassified.

Electron microscopy of serum samples from patients with hepatitis B revealed four types of virus particles:

1. Round particles with a diameter of 15 to 25 nm (20 nm average) that were subsequently found to have a density of 1.20–1.22 g/ml. Negative staining revealed striations in the particles 3 nm apart. The HBsAg was found to be located on the surface of the particles. When precipitin lines of Au antigen (HBsAg) and antibodies that formed in the agar in the Ouchtherlony test were examined by electron microscopy, only 22 nm particles were found.
2. Filamentous particles 22 × 200 nm in size which contain HBsAg on the surface.
3. Particles that contain the small surface antigen. These particles, present in HBsAg positive sera, have a density of 1.29 g/ml and a sedimentation coefficient of 12 S. These particles are present in sera of chronic carriers of the virus and cause in-utero infection of the fetus.
4. The Dane particles. These have a diameter of 42 nm, an outer cover 7 nm in diameter, and an electron dense center of 28 nm, and are probably the whole infectious virus particles (Robinson 1977). HBsAg is located on the surface of the Dane particle. Aggregates of Dane particles are found in the sera of patients. Linear DNA molecules are associated with these particles.

The core of the Dane particles contains a circular DNA molecule 0.78 ± 0.09 nm and a DNA polymerase molecule. In vitro incubation of the viral DNA enzyme complex with the four deoxyribonucleoside triphosphates results in the synthesis of DNA in the form of a linear molecule attached to circular viral DNA.

The 28 nm cores containing the core antigen, HBcAg, can be found without the outer cover in liver tissue obtained from humans and monkeys with hepatitis. Two proteins with molecular weights of 35,000 and 17,000 were found to be associated with the HBcAg.

Antigenic determinants

The composition of the HBsAg present on the surface of the particles is complex and a number of antigens were revealed:

A group-specific antigen, designated *a*, is common to all sera positive for

HBsAg. The particles carry two pairs of antigenic subdeterminants that are type-specific designated *d* or *y* and *w* or *r*. The determinants *d* and *y* are not present in the same serum. A similar situation exists with *w* and *r* antigens.

Four antigenic subtypes were identified: *adw*, *ayw*, *adr*, and *ayr*. These antigenic subtypes are distributed around the world.

Antigen e, which is part of the Dane particle, is found in sera of patients who are positive for HBsAg.

HBsAg from patients' sera

Purified HBsAg contains particles with a diameter of 20 nm composed of proteins, polysaccharides, and lipids. The lipid composition in this antigen is identical to the lipid composition in enveloped virions. Analysis of the proteins revealed the presence of 6 to 9 polypeptides with molecular weights ranging from 97,000 to 23,000, of which two or three are glycoproteins.

Injection of this antigen into rabbits induces the production of antibodies against the group antigen (anti-HBs/*a*) and against the unique antigen (anti-HBs/*d* or *y*). Recently, peptides were isolated from preparations containing HBsAg/*adw* and HBsAg/*ayw*. The use of purified polypeptides for immunization led to the production of antibodies to either *d* or *y*.

Properties of the Dane particle DNA

Two species of DNA molecules were isolated from these particles:

1. Linear DNA molecules, 0.5 to 12 nm in length, associated with the particle, but sensitive to treatment with the enzyme DNase I. The source of these DNA molecules is not known.
2. In the core of the Dane particle a circular DNA molecule 0.78 $\pm$ 0.9 nm equivalent to 1.6 $\times$ 10^6 daltons, is found (Robinson 1977). The total genetic information in the viral DNA is enough to code for a polypeptide of 100,000 daltons. The total molecular weight of the proteins that constitute the Dane particle ranges from 320,000 to 490,000. Thus the viral DNA does not contain enough DNA to code for all the proteins of the Dane particle; a new mechanism for the synthesis of the viral proteins, still unknown, may exist.

In vitro incubation of the Dane particle cores that contain the viral DNA polymerase complex allows synthesis of double-stranded DNA under suitable in vitro conditions. These DNA molecules are synthesized by a mechanism resembling the rolling circle mechanism of DNA synthesis.

Analysis of Dane particle DNA with restriction enzymes showed that it is composed of two single-stranded DNA molecules, both of which are incomplete. One single-stranded molecule is closed circular DNA containing 3,200 nucleotides, and the second, a complementary DNA strand, only 1,600 to 2,800 nucleotides. Under in vitro conditions, the cores synthesize complete circular double-stranded DNA.

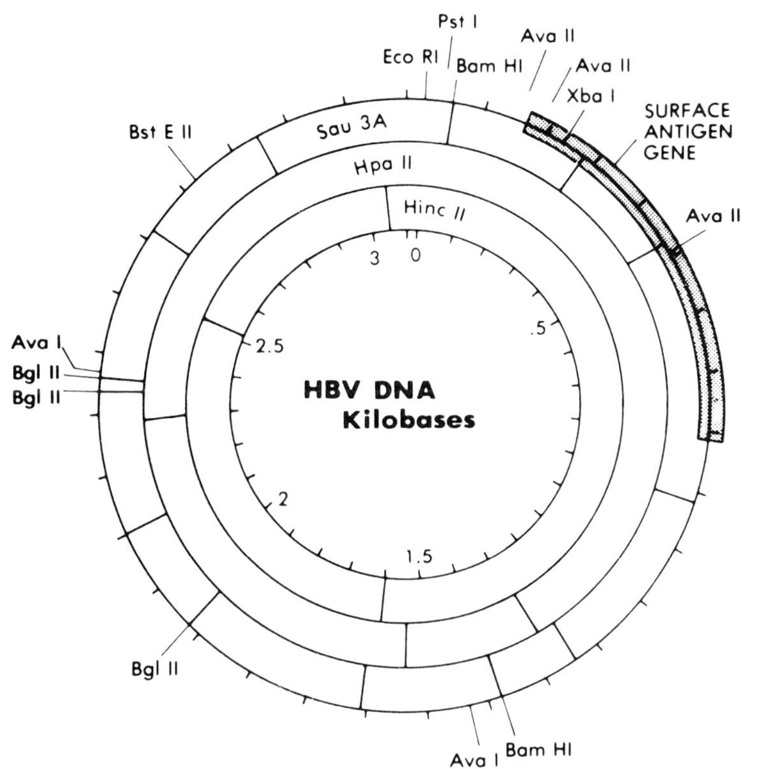

Figure 35. Hepatitis B virus surface antigen gene (shaded area) introduced into the plasmid PHBV-3200 of *E. coli*.
(Valenzuela et al. 1979. Reprinted by permission from *Nature* 280, Fig. 2, p. 816. Copyright © 1979 Macmillan Journals Limited.)

The circular DNA strand can be labeled by the nick-translation procedure, and after cleavage with the restriction enzyme HaeIII (*Haemophilus aegypticus*), the DNA was found to contain 3,880 base pairs. Two additional DNA fragments were discovered after labeling the DNA with ^{32}P and therefore the number of base pairs in the viral DNA must be no less than 4,910. The gene for the surface antigen (HBsAg) has been cloned in a bacterial plasmid (figure 35); thus the viral antigen can be produced by bacterial cells (Valenzuela et al. 1979).

HBsAg is expressed in mammalian cells and bacteria using plasmids with cloned hepatitis B virus DNA

Cloning of the viral DNA in plasmids made it possible to study viral DNA expression. Mouse L cells (TK^-) transformed by a plasmid containing both the HSV-1 thymidine kinase gene and two copies of the HBV genome yielded HBsAg particles of 22 nm. These particles possessed the same characteristics as the human serum particles and were found to produce antibodies to human HBsAg in mice (Dubois et al. 1980).

Viral antigens in liver biopsies

With fluorescent antibodies it was possible to identify the viral antigens present in liver biopsies from persistently infected humans and chimpanzees. Most of the cells have HBsAg in the cytoplasm, some contain HBcAg in the nuclei, and a few have both antigens. The number of infected cells in a liver biopsy ranges from 1 to 100%. All chronic carriers of the virus have Dane particles in their peripheral blood, and many liver cells contain HBcAg.

VACCINATION. HBsAg isolated from the sera of patients was used as an antigen for immunization since this antigen elicits an antibody response in immunized individuals (Buynak et al. 1976).

A formalin-inactivated preparation of purified 20 nm HBsAg particles adsorbed onto an alum adjuvant (prepared by the Merck Institute for Therapeutic Research) was used to immunize humans. Children responded well to the vaccine. Among healthy adults, 95% became positive after three injections (C.E. Stevens and W. Szmuness, Abstracts of the Fifth International Congress of Virology, 1981, p. 180).

Thirteen peptides corresponding to the amino acid sequences predicted from the nucleotide sequence of the HBsAg were synthesized chemically and injected into rabbits. Four of the six soluble peptides that ranged in size from 10 to 34 residues elicited antibodies in the rabbits and these reacted with the native HBsAg molecule (Lerner et al. 1981). These proteins hold promise for vaccine production in the future.

Hepatitis B infection of humans

The hepatitis virus affects the liver hepatocytes leading to:

1. Acute hepatitis that develops into chronic hepatitis and production of antibodies to HBsAg.
2. Acute hepatitis with full recovery. Sera contain HBsAg and antibodies to HBsAg.
3. Chronic hepatitis: HBsAg and anti-HBc antibodies are constantly present.
4. Carrier state: HBsAg and anti-HBc antibodies are constantly present.
5. HBsAg is found in immunodeficient patients.
6. Some patients have serum HBs antibodies without HBsAg.
7. Patients with complexes of antigen-antibody in the serum.

Patients who develop antibodies to HBsAg are capable of rejecting a kidney transplant when the HLA histocompatibility match is not maximal.

MODE OF INFECTION. The disease is transmitted by infusion of blood containing HBsAg. The virus is released in the feces and can infect seronegatives through the respiratory tract or alimentary canal. The virus might also be transmitted by insects such as mosquitoes and bedbugs. It was estimated in the United States that one million people are carriers of HBV. Those who have the viral antigen are also capable of transmitting the virus. It is possible that children may be infected by the virus in-utero (Schweitzer 1975).

Bedbugs were found to mechanically transmit hepatitis B virus to humans in South Africa.

Primary hepatic carcinoma and HBV: integrated viral DNA and viral mRNA

In different parts of the world, patients with liver carcinoma have antibodies to HBsAg and to HBcAg; this led to the idea that HBV may be the cause. The viral DNA was found to be integrated into the DNA of the hepatoma cell, which is further evidence that HBV may be involved in carcinogenesis in humans. A cell line developed from a primary hepatocellular carcinoma in a male from Mozambique, who had serum positive for HBsAg, produces HBsAg in vitro (McNab et al. 1976). The cells were found to contain integrated HBV DNA sequences (Marion et al. 1979; Brechot et al. 1980; Edman et al. 1980). In these cells, three specific virus-coded mRNA species were identified: two were poly(A^+) (21.5 S and 19.5 S), and one was poly($A-$) (27 S) and is larger than the full-length Dane particle DNA, possibly due to cotranscription of the integrated viral DNA and flanking cellular DNA (Chakraborty et al. 1980).

Hepatitis in animals

WOODCHUCK HEPATITIS. A virus similar to human HBV causes hepatitis and hepatomas in woodchucks (Summers et al. 1978). A high incidence of hepatomas in a woodchuck colony in the Philadelphia zoo led to the discovery of this virus which shares many characteristics with human HBV. Cross-serological reactions between the surface and core antigens of these two viruses were demonstrated. The woodchuck hepatitis virus could be a good model for studies on human hepatitis and hepatocellular carcinoma.

Hepatitis in primates

Nonhuman primates and chimpanzees can transmit hepatitis to man, and marmosets can be infected with hepatitis A virus (Deinhardt 1976).

BIBLIOGRAPHY

Blumberg, B.S. Australia antigen and the biology of hepatitis B. *Science 197*:17–25, 1977.

Brechot, C.; Pourcel, C.; Louise, A.; Rain, B.; and Tiollais, P. Presence of integrated hepatitis B virus DNA sequences in cellular DNA of human hepatocellular carcinoma. *Nature 286*:533–535, 1980.

Buynak, E.B.; Roehm, R.R.; Tyrell, A.A.; Bertland, A.U.; Lampson, G.P.; and Hilleman, M.R. Vaccine against human hepatitis B. *JAMA 235*:2832–2834, 1976.

Chakraborty, P.R.; Ruiz-Opazo, N.; Shouval, D.; and Shafritz, D.A. Identification of integrated hepatitis B virus DNA and expression of viral RNA in an HBsAg-producing human hepatocellular carcinoma cell line. *Nature 286*:531–533, 1980.

Deinhardt, F. Hepatitis in primates. *Adv. Virus Res. 20*:113–157, 1976.
Dubois, M.-F.; Pourcel, C.; Rousset, S.; Chany, C.; and Tiollais, P. Excretion of hepatitis B surface antigen particles from mouse cells transformed with cloned viral DNA. *Proc. Natl. Acad. Sci. USA 77*:4549–4553, 1980.
Edman, J.C.; Gray, P.; Valenzuela, P.; Rall, L.B.; and Rutter, W.J. Integration of hepatitis B virus sequences and their expression in human hepatoma cells. *Nature 236*:535–538, 1980.
Lerner, R.A.; Green, N.; Alexander, H.; Liu, F.-T.; Sutcliffe, J.G.; and Shinnick, T.M. Chemically synthesized peptides predicted from the nucleotide sequence of the hepatitis B virus genome elicit antibodies reactive with the native envelope protein of Dane particles. *Proc. Natl. Acad. Sci. USA 78*:3403–3407, 1981.
Macnab, G.M.; Alexander, J.J.; Lecatsas, G.; Bey, E.M.; and Urbanowicz, J.M. Hepatitis B surface antigen produced by a human hepatoma cell line. *Br. J. Cancer 34*:509–515, 1976.
Marion, P.L.; Salazar, F.H.; Alexander, J.J.; and Robinson, W.S. Polypeptides of hepatitis B virus surface antigen produced by a hepatoma cell line. *J. Virol. 32*:796–802, 1979.
Robinson, W.S. The genome of hepatitis B virus. *Annu. Rev. Microbiol. 31*:357–377, 1977.
Schweitzer, I.L. Infection of neonates and infants with the hepatitis B virus. *Prog. Med. Virol. 20*:27–48, 1975.
Summers, J.; Smolec, J.M.; and Snyder, R. A virus similar to human hepatitis B virus associated with hepatitis and hepatoma in woodchucks. *Proc. Natl. Acad. Sci. USA 75*:4533–4537, 1978.
Valenzuela, P.; Gray, P.; Quiroga, M.; Zoldivar, J.; Goodman, H.M.; and Rutter, W.J. Nucleotide sequence of the gene coding for the major protein of hepatitis B virus surface antigen. *Nature 280*:815–819, 1979.

RECOMMENDED READING

Howard, C.R., and Burrell, C.J. Structure and nature of hepatitis B antigen. *Prog. Med. Virol. 22*:36–103, 1976.
Robinson, W.S., and Lutwick, L.I. The virus of hepatitis, type B. *N. Engl. J. Med. 295*:1168–1174; 1232–1236, 1976.
Szmuness, W. Hepatocellular carcinoma and the hepatitis B virus: evidence for a causal association. *Prog. Med. Virol. 24*:40–69, 1978.
Viral Hepatitis (eds. G.N. Kjas, S.N. Cohen, and R. Schmid), Franklin Institute Press, Philadelphia, 1978.
Zuckerman, A.J. Human viral hepatitis. *Pharmacol. Ther. 10*:1–13, 1980.
Zuckerman, A.J. Woodchuck, squirrel and duck hepatitis viruses. *Nature 289*:748–749, 1981.

B. SINGLE-STRANDED DNA VIRUSES

11. PARVOVIRUSES

FAMILY PARVOVIRIDAE

These viruses were found to be associated with adenovirus infected cells. The virions (figure 36) are made up of a nonenveloped isometric capsid with a diameter of 18-26 nm, with icosahedral symmetry, containing 32 capsomeres each with a diameter of 3–4 nm. The DNA genome is single-stranded with a molecular weight ranging from $1.5–2.2 \times 10^6$ and constitutes 20–25% of the virion weight. Half of the virions contain the DNA^+ strand and the other half the DNA^- strand. Cellular functions and a helper virus are required for virus replication to take place in the nucleus of the infected cell.

Genus Parvovirus

Latent rat virus (Kilham virus).

Rodent viruses: MVM, RT, LuIII, TVX, X14, H-1.

There are bovine, porcine, canine, feline panleukopemia, and goose hepatitis parvoviruses. These viruses are competent without a helper virus, and the virions contain DNA^- strands only.

Genus adeno-associated virus

Human and simian adeno-associated virus (AAV) types 1, 2, 3, and 4.

Bovine adeno-associated virus AAV 7.

Avian AAV (AAAV).

Mature virions contain either DNA^+ or DNA^- strands. Replication is dependent on a helper virus.

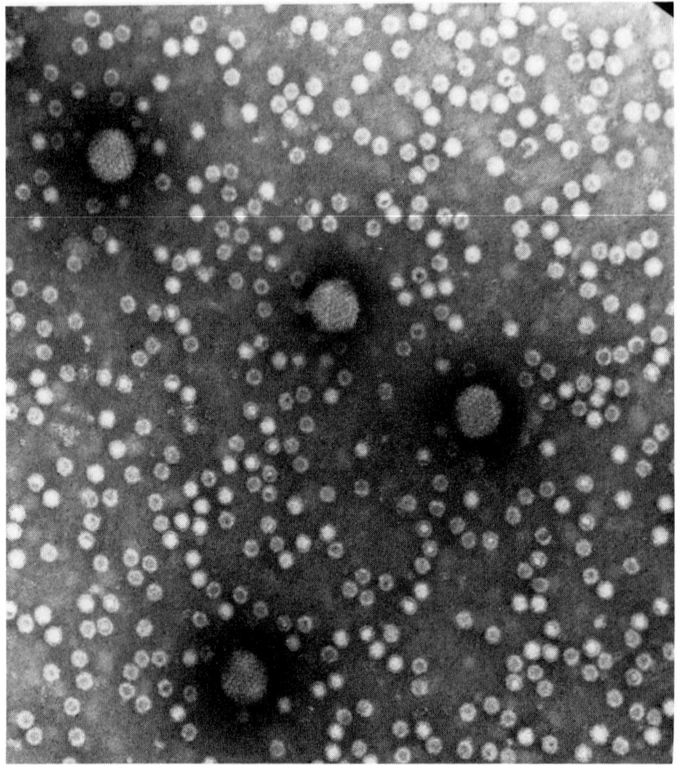

Figure 36. Electron micrograph of parvovirus particles which require adenovirus (large particles) for replication.
(By courtesy of Dr. Daniel Dekegel, Institut Pasteur du Brabant, Brussels, Belgium.)

Genus Densovirus

Genus densoviruses are insect viruses that multiply in tissues of larvae, nymphs, and adults without helper virus. DNA^+ and DNA^- strands are present in the virions.

Molecular aspects of virus replication differs according to dependency on adenovirus

Active, nondefective parvoviruses (e.g., MVM, RV, H-1)

The viral capsid is resistant to chloroform and ether since it does not contain lipids.

The linear single-stranded DNA molecules have molecular weights of 1.5, 1.6, and 1.7 $\times$ 10^6. The viral genome is a noninfectious DNA^- strand.

Three proteins were identified in the viral capsid. In different viruses, the molecular weights of the proteins vary: in RV virus, protein A is 72,000; B is

62,000; and C is 55,000 daltons. In MVM strain, protein A is 92,000; B is 72,000; and C is 69,000 daltons (Clinton and Hayashi 1976).

The total molecular weight of the structural proteins is 233,000, which is about three times the amount of genetic information encoded in the viral DNA. Since it was found that the three polypeptides have a similar organization, it was concluded that the three proteins are cleavage products of a precursor polypeptide coded by one gene. One species of viral mRNA of 9×10^5 daltons is translated into the structural proteins.

DNA polymerase in virions

A DNA polymerase was found in RV but not in H-1 virus. The properties of this enzyme are not known.

Life cycle of an active nondefective virus

The virus replicates in the host cell during the S phase. The particles are adsorbed to the cell membrane, uncoated in the cytoplasm, and replicated in the nucleus. The synthesis of viral DNA starts 8–10 hr after infection and continues until 23 hr postinfection. Two to four hr elapse from the replication of the viral DNA until the first virion progeny appear.

The single-stranded DNA that enters the cells is transformed into linear double-stranded DNA which constitutes the replicative form (RF) of the viral DNA (Singer and Rhode 1977). DNA synthesis most probably occurs by a mechanism of hairpin duplex formation (Tattersall and Ward 1976).

The viral mRNA molecules are transcribed from the double-stranded DNA and are transported from the nucleus to the cytoplasm, the site of protein synthesis. Treatment of infected cells with α-amanitin that inhibits the cellular RNA polymerase II of the host cell inhibited the synthesis of the nuclear viral antigen but did not prevent the synthesis of the cytoplasmic antigen.

Effect of the virus on the host cell

Viruses like RV and MVM effectively inhibit mitosis in the infected cells.

Life cycle of adeno-associated defective virus

The viral DNA

Two classes of virions are produced: Half contain DNA^+ and half contain DNA^- molecules. It is possible to infect cells with the virus in the presence of 5-bromodeoxyuridine (BUdR). BUdR is incorporated into the viral DNA and replaces 70–80% of the thymidines. Centrifugation of the progeny viral DNA in CsCl density gradients revealed that one viral DNA strand has a density of 1.830 g/ml, while the second is lighter, with a density of 1.798 g/ml. The denser DNA is the minus strand. Circular DNA molecules were rarely seen among the linear viral DNA molecules extracted from the virions by osmotic shock. The viral DNA is infectious, provided the cells are infected with adenovirus. The plus and minus DNA strands have complementary sequences in the molecular ends. By self-annealing, circular DNA molecules are formed (Rhode 1977).

The DNAs of AAV-1, 2, and 3 are homologous to each other. No genetic resemblance exists between AAV and the adenoviruses or herpesviruses that serve as helpers for its replication.

Dependency of AAV on a helper virus

AAV replicates and produces virus progeny only in adenovirus infected cells (Hoggan 1970). HSV can only partially replace adenovirus, but in HSV-1 infected cells the viral DNA replicates, the nuclear antigen and empty capsids are made in the nuclei, but infectious virions are not made. Other herpesviruses, like bovine rhinotracheitis, EBV, cytomegalovirus, varicella zoster, and herpesvirus saimiri can serve as helpers for AAV.

Infection of cells with AAV

The adsorption and penetration of AA virions into infected cells do not require the presence of the helper adenovirus. The latent period of the infectious cycle is 17 hr, followed by a 12–15 hr period of viral DNA synthesis. The amount of AAV virions in the infected cells is 10–1,000-fold higher than the number of adeno virions synthesized in the infected cells. Each cell of the African green monkey kidney line produces 2–4 $\times$ 10^5 genomes of AAV-2.

The role of the helper adenovirus

The initiation of mRNA synthesis on the AAV DNA is dependent on the synthesis of adenovirus mRNA which starts 2–3 hr after infection with adenovirus. Infection of adenovirus infected cells with AAV 10 hr after the first infection shortens the lag between the synthesis of AAV mRNA and AAV DNA replication.

Synthesis of AAV mRNA

The mRNA synthesized in AAV-2 infected cells (coinfected with adenovirus-2 or HSV-1) has a molecular weight of 9 $\times$ 10^5 and hybridizes to the DNA strand of AAV DNA. The AAV mRNA present in infected nuclei of KB cells or in the polyribosomes has a sedimentation coefficient of 19–20 S, corresponding to 70% of the viral DNA genome.

Viral proteins

Three viral proteins are made in infected cells: protein A 87,000; protein B 73,000; and protein C 62,000 daltons.

Association of the virus with disease

These viruses were isolated from children suffering from an adenovirus infection (Blacklow et al. 1967; 1968). The nature of association between the two viruses as far as the disease is concerned is not known.

BIBLIOGRAPHY

Blacklow, N.R.; Hoggan, M.D.; Kapikian, A.Z.; Austin, J.B.; and Rowe, W.P. Epidemiology of adenovirus-associated virus infection in a nursery population. *Am. J. Epidemiol.* *88*:368–378, 1968.

Blacklow, N.R.; Hoggan, M.D.; and Rowe, W.P. Isolation of adenovirus-associated viruses from man. *Proc. Natl. Acad. Sci. USA* *58*:1410–1415, 1967.

Clinton, G.M., and Hayashi, M. The parvovirus MVM: A comparison of heavy and light particle infectivity and their density conversion in vitro. *Virology* *74*:57–63, 1976.

Hoggan, M.D. Adenovirus-associated viruses. *Prog. Med. Virol.* *12*:211–239, 1970.

Rhode, S.L. Replication process of parvovirus H-1. X. Physical mapping studies of the H-1 genome. *J. Virol.* *22*:446–458, 1977.

Singer, I.I., and Rhode, S.L. Replication process of the parvovirus H-1. VII. Electron microscopy of replicative-form DNA synthesis. *J. Virol.* *21*:713–723, 1977.

Tattersall, P., and Ward, P. Rolling hairpin model for replication of parvovirus and linear chromosomal DNA. *Nature* *263*:106–109, 1976.

RECOMMENDED READING

Berns, K.I.; Hauswirth, W.W.; Fife, K.H.; and Lusky, E. Adeno-associated virus DNA replication. *Cold Spring Harbor Symp. Quant. Biol.* *43*:781–787, 1978.

Kurstak, E. Small DNA densonucleosis virus (DNV). *Adv. Virus Res.* *17*:207–241, 1972.

Rose, J.A. Parvovirus Reproduction. *In: Comprehensive Virology* (H. Fraenkel-Conrat and R.R. Wagner, eds.) Plenum Press, New York, 1974, Vol. 3, pp. 1–61.

Siegel, G. Parvoviruses as contaminants of permanent cell lines. V. The nucleic acid of KBSH-virus. *Arch. Ges. Virusforsch.* *37*:267–274, 1972.

Ward, D.C., and Tattersall, P. (eds.) *Replication of Mammalian Parvovirus.* Cold Spring Harbor Laboratory, Cold Spring Harbor, New York, 1978.

C. RNA MINUS VIRUSES

12. DOUBLE-STRANDED RNA VIRUSES MADE FROM SINGLE-STRANDED RNA: REOVIRUSES

FAMILY REOVIRIDAE

These include Reo-, Orbi- and Rota-viruses. The virus (no lipoprotein envelope) has an isometric capsid 60–80 nm in diameter made of two layers of capsomeres and arranged in icosahedral symmetry. The virions contain the RNA-dependent RNA polymerase and a double-stranded RNA genome fragmented into ten pieces, each containing one viral gene. The total molecular weight of the RNA fragments ranges from $10–16 \times 10^6$.

Genus Reovirus

Human reovirus types 1, 2, and 3 (figure 37). These viruses have a common complement fixing antigen.
Simian, canine, and avian reoviruses.

Genus Orbivirus

Virus replicates in vertebrates and in insects. The genus includes 38 virus strains. The virion has a diameter of 55–65 nm.

Blue tongue virus.
African horse sickness virus.
Colorado tick fever virus.

Genus Rotavirus (figure 38)

Human infantile enteritis.
Epizootic diarrhea of infant mice.

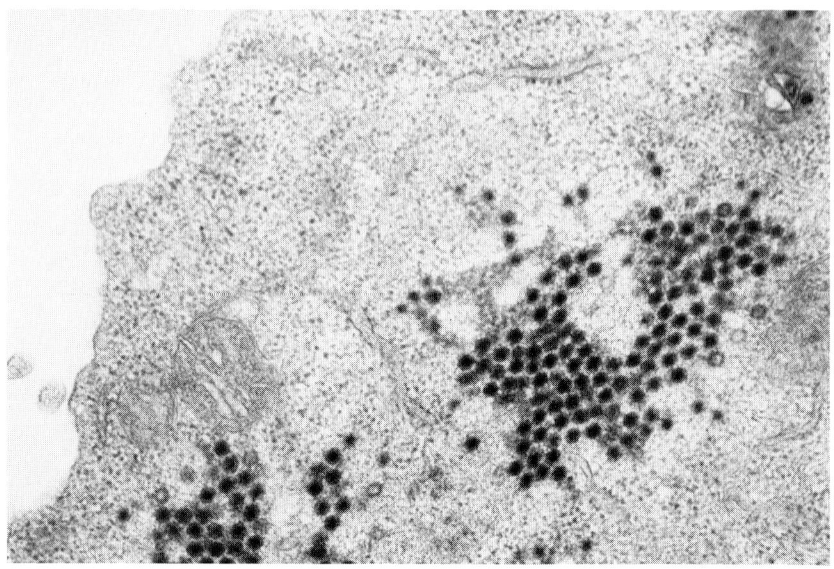

Figure 37. Electron micrograph of reovirus infected cells. Clusters of reovirions are found in the cell cytoplasm.
(By courtesy of Dr. Daniel Dekegel, Institut Pasteur du Brabant, Brussels, Belgium.)

Nebraska calf scours.
Pig diarrhea.
Simian virus SA11.
There are also plant reoviruses (phytoreovirus and fijivirus).

Molecular aspects of reovirus replication

Studies on the replication of reoviruses were done mainly with reovirus 3 which replicates in mouse L cells cultured in suspension. The virions adsorb and penetrate into the cells, and the outer capsid of the virion is removed in the uncoating process; the RNA-dependent RNA polymerase present in the virions is activated, and viral mRNA is synthesized. A latent period of 6 hr is followed by the synthesis of virus progeny that reaches a maximum after 18 hr. At this stage, every infected cell produces 1000 PFU of the new virus progeny (equivalent to 20,000 virions/cell). A number of specific functions have been assigned to the viral hemagglutinin which plays a central role in virus-cell interactions (Fields et al. 1980). After entry into the host cell, the hemagglutinin (encoded by the S1 double-stranded RNA segment) is responsible for recognition by both humoral and cellular components of host immunity. The hemagglutinin also recognizes different neural cells and is thus the prime determinant of the pattern of neurovirulence of the virus.

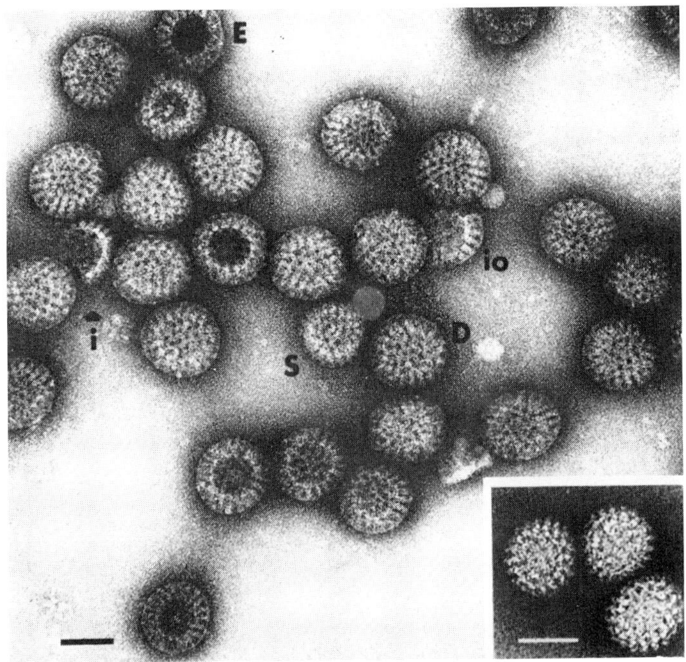

Figure 38. Electron micrograph of a negatively stained preparation of human rotavirus. D: double-shelled particles; S: single-shelled particles; E: empty capsids; i: fragment of inner shell; io: fragments of a combination of inner and outer shell. Inset: single-shelled particles obtained by treatment of the virus preparation with 100 μg/ml of SDS, immediately prior to processing for E.M. Bars: 50 nm.
(Esparza and Gil 1978. Reprinted by permission from *Virology* 91, Fig. 1, p. 143.)

Synthesis of viral mRNA

The ten species of dsRNA molecules that comprise the reovirus genome fall into three size classes: three L species (mol. wt. 2.5–2.7 × 10^6); three M species (mol. wt. 1.2–1.4 × 10^6); and four S species (mol. wt. 0.6–0.8 × 10^6) (Bellamy and Joklik 1967). Each dsRNA fragment of the viral genome is transcribed in its entirety into single-stranded RNA by the viral RNA-dependent RNA polymerase present within the viral core, forming 10 mRNA species. The 10 viral dsRNA fragments have the same polarity of transcription, and no hybrids between mRNA molecules of the different viral genes are formed. The viral mRNA is polyadenylated at the 3′ end and is translated by the cellular ribosomes in the cytoplasm. The 11 reovirus-coded polypeptides that are known consist of three λ polypeptides coded by the three L segments; four μ polypeptides coded by the three M segments; and four σ polypeptides coded by the four S segments (Both et al. 1975). The specific polypeptides encoded by each of the 10 segments of reovirus double-stranded RNA have been

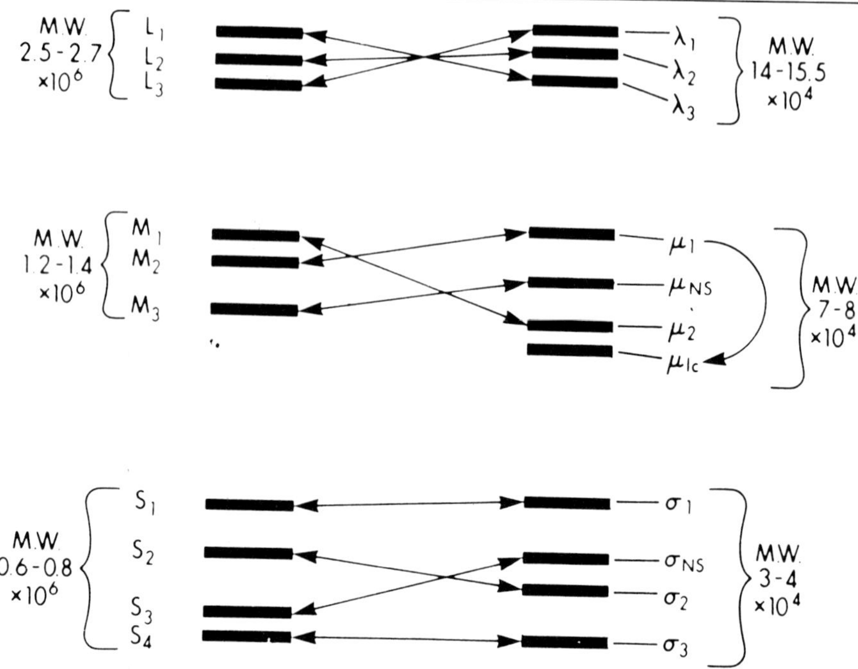

Figure 39. RNA coding assignments for reovirus type 3, strain Dearing. (McCrae and Joklik 1978. Reprinted by permission from *Virology* 89, Fig. 18, p. 591.)

identified by McCrae and Joklik (1978) using wheat germ ribosomes for mRNA translation in vitro. Each viral protein can be correlated by its molecular weight with the gene coding for it, as shown in figure 39. The large mRNA species synthesize the longer polypeptide chains.

Synthesis of the viral double-stranded RNA molecules

The viral mRNA species are the templates for the viral RNA. At the time of virion formation, the ten mRNA species are inserted into the newly made viral cores [after cleavage of the poly(A) sequences] together with the viral RNA-dependent RNA polymerase. Inside the viral core, the RNA polymerase synthesizes the RNA^+ complementary RNA strand for each RNA species, and the double-stranded RNA molecules of the genome are made. The core is coated with the second capsomere layer, and the mature virions are formed.

GENUS ORBIVIRUS. These viruses resemble the members of the genus reovirus in their molecular properties.

Diseases caused by members of the reovirus family

ROTAVIRUSES. These viruses are a major cause of gastroenteritis in children. The virus is excreted in the feces (Esparza and Gil 1978).

COLORADO TICK FEVER VIRUS. Colorado tick fever virus is transmitted by orbivirus infected ticks. The disease is manifested by high fever, head, eye, and back pains, as well as leukopenia.

REOVIRUS 1, 2, AND 3. These viruses are excreted in the feces but are not associated with a defined disease syndrome.

BIBLIOGRAPHY

Bellamy, A.R., and Joklik, W.K. Studies on reovirus RNA. II. Characterization of reovirus messenger RNA and of the genome RNA segments from which it is transcribed. *J. Mol. Biol. 29*:19–26, 1967.

Both, G.W.; Lavi, S.; and Shatkin, A.J. Synthesis of all the gene products of the reovirus genome in-vivo and in-vitro. *Cell 4*:173–180, 1975.

Esparza, J., and Gil, F. A study on the ultrastructure of human rotavirus. *Virology 91,* 141–150, 1978.

Fields, B.N.; Weiner, H.L.; Drayna, D.T.; and Sharpe, A.H. The role of the reovirus hemagglutinin in viral virulence. *Ann. NY Acad. Sci. 354*:125–134, 1980.

Laemmli, U.K. Cleavage of structural proteins during the assembly of the head of bacteriophage T4. *Nature 227*:680–685, 1970.

McCrae, M.A., and Joklik, W.K. The nature of the polypeptide encoded by each of the ten double-stranded RNA segments of reovirus type 3. *Virology 89*:578–593, 1978.

RECOMMENDED READING

Gorman, B.M. Review article: Variation in orbiviruses. *J. Gen. Virol. 44*:1–15, 1979.

Gross, R.K., and Fields, B.N. Genetics of reoviruses. *In: Comprehensive Virology* (H. Fraenkel-Conrat and R.R. Wagner, eds.) Plenum Press, New York, 1977, Vol. 9, pp. 291–340.

Joklik, W.K. Reproduction of reoviridae. *In: Comprehensive Virology* (H. Fraenkel-Conrat and R.R. Wagner, eds.) Plenum Press, New York, 1974, Vol. 2, pp. 231–334.

Joklik, W.K. The structure and function of the reovirus genome. *Ann. NY Acad. Sci. 354*:107–124, 1980.

McNulty, M.S. Rotaviruses. *J. Gen. Virol. 40*:1–18, 1978.

13. SINGLE-STRANDED RNA MINUS VIRUSES WITH AN INTACT GENOME: RHABDOVIRUSES

FAMILY RHABDOVIRIDAE

The virions are bullet-shaped, 130–300 nm long, and 70 nm wide with a lipoprotein envelope containing peplomeres that cover an elongated nucleocapsid with helical symmetry. The virion contains five major proteins and an enzyme, the RNA-dependent RNA polymerase. The RNA genome is linear and single-stranded, with a molecular weight of $3.5–4.6 \times 10^6$.

Genus Vesiculovirus

Vesicular stomatitis virus (figure 40) replicates in vertebrates and invertebrates.
Chandipura virus has been isolated from humans.
Kern Canyon virus has been isolated from a bat.

Genus Lyssavirus

Rabies virus } have been isolated from humans
Duvenhage virus }

Genus Sigmavirus

Drosophila σ virus is responsible for CO_2 sensitivity in Drosophila.

Plant rhabdoviruses (ungrouped).

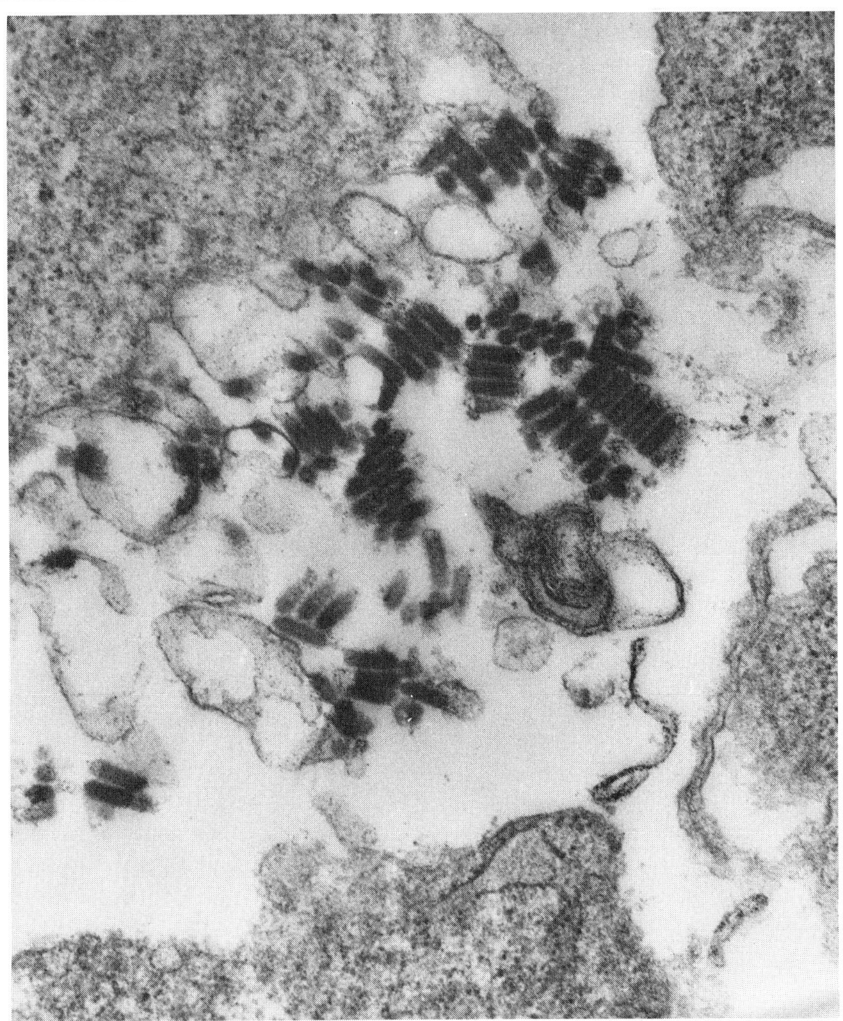

Figure 40. Electron micrograph of vesicular stomatitis virions in a cell section and their ribonucleoprotein complexes released from damaged virions.
(By courtesy of Dr. Daniel Dekegel, Institut Pasteur du Brabant, Brussels, Belgium.)

RABIES VIRUS

The virion and its properties

The virions have a typical rhabdovirus structure (figure 41) and are 175 nm long and 75 nm wide. The protein peplomeres on the envelope are capable of agglutinating red blood cells. Inside the virion, a single-stranded RNA genome is arranged in a helical nucleocapsid.

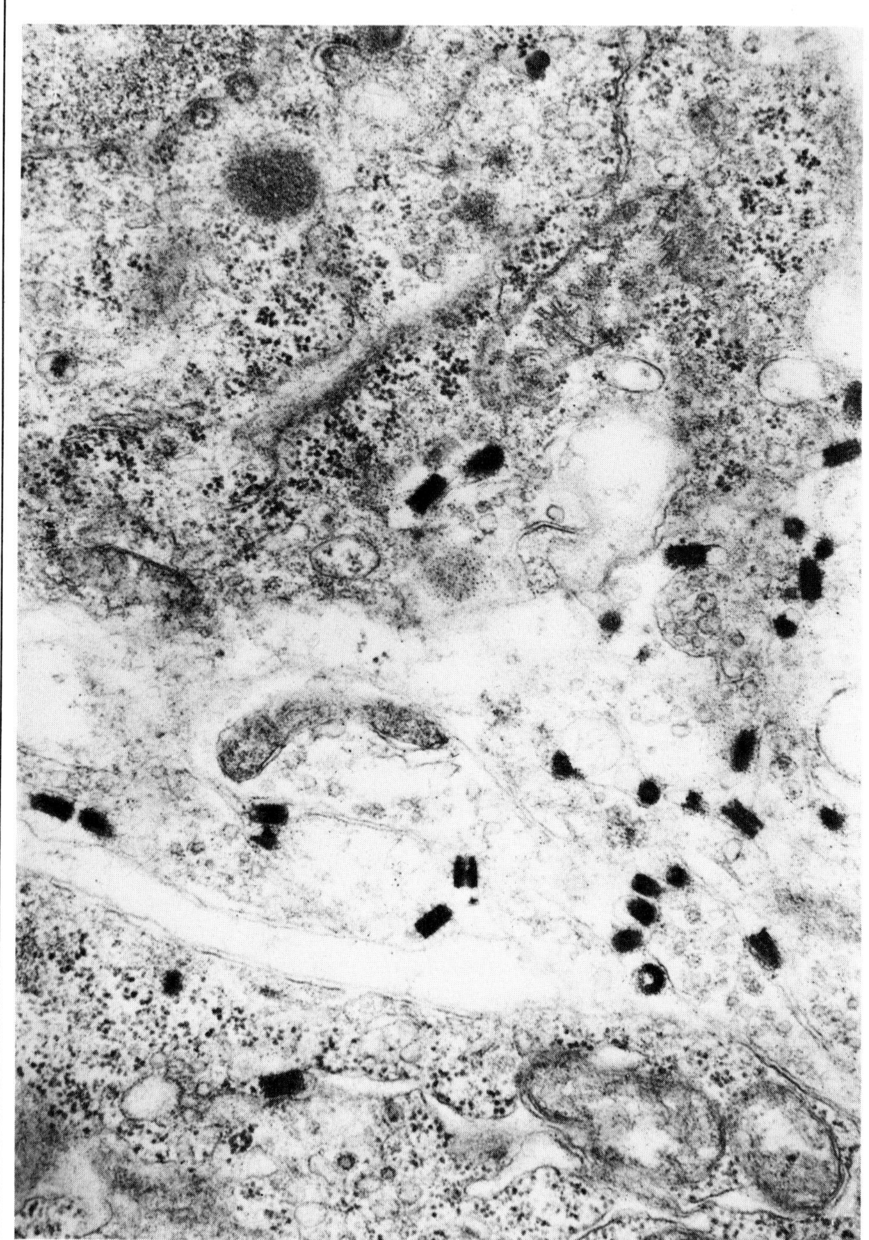

Figure 41. Electron micrograph of rabies virus in the cytoplasm of infected cells. (By courtesy of Dr. Daniel Dekegel, Institut Pasteur du Brabant, Brùssels, Belgium.)

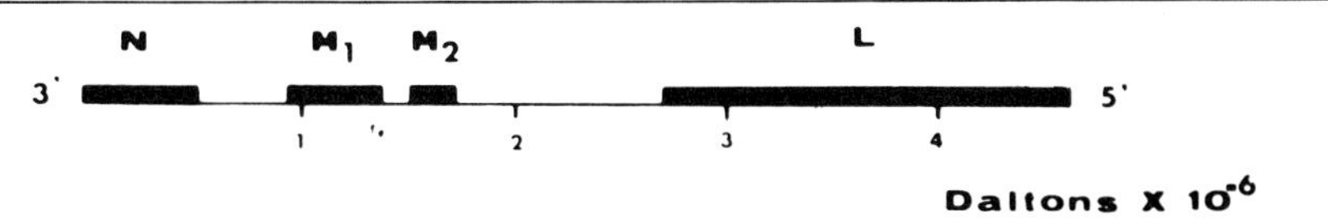

Figure 42. Tentative map of the rabies genome. The target sizes of proteins N, M_1, M_2, and L are shown assuming a genome size of 4.6×10^6 daltons. The four cistrons are thus assumed to terminate at 5.1×10^5, 1.4×10^6, 1.7×10^6, and 4.6×10^6 daltons.
(Flamand and Delagneau, 1978. Reprinted by permission from *J. Virol.* 28, Fig. 5, p. 523.)

Table 3. Viral proteins found in rabies

Protein	Property	Molecular weight $\times 10^3$	Number of molecules per virion
G	Glycoprotein	80	1,783
N	Nucleoprotein	60	1,713
M_1	Membrane protein	45	789
M_2	Membrane protein	20	1,661
L	Associated with transcriptase activity	190	—
NS	Minor nucleocapsid protein	55	76

(After Sokol et al. 1971; Flamand and Delagneau 1978.)

The virions contain an RNA minus genome that is not infectious and that is transcribed by the RNA polymerase. This enzyme has been demonstrated in purified virions (Kawai 1977). The molecular weight of the RNA genome is 4.6×10^6 and it contains four cistrons. A possible map of the rabies genome is presented in figure 42 (Flamand and Delagneau 1978).

The RNA molecule is associated with a phosphorylated nucleoprotein (N) in the nucleocapsid which is surrounded by a viral envelope containing two nonglycosylated membrane proteins (M_1 and M_2) and one glycoprotein (G). The high molecular weight L protein found in both virions and infected cells is probably associated with transcriptase activity (table 3).

Infectivity of rabies virus

The virus can infect all warm-blooded animals such as foxes, dogs, cats, bats, and so on. Vampire bats were found to transmit rabies virus by biting.

Rabies in humans

Humans are infected when they are bitten by rabid animals previously infected with the virus; the animals secrete rabies virus in their saliva. Symptoms are characterized by opisthotonic posturing, rigidity, and hydrophobia (fear of water), followed by paralysis and death. The incubation period in humans is usually several weeks or months but may be as long as a year. The closer the bite is to the CNS, the shorter the incubation period. The rabies virus penetrates from the area of the bite into a nerve and travels along axons to the brain where it infects nerve cells only.

Laboratory diagnosis

Touch preparations from the brain of a suspected rabid animal are stained with fluorescein-conjugated antibodies prepared against rabies virions. Histological preparations from the brain are used to identify intracytoplasmic inclusions (Negri bodies). Hair follicles can also be removed and stained by immunofluorescence for rabies virus. A suspension of the brain is also injected into mice intracerebrally, and the mice are observed for three weeks. Those that die are autopsied (or, if no mice die, all are sacrificed and autopsied), and the brains are studied for the presence of Negri bodies and rabies virus antigens.

Epidemiology of rabies

The virus is usually present in the saliva of rabid dogs. Some rabid animals, however, although infected with the virus, do not have the virus in the cells of the salivary glands and still might die of rabies. The virus is spread to domestic animals by wild animals (e.g., foxes in Europe) (see chapter 1).

Immunization of humans and animals

Rabies virus is now grown in human diploid cells under in vitro conditions for the production of a vaccine after inactivation of the virus with formaldehyde (Wiktor et al. 1964). Three consecutive intradermal injections of the vaccine produce lasting immunity. The vaccine is given to immunize individuals bitten by an animal suspected to be rabid (see chapter 24). A live attenuated virus vaccine named *Kelev* was developed by Kumarov for the immunization of dogs and cattle.

GENUS VESICULOVIRUS

Vesicular stomatitis virus (VSV)

Structure of infectious B virions

The virions are bullet-shaped, 180 nm long, and 65 nm wide, with one round end. The virion membrane covers a ribonucleocapsid that contains the viral RNA genome of $3.1–4.0 \times 10^6$ daltons. The ribonucleocapsid has a helical form with 35 ± 1 turns. A virion-associated RNA polymerase (Baltimore et al. 1970) transcribes the viral RNA into five monocistronic mRNA species that code for the fine structural proteins of the virus (Banerjee et al. 1977). The proteins N, L, and NS are associated with the viral RNA in the nucleocapsid that is infectious. The matrix protein (M) surrounds the nucleocapsid that is enclosed in an envelope into which glycoprotein molecules (G) are inserted as surface projections that act as a type-specific immunizing antigen (table 4).

Table 4. The vesicular stomatitis virus proteins

Protein	Function	Molecular weight $\times 10^3$	Number of molecules per virion
G	Glycoprotein (the membrane peplomeres)	170	
M	Membrane protein	29	
N	Nucleocapsid	50	2,300
NS	Nucleocapsid (phosphoprotein)	40	230
L	RNA polymerase	170	60

(After Emerson 1976.)

Additional enzyme activities found in the ribonucleoprotein core include a guanyltransferase, methyltransferases, and a poly(A) polymerase.

Defective truncated (T) virions

Infection of cells at consecutive passage levels at a high multiplicity of infection leads to the appearance of interfering (or defective) virus particles (Pringle 1975). These truncated (T) virions are shorter than the B virions (65 nm long) and contain only one-third of the viral RNA genome (1.2–1.3 $\times 10^6$ daltons).

Replicative cycle of the virus

The virion and cell membranes interact and fuse, and the nucleocapsid enters into the cytoplasm. After a latent period of two hr, the synthesis of the virions starts, reaching a maximum eight hr after infection. The RNA polymerase in the nucleocapsid is activated and RNA^+ molecules are made that are polyadenylated and capped prior to serving as mRNA. The replicative intermediate of the viral RNA has a sedimentation coefficient of 25–35 S, while the virion RNA has a sedimentation coefficient of 42 S. The viral mRNA can be resolved into 28 S molecules and 12–16 S molecules. The 28 S mRNA is translated into the L protein and the 12–16 S mRNA into N, NS, and M proteins. Electrophoresis of the viral mRNA revealed three species in the polyribosomes with molecular weights of 0.75×10^6, 0.59×10^6, and 0.35×10^6 daltons. A model for the synthesis of viral mRNA is presented in figure 43 (Testa et al. 1980). The nucleotide sequence in the mRNA of VSV that allows interaction with the ribosomes is presented in figure 44 (Rose 1978).

Synthesis of virions

The virions are formed by a budding process when the ribonucleoprotein interacts with the cell membrane into which the protein is inserted.

Pseudotypes

Superinfection with VSV of cells latently infected with enveloped RNA or DNA viruses expressing surface glycoproteins may lead to the appearance of

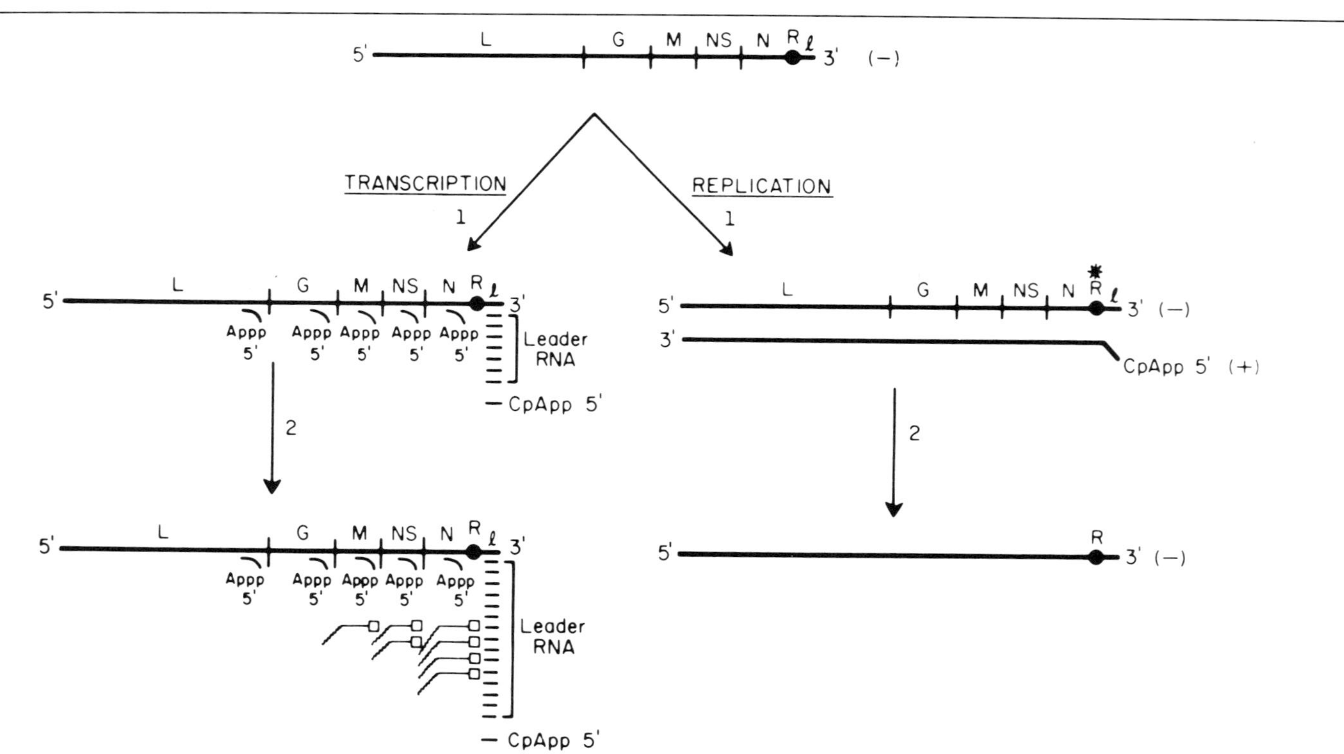

Figure 43. A model for the biosynthesis of VSV mRNA species in vitro. According to this tentative model, the virion-associated RNA polymerase initiates transcription at the 3′-terminus of the genome with the synthesis of a small leader RNA molecule followed by multiple initiations at different promotor sites on the RNA genome. Elongation and completion of the individual mRNAs is dependent on prior transcription of the 3′ proximal genes.

(Testa et al., 1980. Reprinted by permission from *Cell* Vol 21, Fig. 8, p. 274. Copyright © 1980 by MIT)

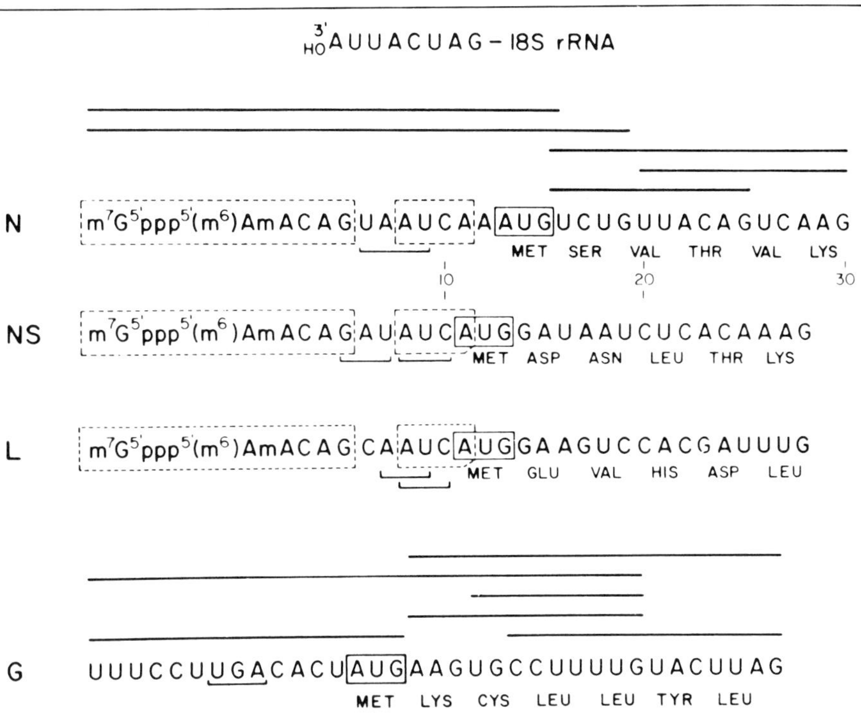

Figure 44. Nucleotide sequences of ribosome-binding sites from the VSV mRNAs specifying the N, NS, L, and G proteins. The T_1 partial products used to establish the sequences of spots 1 and 6 (N and G mRNA sites) are indicated by the solid lines above the sequences. The figure shows the regions of homology (dashed boxes) between N, NS, and L sequences complementary to the 3′-end of 18S ribosomal RNA (brackets under sequences).
(Rose, 1978. Reprinted by permission from *Negative Strand Viruses and the Host Cell*. B.W.J. Mahy and R.D. Barry, eds. Copyright by Academic Press Inc. [London] Ltd. 1978, Fig. 4, p. 55.)

pseudotype virions that contain the genome of VSV and the coat protein of the second virus. These pseudotypes can be recognized by plaque reduction after neutralization with antisera to both viruses. This technique is used to search for the presence of an additional virus in the cells used for infection with VSV (Pringle 1975).

PIKE FRY RHABDOVIRUS (RED DISEASE OF PIKE). Pike fry virus has a typical rhabdovirus morphology. The virions contain an RNA minus genome of 4 × 10^6 daltons.

BIBLIOGRAPHY

Baltimore, D.; Huang, A.S.; and Stampfer, M. Ribonucleic acid synthesis of vesicular stomatitis virus. II. An RNA polymerase in the virion. *Proc. Natl. Acad. Sci. USA 66*:572–576, 1970.

Banerjee, A.K.; Abraham, G.; and Colonno, R.J. Vesicular stomatitis virus: mode of transcription. *J. Gen. Virol. 34*:1–8, 1977.

Emerson, S.U. Vesicular stomatitis virus: structure and function of virion components. *Curr. Top. Microbiol. Immunol.* *73*:1–34, 1976.

Flamand, A., and Delagneau, J.F. Transcriptional mapping of rabies virus in vivo. *J. Virol.* *28*:518–523, 1978.

Kawai, A. Transcriptase activity associated with rabies virion. *J. Virol.* *24*:826–835, 1977.

Pringle, C.R. Conditional lethal mutants of vesicular stomatitis virus. *Curr. Top. Microbiol. Immunol.* *69*:85–116, 1975.

Rose, J.K. Ribosome recognition sites in vesicular stomatitis virus messenger RNA. *In: Negative Strand Viruses and the Host Cell* (B.W.J. Mahy, and R.D. Barry, eds.), Academic Press, N.Y., 1978, pp. 47–60.

Sokol, F.; Stancek, D.; and Koprowski, H. Structural proteins of rabies virus. *J. Virol.* 7:241–249, 1971.

Testa, D.; Chandra, P.K., and Banerjee, A.K. Unique mode of transcription in-vitro by vesicular stomatitis virus. *Cell* *21*:267–275, 1980.

Wiktor, T.J.; Fernandes, M.U.; and Koprowski, H. Cultivation of rabies virus in human diploid cell strain W1-38. *J. Immunol.* *93*:353–366, 1964.

RECOMMENDED READING

Ball, L.A. and Wertz, G.W. VSV RNA synthesis: how can you be positive? *Cell* *26*:143–144, 1981.

Huang, A.S. Viral pathogenesis and molecular biology. *Bacteriol. Rev.* *41*:811–821, 1977.

Pringle, C.R. Genetics of rhabdoviruses. *In: Comprehensive Virology* (H. Fraenkel-Conrat and R.R. Wagner, eds.), Plenum Press, New York, 1977, Vol. 9, pp. 239–290.

Schneider, L.G., and Diringer, H. Structure and molecular biology of rabies virus. *Curr. Top. Microbiol. Immunol.* *75*:153–180, 1977.

Toma, B., and Andral, L. Epidemiology of fox rabies. *Adv. Virus Res.* *21*:1–36, 1977.

Wagner, R.R. Reproduction of rhabdoviruses. *In: Comprehensive Virology* (H. Fraenkel-Conrat and R.R. Wagner, eds.), Plenum Press, New York, 1975, Vol. 4, pp. 1–94.

14. SINGLE-STRANDED RNA MINUS VIRUSES WITH AN INTACT GENOME: PARAMYXOVIRUSES

FAMILY PARAMYXOVIRIDAE

The virions are round with a diameter of 150 nm, but there are also pleomorphic particles. The RNA is a single minus strand with a molecular weight of $5–8 \times 10^6$. The virions have a ribonucleocapsid within a lipid envelope and are genetically stable.

Genus Paramyxovirus

Newcastle disease virus (NDV) infects chickens (figure 45).
Mumps virus infects humans.
Parainfluenza virus (Sendai virus; figure 46) infects humans and mice.
Parainfluenza virus 2 infects humans, mice, dogs, bats, and birds.
Parainfluenza virus 3 infects humans, horses, and cattle.

Turkey parainfluenza virus Yucaipa virus Finch and parrot paramyxovirus	} infect birds.

Genus Morbillivirus (measles: distemper group)

Measles virus (figure 47) infects humans.
Canine distemper virus infects dogs and cats.
Rinderpest virus infects cattle.

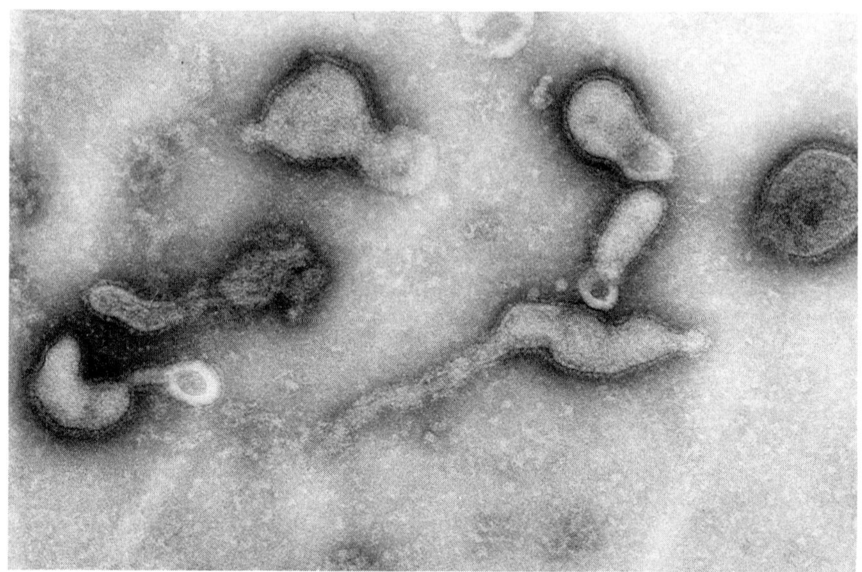

Figure 45. Electron micrograph (negative staining) of Newcastle disease virus (× 120,000). (By courtesy of Dr. Daniel Dekegel, Institut du Brabant, Brussels, Belgium.)

Pneumovirus (respiratory syncytial virus subgroup)

Respiratory syncytial virus infects humans.
Bovine respiratory syncytial virus infects cattle.
Pneumonia virus of mice.

VIRAL PROTEINS AND RNA

There are two types of spikes on the envelope of paramyxoviruses. Both are glycoproteins: Type HN has hemagglutinating and neuraminidase activity; type F induces cell fusion and hemolysis. There is a larger precursor protein (Fo) to glycoprotein F that has been observed with Sendai virus and with NDV. The M protein forms part of the virus envelope (table 5).

The molecular weight of the viral RNA is $5–8 \times 10^6$, and in the virion it is associated with the P (RNA polymerase) and NP proteins forming the helical nucleocapsid, 17–18 nm long and 5 nm wide.

Adsorption and penetration of virions into the cells

The virions attach to neuraminic acid-containing receptors (HN glycoprotein) on the host cell membrane. After adsorption, the precursor Fo protein is cleaved and fusion of the virion with the cell membrane is induced by the F glycoprotein. A general feature of paramyxoviruses is that the precursor glycoprotein (Fo) undergoes proteolytic cleavage to yield two disulfide-bonded

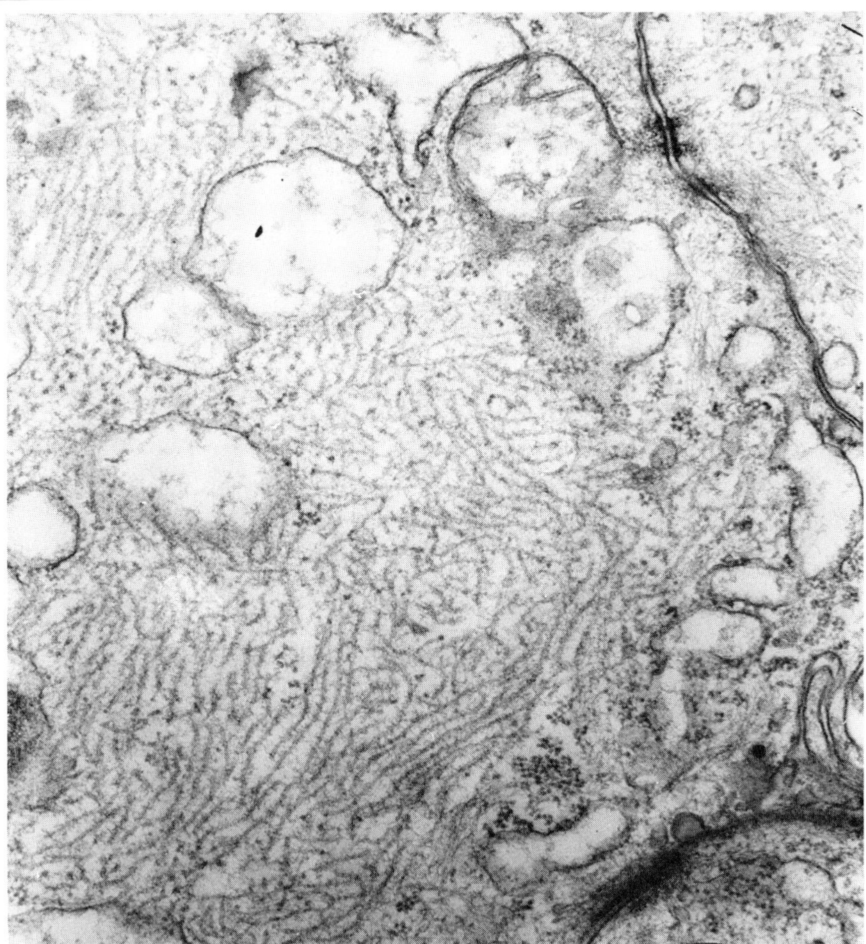

Figure 46. Electron micrograph of Sendai virus infected cell (× 40,000). Note the viral ribonucleoprotein complexes in the cell cytoplasm.
(By courtesy of Dr. Daniel Dekegel, Institut Pasteur du Brabant, Brussels, Belgium.)

glycoproteins (F_1 and F_2). Activation of cell fusion, hemolysis, and the initiation of infection is the result of a cleavage-induced conformational change in which the two polypeptides form an active complex (the F protein) (Scheid and Choppin 1977). It should be noted that cleavage of a precursor (HNo) to the HN protein is also a precondition for its biological activity (Nagai and Klenk 1977).

After entry of the nucleocapsid into the cytoplasm, the viral RNA-dependent RNA polymerase is activated and transcribes the viral RNA minus genome into RNA plus mRNA.

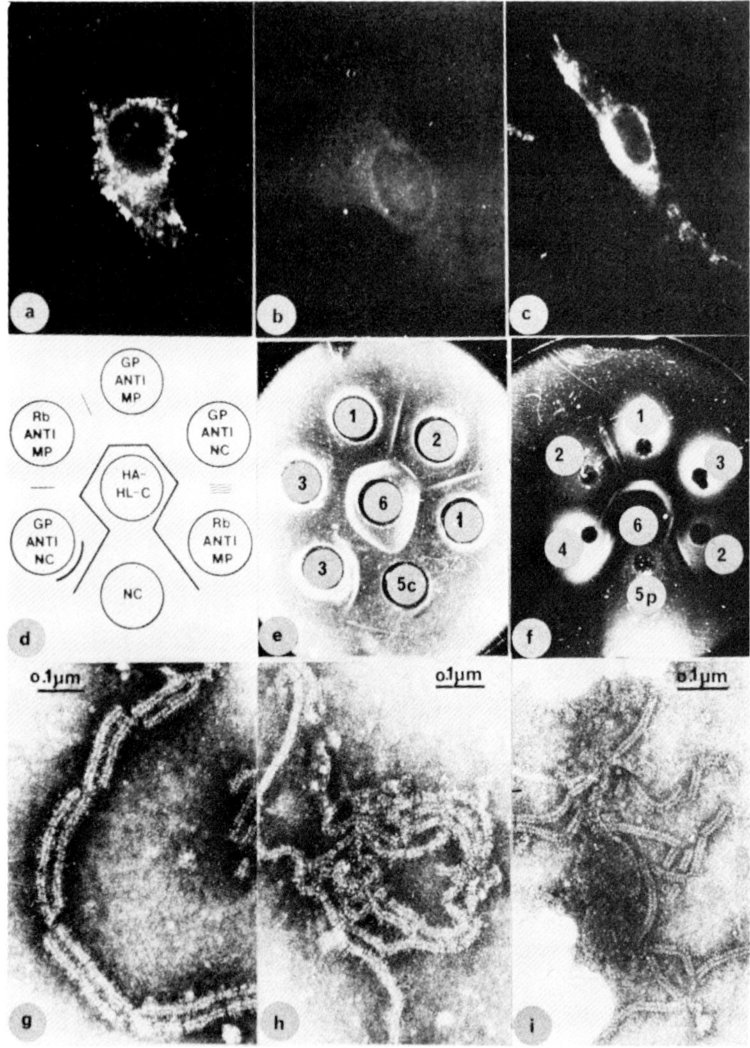

Figure 47. Antigens of measles virus. a–c: Indirect immunofluorescent staining of HEP_2 cells infected with measles virus; a—unfixed cells; b—acetone-fixed cells stained with anti-HL serum and FITC-conjugated anti-IgG; c—cells fixed by drying and stained similarly; d–f: Immunodiffusion reactions of HA-HL complex and unpurified nucleocapsid (NC); 1, guinea-pig anti-HA; 2, rabbit anti-MP; 3, guinea-pig anti-HL; 4, guinea-pig anti-NC; 5c, crude preparation of NC treated with SDS; 5p, purified preparation of NC treated with SDS; 6, Tween-20 treated HA; g–i, electron micrographs of NC treated with antisera; equal volumes of NC and antibodies were incubated 1 hr at 37°C and the product applied to carbon-coated grids with 2% phosphotungstate contrast stain; g-NC with guinea-pig anti-NC; h-NC with unabsorbed rabbit anti-MP; i-NC with absorbed rabbit anti-MP.

(Fraser et al. 1978. Reprinted by permission from *Negative Strand Viruses and the Host Cell*. B.W.J. Mahy and R.D. Barry, eds. Copyright by Academic Press Inc. [London] Ltd. 1978, Fig. 1, p. 778.)

Table 5. Proteins present in the paramyxovirus

Protein	Molecular weight ($\times 10^3$)	Function
P	69	RNA polymerase
NP	56–61	In the nucleocapsid
M	38–41	Membrane protein
F	53–56	Cell fusion glycoprotein
Fo	65	Precursor to F glycoprotein
HN	67–74	Hemagglutinin and neuraminidase (glycoprotein)

Viral RNA species in the infected cells

The virion genomic RNA has a sedimentation coefficient of 50S in sucrose gradients, but in the cytoplasm of infected cells, viral RNA with sedimentation coefficients of 18, 22, and 35 S are present. The 35 S molecule contains 70% of the genetic information of the viral RNA. The 18 S RNA molecules have a molecular weight of $5.5–15 \times 10^5$. Each of these RNA species seems to contain genetic information for one viral gene. Although not yet proven, it appears that, even though the viral genome consists of one RNA molecule that replicates in the cytoplasm, each viral gene operates by separately producing the mRNA species. The nature of the enzyme that synthesizes the new progeny of viral RNA minus genomes is not known.

Synthesis of viral proteins

Arginine must be present in the medium of cells infected with NDV for the synthesis of virions. In the absence of arginine in the culture medium, the nucleocapsids, hemagglutinin, and neuraminidase are synthesized, but the process of budding of the nucleocapsids through the cell membrane is prevented. As a result, red blood cells cannot adsorb to the arginine-deprived infected cells and the hemadsorption test is negative. Addition of arginine to the medium results in the formation of complete virions.

Defective virions

Incomplete virions are formed in infected cells during the course of virus synthesis. Defective Sendai virions were found to contain a short RNA genome with a sedimentation coefficient of 19–24 S. The complete RNA genome has a sedimentation coefficient of 50 S. Large quantities of defective virus are produced in chick embryos or cell cultures infected at a high multiplicity of infection, and these particles interfere with the replication of the residual infectious virus.

Persistent infection in cultured cells

Paramyxoviruses infect only some of the cells in culture, and the virus can persist throughout a number of consecutive cell passages in vitro. This phenomenon was described with Sendai, mumps, NDV, and simian virus 5 (SV5) viruses.

Paramyxovirus mutants

Plaque morphology mutants

PLAQUE SIZE MUTANTS. In uncloned virus stocks, it is possible to detect plaques that differ in size. Purified virus clones have been obtained with distinct plaque morphology. However, in NDV, the plaque size property is unstable.

MUTANTS CAUSING UPTAKE OF THE STAIN NEUTRAL RED BY INFECTED CELLS. Wild type virus infected cells that form the plaque do not incorporate the dye neutral red, and therefore the plaque can be seen as a transparent region in the cell monolayer that stains red with the dye. Virus mutants were isolated from the wild type stocks that caused infected cells to incorporate the dye and appear as red plaques.

MUTANTS WITH OPAQUE PLAQUES. A spontaneous mutant that gave opaque plaques was described for NDV.

Ts mutants

Five complementation groups (A–E) were found among the ts mutants of NDV. From Sendai virus, mutagenized with 5-fluorouracil (5FU), seven complementation groups of ts mutants were obtained. Mutants of measles virus were divided into four complementation groups.

DISEASES CAUSED BY PARAMYXOVIRUSES

Mumps in humans

Mumps virus infects children mainly and affects the salivary glands, especially the parotid gland. The virus can also cause meningoencephalitis, orchitis, oophoritis, and pancreatitis.

The virus replicates in the pharynx of the infected child and then spreads to the salivary glands. The incubation period is 16–18 days, and the virus can be isolated from the saliva or blood at the onset of symptoms as well as from the urine. After recovery from the virus infection, the patient has lifelong immunity. A live attenuated virus vaccine is available for immunization of children (chapter 24).

Parainfluenza virus infections of humans

These viruses cause croup (laryngo-tracheo-bronchitis) and can be isolated in cell cultures from throat washings of sick children. The presence of the virus in the infected cultures can be detected by hemadsorption of red blood cells to the infected cells.

Newcastle disease in chickens

This virus infection appears as an epidemic in chicken flocks. The virulence of the virus strains determines the severity of the epidemic. The virus spreads by an airborne infection and infects healthy chickens via the respiratory tract. It also infects the CNS, causing a meningoencephalomyelitis.

NDV can cause conjunctivitis in humans that heals spontaneously after 3–4 days.

MORBILLIVIRUSES

These viruses lack neuraminidase on the virion envelope and are therefore unable to adsorb to neuraminic acid receptors. Measles virus is an exception; it can hemagglutinate red blood cells.

Measles virus in humans

The virions have a pleomorphic shape ranging from 120–270 nm in diameter. The nucleocapsid has a diameter of 17–18 nm with a central core of 5 nm. The viral RNA has a molecular weight of 6.2×10^6 and a sedimentation coefficient of 52 S.

One of the virion proteins has a molecular weight of 60,000 and belongs to the nucleocapsid (figure 46). This protein is cleaved by the proteolytic enzymes in the virion envelope into two proteins of 38 and 24×10^3 daltons. A 70,000 dalton protein is also present in the nucleocapsid (Mountcastle and Choppin 1977) and is probably the viral RNA polymerase (Seifried et al. 1978). The largest polypeptide (probably the HA protein of the envelope) has a molecular weight of 79,000. Other envelope proteins are the M protein of 36,000 daltons and two proteins of 40,000 and 20,000 daltons which are probably the F_1 and F_2 proteins, respectively (Tyrrell and Norrby 1978). Measles virus fuses cells less well than Sendai virus. Various properties of measles virus are illustrated in figure 47.

The virus spreads as an airborne infection in droplets from the mouth. The incubation period in the infected individual is 9–12 days, and the disease is accompanied by fever, cough, conjunctivitis, and macules (Koplik spots) on the mucosa. Three days later, a red rash appears on the face and head. In 10–20% of infected children, otitis media appears, followed by pneumonia and a secondary bacterial infection. Encephalitis appears in 1 out of 1,000 children infected with measles virus. The disease is lethal in about 15% of the children who develop encephalitis. Measles infection of the brain causes progressive damage, including epilepsy and changes in personality. Subacute sclerosing panencephalitis (SSPE) is a rare disease in children and young adults which appears many years after the childhood measles virus infection.

A live attenuated measles vaccine is available for the immunization of children (chapter 24).

Hecht giant cell pneumonia

A lethal pneumonia characterized by multinuclear giant cells that appear in the lungs was reported in children with malignant diseases and in children after measles.

Damage to the immune system

In lethal infections of children with measles virus, destruction of the cortex of the thymus was noted. The reason for this damage is not known, but measles virus can infect lymphocytes in vitro.

Subacute sclerosing panencephalitis (SSPE)

This is a progressive and fatal CNS infection in children and young adults, initially characterized by loss of intellectual faculties, followed later by paralysis, myoclonic seizures, coma, and death. Pathological changes are those of a diffuse (pan-) encephalitis in both the gray and white matter of the brain, areas of demyelination, striking astrocytosis, inflammation, and neuronal loss. Cowdry A inclusion bodies found in neurons and glial cells contain measles virus nucleocapsids. In the serum and cerebrospinal fluid, a high titer of antibodies to measles virus is found. Measles virus antigens are also found by immunofluorescence and immunoperoxidase techniques in brain tissue from SSPE patients. Cocultivation or cell fusion of SSPE brain cells with permissive cells resulted in the isolation of measles virus (Horta-Barbosa et al. 1969). The virus has been demonstrated in lymph nodes of SSPE patients (Horta-Barbosa et al. 1971). About 50% of the SSPE patients were found to have had measles before the age of two and the SSPE started at ten years of age.

Measles virus isolated from SSPE brains differs from the virus that causes conventional measles in children. The RNA of the SSPE virus has only 60% homology with the 18S RNA of measles virus, indicating only partial relatedness.

Respiratory syncytical virus (RSV)

This virus causes acute infections of the lower respiratory tract and epidemics of broncheolitis in the winter. About 25% of childhood pneumonia in early infancy is caused by RSV. The virus infection in newborn babies can be lethal. No vaccine is available.

BIBLIOGRAPHY

Frazer, K.B.; Gharpure, M.; Shirodaria, P.V.; Armstrong, M.A.; Moore, A.; and Dermott, E. Distribution and relationships of antigens of measles virus membrane complex in the infected cell and the nature of the haemolysin antigen. *In: Negative Strand Viruses and the Host Cell* (B.W.J. Mahy and R.D. Barry, eds.), Academic Press, London, New York, San Francisco, 1978, pp. 771–780.

Horta-Barbosa, L.; Fuccillo, D.A.; Sever, J.; and Zerman, W. Subacute sclerosing panencephalitis. Isolation of measles virus from a brain biopsy. *Nature 221*:974, 1969.

Horta-Barbosa, L.; Hamilton, R.; Wittig, B.; Fuccillo, D.A.; and Sever, J.L. Subacute sclerosing panencephalitis: Isolation of suppressed measles virus from lymph node biopsies. *Science 173*:840–841, 1971.

Mountcastle, W.E., and Choppin, P.W. A comparison of the polypeptides of four measles virus strains. *Virology 78*:463–474, 1977.

Nagai, Y., and Klenk, H.-D. Activation of precursors to both glycoproteins of Newcastle disease virus by proteolytic cleavage. *Virology* 77:125–134, 1977.

Scheid, A., and Choppin, P.W. Two disulfide-linked chains constitute the active F protein of paramyxoviruses. *Virology 80*:54–66, 1977.

Seifried, A.S.; Albrecht, P.; and Milstein, J.B. Characterization of an RNA-dependent RNA polymerase activity associated with measles virus. *J. Virol. 25*:781–787, 1978.

Tyrrell, D.J., and Norrby, E. Structural polypeptides of measles virus. *J. Gen. Virol. 39*:219–229, 1978.

RECOMMENDED READING

Bratt, M.A., and Hightower, L.E. Genetics and paragenetic phenomena of paramyxoviruses. *In: Comprehensive Virology* (H. Fraenkel-Conrat and R.R. Wagner, eds.), Plenum Press, New York, 1977, Vol. 9, pp. 457–534.

Choppin, P.W., and Compans, R.W. Reproduction of paramyxoviruses. *In: Comprehensive Virology* (H. Fraenkel-Conrat and R.R. Wagner, eds.), Plenum Press, New York, 1975, Vol. 4, pp. 95–178.

Ishida, N., and Homma, M. Sendai Virus. *Adv. Virus Res. 23*:349–383, 1978.

Kingsbury, D.W. Paramyxovirus replication. *Curr. Top. Microbiol. Immunol. 59*:1–34, 1972.

Lucas, A.; Coulter, M.; Anderson, R.; Dales, S.; and Flintoff, W. In vivo and in vitro models for demyelinating diseases. II. Persistence and host-regulated thermosensitivity in cells of neural derivation infected with mouse hepatitis and measles viruses. *Virology 88*:325–337, 1978.

15. SINGLE-STRANDED RNA MINUS VIRUSES WITH A FRAGMENTED LINEAR GENOME: ORTHOMYXOVIRUSES

FAMILY ORTHOMYXOVIRIDAE

These virions (figure 48) are enveloped with a lipid membrane, are round or elongated in shape, with a diameter of 80–120 nm. The viral proteins, hemagglutinin (HA) and neuraminidase (NA) are in the envelope (figures 49 and 50). The nucleocapsid contains the RNA-dependent RNA polymerase, and a single-stranded RNA genome is made up of eight distinct species with a total molecular weight of between $4–5 \times 10^6$. Each RNA molecule comprises one gene. Genes are exchanged when recombination between different virus strains takes place.

Genus influenzavirus

Influenza A, B, and C viruses

Human influenza virus A strains that cause epidemics or pandemics are designated according to the antigenicity of the hemagglutinin or the neuraminidase. There are also influenza virus strains that belong to groups B and C found only in man, but these are not associated with pandemics.

The group A viruses are designated according to the antigenicity of the HA and NA antigens and the animal source of the virus. The H (hemagglutinin) types include HO through H3 (human), HSW1 (swine), Heq1 and Heq2 (equine), HAV1 through HAV8 (avian), and the N (neuraminidase) types

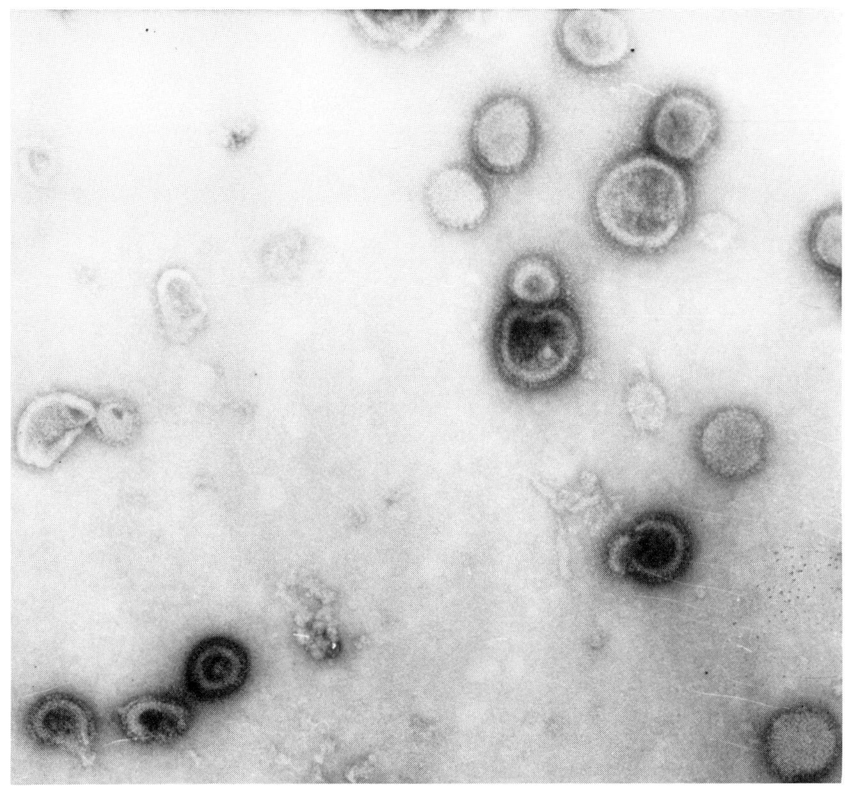

Figure 48. Electron micrograph (negative stain) of influenza virions PR 8 strain. (By courtesy of Dr. Daniel Dekegel, Institut Pasteur du Brabant, Brussels, Belgium.)

include N1 and N2 (human), Neq1 and Neq2 (equine), and Nav1 through Nav5 (avian). These surface antigens may be shared to give the following strains: Hsw1N1 (swine), Heq2Neq1, Heq2Neq2 (equine), and Hav1Neq1, Hav2Neq1, Hav3Nav1, Hav4Nav1, Hav5Nav2, Hav6N2, Hav7Neq2, Hav8-Nav4 (fowl plague virus: respiratory infections in birds).

Melnick (1980) reported on the regrouping of the hemagglutinin and neuraminidase subtypes of influenza virus: HO, H1 and HSW1 are named H1; H2 is unchanged; new H3 replaces H3, Heq2, Hav7; new H4 instead of Hav4; new H5 replaces Hav5; H6 replaces Hav6; H7 replaces Heq1, Hav1; H8 replaces Hav8; H9 replaces Hav9; H10 replaces Hav2; and H11 replaces Hav3.

N1 and N2 remain unchanged; N3 replaces Nav2, Nav3 and Nav6; N4 replaces Nav4; N5 replaces Nav5; N6 replaces Nav1; N7 replaces Neq1; and N8 replaces Neq2.

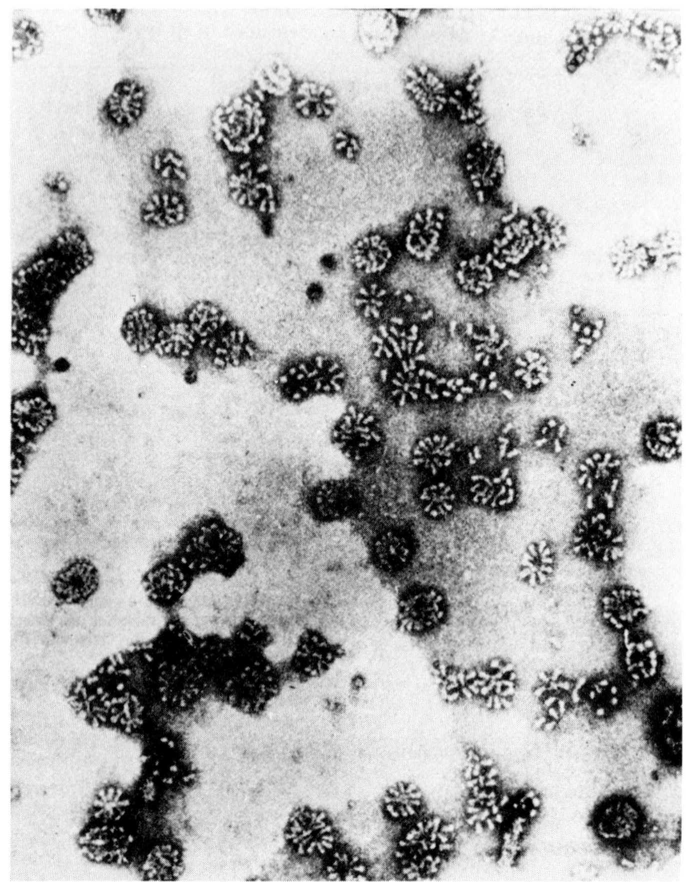

Figure 49. Electron micrograph (negative stain) of purified hemagglutinin and neuraminidase subunits, aggregated into clusters by binding of hydrophobic regions (× 250,000).
(Laver 1976. Reprinted by permission from *Influenza: Virus, Vaccines, Strategy*. P. Selby, ed. Copyright by Academic Press Inc. [London] Ltd. 1976, Fig. 3, p. 167.)

Viral proteins

Each of the eight RNA fragments constituting the viral genome codes for a protein as shown in table 6 (Laver 1973; Palese and Schulman 1976; Ritchey et al. 1976 *a,b*; Palese et al. 1977; Scholtissek 1978). The neuraminidase gene was identified as the sixth RNA segment of A/PR8 virus and as the fifth segment of A/Hong Kong virus, whereas the gene for the nucleoproteins was identified as segment 5 of PR8 virus and segment 6 of Hong Kong virus, respectively. The HA glycoprotein that is synthesized as the product of the fourth gene segment is cleaved into two polypeptides HA1 and HA2 which are linked by a disulphide bond. The virions contain the RNA-dependent RNA polymerase that is

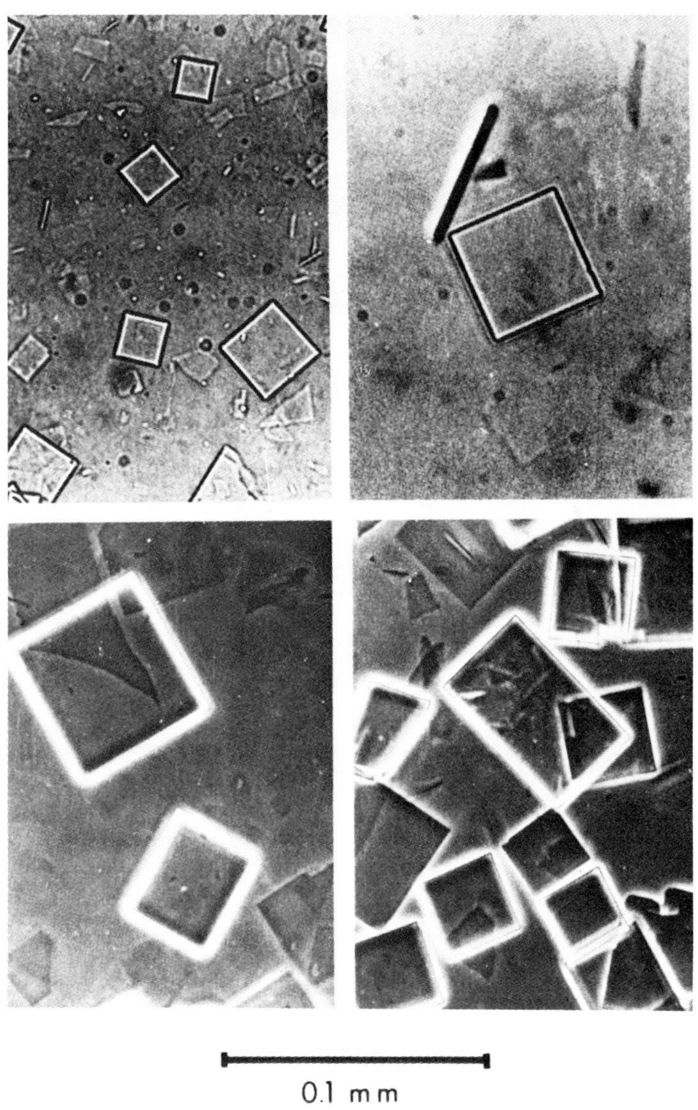

Figure 50. Crystals of purified neuraminidase.
(Laver, 1978. Reprinted by permission from *Virology* 86, Fig. 2, p. 82.)

Table 6. Proteins coded by various RNA species of influenza virus

RNA segment	Molecular weight ($\times 10^4$)	Protein	Molecular weight ($\times 10^3$)	Function
1	98–115	P3	100	Internal protein
2	95–99	P1	89–96	Internal protein
3	93–98	P2	80–83	Internal protein
4	73–85	HA	75–80	Hemagglutinin
5	64–70	NP	53–60	Nucleoprotein
6	57–64	NA	55–70	Neuraminidase
7	31–47	M	21–27	In the inner part of the membrane
8	24–39	NS	23–25	Nonstructural protein

associated with three proteins in the ribonucleoprotein (RNP) complex. The cellular RNA polymerase II is essential for the infectious process in influenza-infected cells.

Structure and organization of the influenza virus hemagglutinin gene

The complete nucleotide sequence of the HA gene was obtained by Porter and associates (1979), using polyadenylation of the eight RNA components of fowl plague virus (FPV) followed by incubation with the RNA-dependent DNA polymerase (reverse transcriptase), using an oligo(dT) primer. Under these conditions, full-length complementary double-stranded DNA copies (cDNA) of the eight viral genes were obtained. The viral cDNA molecules, treated with edonuclease S1 to remove ssDNA molecules, were ligated to the DNA of the plasmid pBR322 cut with *Hind*III using the *Hind*III linker method. The DNA mixture was used to transform *E. coli* strain K12HB101, and 13 colonies of bacteria containing the plasmid were established. One plasmid pBR322-FPV4-10 contained a large insert of about 1,700 nucleotides. The plasmid was hybridized to a ^{32}P-labeled gene 4 probe. The DNA insert in the plasmid was shown to be a nearly full-length copy of the HA gene.

Adsorption, penetration, and uncoating of viral RNA

The virions adsorb to neuraminic acid-containing receptors on the surface of the cell membrane. The viral neuraminidase cleaves the cellular receptor and the viral envelope fuses with the cell membrane so that the viral RNA is released into the cell cytoplasm.

Viral RNA synthesis

The initiation of RNA synthesis within the viral RNP is dependent on the host cell. In α-amanitin treated cells in which the cellular RNA polymerase II is inhibited, synthesis of viral RNA is prevented. Similarly, treatment of infected cells with actinomycin D, which binds to cellular DNA and inhibits transcription, prevents the synthesis of viral RNA. It was suggested that cap structures of the cellular mRNAs are cleaved and used as primers for the initiation of the viral RNA-dependent RNA polymerase. The virion enzyme is stimulated to initiate RNA synthesis by the dinucleotide primers ApG or GpG (Bouloy et al. 1978; Skehel and Hay 1978).

The synthesis of viral RNA^{+} molecules that serve as mRNA occurs in the RNP complex. The viral mRNAs released from the RNP complex are translated by the cellular ribosomes. New viral RNA minus molecules are synthesized by the RNA-dependent RNA polymerases that use the viral RNA^{+} molecules as templates.

Virion assembly

The newly synthesized viral proteins interact with the progeny viral RNA^{-} molecules to form the RNP complex. The mechanism that allows assortment of the eight fragments into one viral RNP complex is not yet known. The viral glycoproteins are inserted into the cytoplasmic membrane and, in due course, into the outer membrane of the cell. The viral RNPs interact with the cell membrane at sites where the viral M protein molecules are inserted. Attachment of the RNP to the cell membrane causes the latter to undergo a budding process whereby the RNP is enveloped by the cell membrane and released from the infected cell.

The viral HA, a glycoprotein situated in the virion envelope arranged as spikes on the virus surface, is a trimer of about 250,000 daltons. The monomers of HA are made in infected cells as precursor polypeptides with the addition of 18 amino acids at the N terminus. During maturation and virus assembly, the precursor is cleaved to HA1 and HA2 polypeptides which are linked by disulfide bridges. Cleavage of the HA glycoprotein is necessary for the infectivity of influenza virions (Klenk et al. 1975).

Incomplete defective virions: von Magnus effect

Infection of chick embryos with large amounts of influenza virus leads to the synthesis of defective virus progeny that interferes with the infectious process. The reduction in infectivity caused by such virus particles is called the von Magnus effect. Analysis of virion RNA revealed that one RNA gene is missing, thus leading to incomplete (defective) virus.

Plaque assay for influenza virus

Influenza virus can form plaques on monolayer cultures of chick embryo fibroblasts only if the proteolytic enzyme trypsin is added to the culture medium. The enzyme cleaves a protein in the virion envelope and allows the virus particles released from the cells to infect neighboring cells (figure 51).

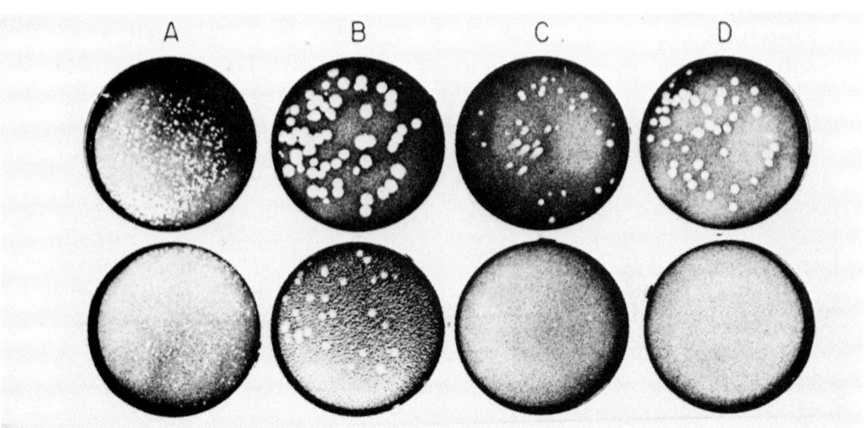

Figure 51. Trypsin dependent plaque formation by WSN(HA) Turkey/Ontario (NA) recombinants in chick embryo fibroblasts.
Upper row: 5 μg trypsin/ml. Lower row: No trypsin
A — Turkey/Ontario
B — WSN
C — Recombinant containing all RNA segments from turkey/Ontario except that coding for the WSN hemagglutinin
D — Recombinant containing all RNA segments from WSN except that coding for the turkey/Ontario neuraminidase
(Bean and Webster 1978. Reprinted by permission from *Negative Strand Viruses and the Host Cell*. B.W.J. Mahy and R.D. Barry, eds. Copyright by Academic Press Inc. [London] Ltd. 1978, Fig. 3, p. 689.)

Influenza virus recombinants

Infection of the same cell with two influenza viruses leads to the appearance of virus recombinants that contain RNA genes from both parent viruses. An analysis of the RNAs in a recombinant derived from a dual infection with influenza virus strains PR8 and HK is presented in figure 52. Table 7 shows the relationship between the amount of virus obtained in embryonated eggs and the gene arrangement in the recombinant viruses.

Mutants of orthomyxoviruses

Ts mutants of influenza viruses were found to belong to seven or eight recombination-complementation groups. The hypothesis is that each recombination group can be equated with any individual gene (Webster and Bean 1978).

Influenza in man

The virus infection starts with an upper respiratory tract infection and leads to coughing, nasal excretions, and, in some cases, pneumonia. Different virus strains differ in virulence (table 8).

Influenza persists as an uncontrollable infection in man because of the changes that the HA and NA antigens undergo. This is referred to as antigenic

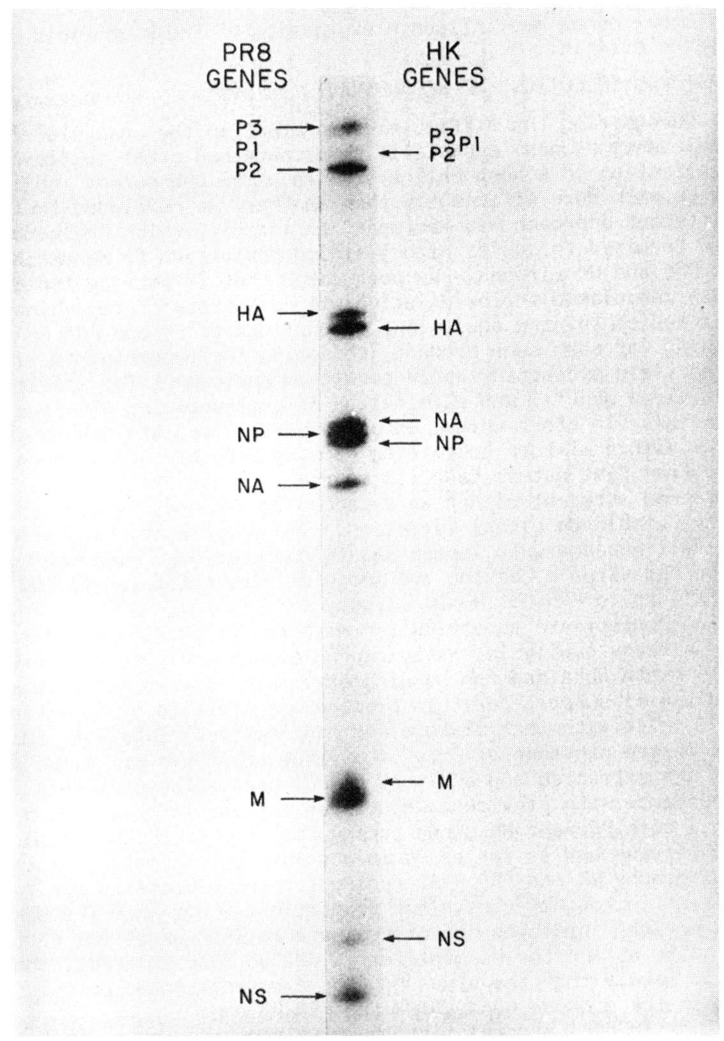

Figure 52. Polyacrylamide gel analysis of the RNAs of a recombinant mixture obtained 8 hrs after mixed infection of MDCK cells with influenza A/HK/8/68 and A/PR/8/34 viruses. Arrows on the left indicate PR8 virus RNAs in the mixture; arrows on the right indicate HK virus RNAs.

(Schulman and Palese, 1978. Reprinted by permission from *Negative Strand Viruses and the Host Cell*. B.W.J. Mahy and R.D. Barry, eds. Copyright by Academic Press Inc. [London] Ltd. 1978, Fig. 1, p. 668.)

Table 7. Relationship of yield in embryonated eggs and gene constellation of recombinants of PR8 and HK viruses[a]

	Gene derivation[b]								
Recombinant	P1	P2	P3	HA	NA	NP	M	NS	HA titer
1	P	P	P	P	H	P	P	P	16,384
2	H	H	H	P	H	P	P	P	8,192
3	P	P	P	P	H	P	P	P	8,192
4	P	P	P	H	P	P	P	P	8,192
5	P	P	P	H	P	P	P	P	8,192
6	H	H	P	H	P	P	P	H	4,096
7	H	H	H	H	H	H	H	P	256
8	H	H	H	P	P	P	H	P	512
9	H	P	H	P	H	P	P	P	256
10	H	H	H	P	H	H	H	H	256
11	H	H	P	P	H	H	P	P	512
12	P	P	P	H	P	P	H	P	128
13	H	H	P	H	P	H	H	P	32
14	H	P	H	H	P	P	H	P	256

(From Schulman and Palese 1978. Reprinted by permission from *Negative Strand Viruses and the Host Cell.* B.W.J. Mahy and R.D. Barry, eds. Copyright by Academic Press, Inc. London, Ltd., 1978, Table 2, p. 666.)

[a]All cloned recombinants derived from a single mixture in bovine kidney cells.

[b]H = gene from HK virus; P = gene from PR8 virus.

Table 8. Influenza A viruses (or comparable serotypes) of defined human virulence used in the study or from which recombinants were made

Name	Surface antigens	Origin and passage	Whether tested in human trials	Known or presumed virulence
A/PR/8/34	H0 N1	Unrecorded passages	Yes	0
A/Okuda/57	H2 N2	280 egg passes	No	±
A/Hong Kong/1/68	H3 N2	Few egg passes	Yes	+ + +
A/England/939/69	H3 N2	Isolate from clinical influenza	No	+ + +
A/England/878/69	H3 N2	Few egg passes	Yes	+ + +
A/England/42/72	H3 N2	" " "	Yes	+
A/Port Chalmers/1/73	H3 N2	" " "	Yes	+ + +
A/Finland/4/74	H3 N2	" " "	Yes	+ + +
A/Scotland/840/74	H3 N2	" " "	Yes	+ + +
A/Victoria/3/75	H3 N2	" " "	No	+ + +

(From Beare, et al. 1978. Reprinted by permission from *Negative Strand Viruses and the Host Cell*. B.W.J. Mahy and R.D. Barry, eds. Copyright by Academic Press, Inc. London, Ltd., 1978, Table 1, p. 747.)
Note: Testing in human trials consisted of inoculation into antibody-free volunteers and observation for clinical effects, virus excretions and antibody rises. Virulence: + + + = capable of causing influenzal symptoms; + + = local symptoms and constitutional symptoms; + = local symptoms only; ± = overattenuation; 0 = non-infectious. Only mild effects were observed with wild A/England/42/72, but it was difficult to find antibody-free volunteers.

drift and in type A influenza virus it is believed to be due to selection, in an immune population, of mutant virus particles with altered antigenic determinants on the HA proteins (see chapter 25). Changes in the amino acid sequences of the HA of influenza A/Hong Kong virus over a period of nine years (1968–1977) have been demonstrated (Laver et al. 1980).

Influenza virus vaccines

The available killed virus vaccines contain the currently known epidemic-causing influenza virus strains. When a new influenza virus strain appears and threatens to develop into an epidemic, a new vaccine must be prepared, but influenza virus freshly isolated from man does not grow well in embryonated eggs. To facilitate production of the new virus strains in large quantities, recombinants between the new virus and an influenza strain capable of replication in embryonated eggs are made. A recombinant virus is selected that contains the HA or NA of the new influenza mutant and all the other genes of the adapted virus strain (Murphy et al. 1978). The recombinant virus is further propagated in embryonated eggs and used for the production of inactivated influenza virus vaccine (see chapter 24).

A new approach to vaccine production: cloning of the viral HA gene in bacterial plasmids.

Genetic engineering techniques can be used to clone any viral gene in bacterial plasmids (see chapter 7). Expression of the viral gene in bacteria into which the cloned plasmids have been inserted provides a method of large-scale produc-

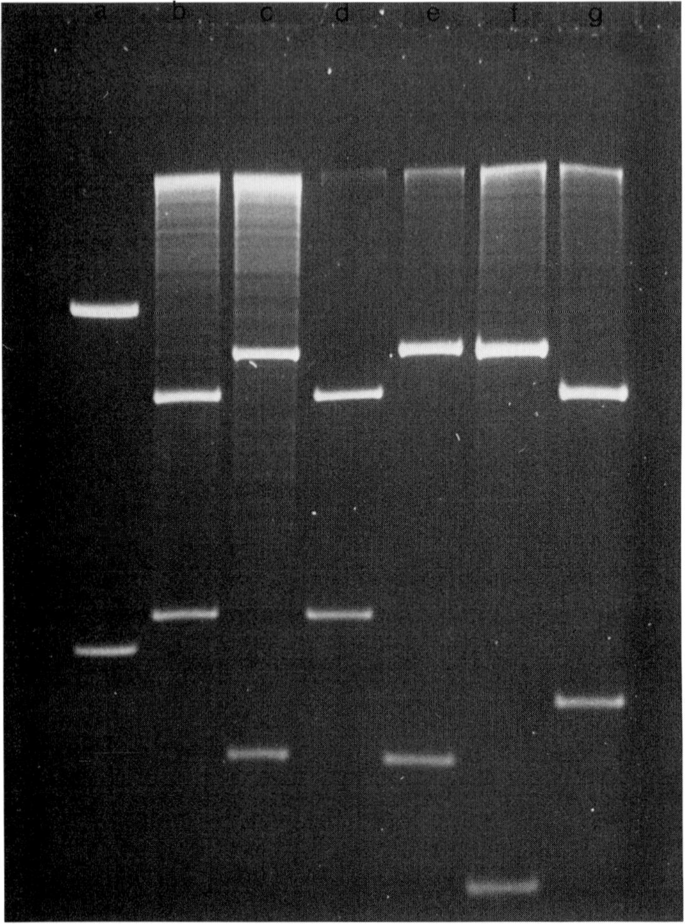

Figure 53. Orientation of FPV gene in plasmids pWT111, pWT121, and pBR322. A. Cleavage of plasmid DNA with restriction enzymes and electrophoresis in agarose gels. Lane (a) contains PM-2 DNA digested with HindIII. The other lanes contain PstI digests of various plasmids. B. Structure of pWT121/FPV411(R). The plasmid shown contains the FPV-HA gene cloned at the *Hin*dIII site in the R orientation.

(Emtage et al. 1980. Reprinted by permission from *Nature* 283, Fig. 2, p. 172. Copyright © 1980 Macmillan Journals Limited.)

tion of viral antigens. The bacteria produce the viral antigen as part of their cellular proteins; the amount of the viral protein produced is 1–2% of the total bacterial proteins. The viral protein can be purified by chromatographic methods (e.g., columns containing specific antibodies to the viral antigen). These methods could eliminate the need for cell culture and virus preparation on a large scale for vaccine production. The bacterial fermentation unit could have a library of bacterial strains with plasmids containing selected viral genes from a number of viruses for the preparation of antigens as the need arises.

An example of the potential for such technology was recently reported

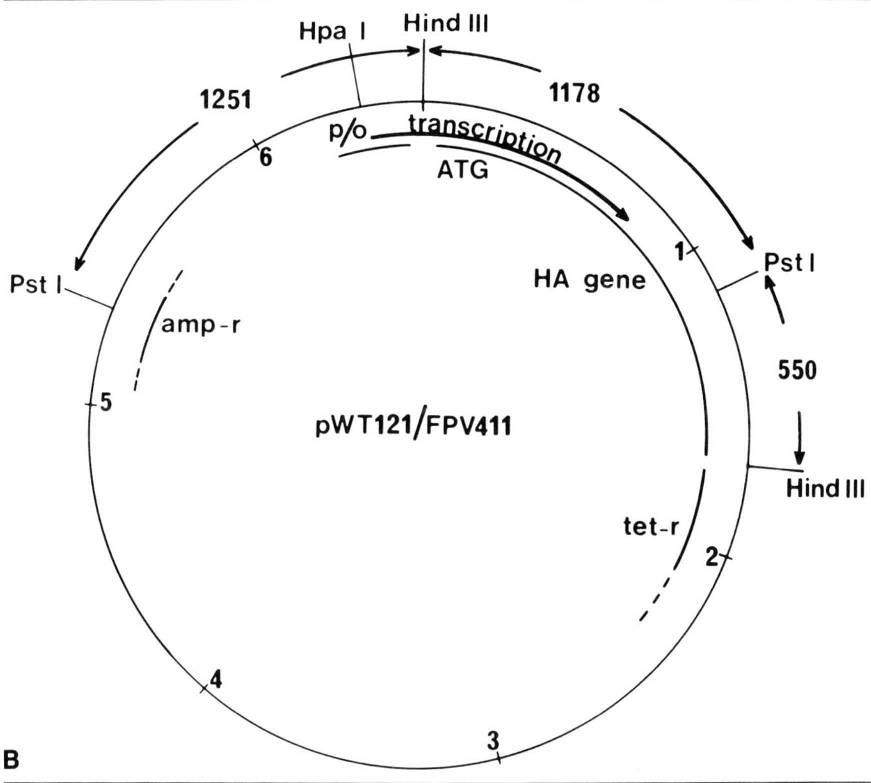

(Emtage et al. 1980). The gene sequence for the hemagglutinin of fowl plague virus was cloned in the pBR322 plasmid and was inserted into a bacterium (figure 53). The viral HA gene was expressed in the bacteria and corresponded to 0.75% of the bacterial proteins. The viral HA thus produced was antigenic and could be detected by specific antibodies.

AMANTADINE: A POSSIBLE PROPHYLACTIC DRUG. Amantadine HCl was found to be effective in the prophylaxis of influenza A but was ineffective against influenza B. The drug is not really effective in curing influenza. This drug is also useful in the treatment of Parkinson's disease.

SWINE INFLUENZA VIRUS AND HUMAN INFLUENZA. A virus that resembled swine influenza virus was isolated from a number of fatal influenza cases in the United States. The results of the program to immunize the whole population in the United States are discussed in chapter 24.

BIBLIOGRAPHY

Bean, W.J., and Webster, R.G. Phenotypic properties associated with influenza genome segments. *In: Negative Strand Viruses and the Host Cell* (B.W.J. Mahy and R.D. Barry, eds.), Academic Press, London, New York, San Francisco, 1978, pp. 685–692.

Beare, A.S. Recombinant influenza viruses from parents of known virulence for man. *In: Negative Strand Viruses and the Host Cell* (B.W.J. Mahy and R.D. Barry, eds.), Academic Press, London, New York, San Francisco, 1978, pp. 745–755.

Bouloy, M.; Plotch, S.J.; and Krug, R.M. Globin mRNAs are primers for the transcription of influenza viral RNA in vitro. *Proc. Natl. Acad. Sci. USA 75*:4886–4890, 1978.

Emtage, J.S.; Tacon, W.C.A.; Catlin, G.H.; Jenkins, B.; Porter, A.G.; and Carey, N.H. Influenza antigenic determinants are expressed from haemagglutinin genes cloned in *Escherichia coli. Nature 283*:171–174, 1980.

Klenk, H.-D.; Rolt, R.; Orlich, M.; and Blödorn, J. Activation of influenza A viruses by trypsin treatment. *Virology 68*:426–434, 1975.

Laver, W.G. Crystallization and peptide maps of neuraminidase "heads" from H2N2 and H3N2 influenza virus strains. *Virology 86*:78–87, 1978.

Laver, W.G. Immune response to a subunit vaccine. *In: Influenza: Virus, Vaccines and Strategy* (P. Selby, ed.), Academic Press, London, New York, 1976, pp. 163–175.

Laver, W.G. The polypeptides of influenza viruses. *Adv. Virus Res. 18*:57–104, 1973.

Laver, W.G.; Air, G. M.; Dopheide, T.A.; and Ward, C.W. Amino acid sequence changes in the haemagglutinin of A/Hong Kong (H3N2) influenza virus during the period 1968–77. *Nature 283*:454–457, 1980.

Melnick, J. Taxonomy of Viruses 1980. *Prog. Med. Virol. 26*:214–232, 1980.

Murphy, B.R.; Wood, F.T.; Massicot, J.G.; and Chanock, R.M. Temperature-sensitive mutants of influenza virus. XVI. Transfer of two ts lesions present in the Udorn/72-ts-1 A2 donor virus to the Victoria/3/75 wild-type virus. *Virology* 88:244–251, 1978.

Palese, P.; Ritchey, M.B.; and Schulman, J.L. Mapping of the influenza virus genome. II. Identification of the P1, P2 and P3 genes. *Virology 76*:114–121, 1977.

Palese, P., and Schulman, J.L. Mapping of the influenza virus genome: identification of the hemagglutinin and the neuraminidase genes. *Proc. Natl. Acad. Sci. USA* 73:2142–2146, 1976.

Porter, A.G.; Barker, C.; Carey, N.H.; Hallewell, R.A.; Threlfall, G.; and Emtage, J.S. Complete nucleotide sequence of an influenza virus haemagglutinin gene from cloned DNA. *Nature 282*:471–477, 1979.

Ritchey, M.B.; Palese, P.; and Kilbourne, E.D. RNAs of influenza A, B and C viruses. *J. Virol.* 18:738–744, 1976*a*.

Ritchey, M.B.; Palese, P.; and Schulman, J.L. Mapping of influenza virus genome. III. Identification of genes coding for nucleoprotein membrane protein and nonstructural protein. *J. Virol.* 20:307–313, 1976*b*.

Scholtissek, C. The genome of influenza virus. *Curr. Top. Microbiol. Immunol. 80*:139–169, 1978.

Schulman, J.L., and Palese, P. Biological properties of recombinants of influenza A/Hong Kong and A/PR8 viruses: effects of genes for matrix protein and nucleoprotein on virus yield in embryonated eggs. *In: Negative Strand Viruses and the Host Cell* (B.W.J. Mahy and R.D. Barry, eds.), Academic Press, London, New York, San Francisco, 1978, pp. 663–674.

Skehel, J.J., and Hay, A.J. Influenza virus transcription. *J. Gen. Virol. 39*:1–8, 1978.

Webster, R.G., and Bean, W.J. Genetics of influenza virus. *Annu. Rev. Genet. 12*:415–431, 1978.

RECOMMENDED READING

Beare, A.S. Live viruses for immunization against influenza. *Prog. Med. Virol. 20*:49–83, 1975.

Compans, R.S., and Choppin, P.W. Reproduction of myxoviruses. *In: Comprehensive Virology* (H. Fraenkel-Conrat and R.R. Wagner, eds.), Plenum Press, New York, 1975, Vol. 4, pp. 179–252.

Hightower, L.E., and Bratt, M.A. Genetics of orthomyxoviruses. *In: Comprehensive Virology* (H. Fraenkel-Conrat and R.R. Wagner, eds.), Plenum Press, New York, 1977, Vol. 9, pp. 535–598.

Kilbourne, E.D. (ed.) *The Influenza Viruses and Influenza.* Academic Press, New York, 1975.

Kilbourne, E.D. Molecular epidemiology—influenza as archetype. *Harvey Lect.* 1977–1978, 225–258, 1979.

Melnick, J.L. Viral vaccines. *Prog. Med. Virol. 23*:158–195, 1977.

Sabin, A.B. Mortality from pneumonia and risk conditions during influenza epidemics. High influenza morbidity during nonepidemic years. *JAMA* 237:2823–2828, 1977.

Schulze, I.T. Structure of the influenza virion. *Adv. Virus Res. 18*:1–56, 1973.

Seal, J.R.; Sencer, D.J.; and Meyer, H.M. A status report on national immunization. *J. Infect. Dis. 133*:715–720, 1976.

Selby, P. (ed.). *Influenza.* Sandoz Institute for Health and Socio-Economic Studies. Academic Press, New York, 1976.

Webster, R.G. On the origin of pandemic influenza viruses. *Curr. Top. Microbiol. Immunol. 59*:75–106, 1972.

16. SINGLE-STRANDED RNA MINUS VIRUSES WITH A FRAGMENTED LINEAR GENOME: ARENAVIRUSES

FAMILY ARENAVIRIDAE

The virions are round or pleomorphic, with a diameter of 50–300 nm, and have an outer lipid envelope with numerous surface projections or spikes. A special feature of these virions is that they contain cellular ribosomes that are enveloped together with the viral components. Three proteins are present in the virions. The RNA genome is composed of five single-stranded RNA minus molecules: two virus-specific molecules and three of cell origin. There is a virion-associated transcriptase, as well as poly(U) and poly(A) RNA polymerases associated with ribosome fractions of detergent-disrupted virus. Virus particles are formed by budding through the cell membrane.

Genus Arenavirus (LCM virus group)

Lymphocytic choriomeningitis (LCM: the prototype, worldwide distribution)
Lassa (West Africa)

Tacaribe complex includes:

Tacaribe (Trinidad)
Tamiami (Florida)
Machupo (Bolivian hemorrhagic fever)
Junin (Argentinian hemorrhagic fever)
Pichinde (Colombia)
Amapari (Brazil)

Parana (Paraguay)
Latino (Bolivia)

The name *arenavirus* is derived from the Latin *arenosus,* meaning sandy—the description of the morphology of the virions in the electron microscope.

Tacaribe virus, which was isolated from bats, does not cause any disease in man.

Viral RNA and proteins

Comparison of three viruses—Pichinde, Tacaribe, and Tamiami—revealed that the viral RNA is made of one large (L) RNA molecule of 3.2×10^6 daltons (31S) and one small (S) molecule of 1.6×10^6 daltons (22S) (Vezza et al. 1978). The additional RNA molecules (28 and 18S rRNA and 4-6S) are derived from the cellular ribosomes which are incorporated into the virions.

Tacaribe and Tamiami virions have a major N protein in the nucleocapsid with a molecular weight of 68,000 and 66,000, respectively, and a single glycoprotein G in the envelope of 42,000 and 44,000 daltons, respectively, as well as some minor protein species (Gard et al. 1977). In the Pichinde virions, in addition to the major N protein of 66,000 daltons, there are also G1 (64,000 daltons) and G2 (38,000 daltons) external glycoproteins, as well as minor proteins. LCM virions contain the N protein (63,000 daltons) and the GP1 (54,000 daltons) and GP2 (35,000 daltons) glycoproteins (Buchheimer et al. 1978). The viral RNA is synthesized in the cytoplasm of the infected cells where a nonstructural protein (GP-C) of 74–75,000 daltons was found in LCM virus preparations. This protein may be a precursor to the G1 and G2 proteins.

Virus replication is dependent on the cellular RNA polymerase II.

Defective interfering virus

Infection of the hamster cell-line BHK-21 with LCM virus leads to the appearance of noninfectious defective virions. In these particles, the viral RNA is shorter than the wild type viral RNA, and the virions do not contain ribosomes (Pederson 1978).

Virus recombination

High-frequency genetic recombination occurs with ts mutants of Pichinde virus. This is a characteristic of RNA viruses with a segmented genome such as bunyaviruses, influenzaviruses, and reoviruses (Vezza and Bishop 1977).

DISEASES CAUSED BY ARENAVIRUSES

Three of the ten known arenaviruses cause disease in humans with a high mortality rate.

Lymphocytic choriomeningitis

LCM virus is the most extensively studied of the arenaviruses. The domestic mouse is the natural host of LCM virus. Mice are usually infected shortly after birth and become lifelong carriers of LCM virus (reviewed by Hotchin 1971; Lehmann-Grube 1971). Considerable amounts of the virus are found in all organs, including the blood. Many of these asymptomatic LCM-carriers develop a progressive *runting syndrome* before or at one year of age which has subsequently been attributed to chronic glomerulonephritis from local deposition of circulating LCM virus antibody complexes (Oldstone and Dixon 1970). Humans are infected with LCM virus mostly in the late autumn and early winter when mice excreting the virus in their urine come indoors.

Pet hamsters have been shown to be a source for human infection with LCM virus (Hirsch et al. 1974). LCM virus usually produces a self-limiting pneumonitis but in rare instances spreads to the CNS to produce aseptic meningitis.

Acute LCM can be produced in weanling and adult mice, but not in newborns, by intracerebral inoculation of virus. This increased susceptibility of older mice to intracerebral infection with LCM virus, combined with the ability of mice infected during gestation or within 24 hours of birth to survive despite widespread systemic infection, was presumptive evidence that acute LCM disease was immunopathologic.

Multiple investigations demonstrated that acute LCM might be prevented by various immunosuppressive regimens, even though virus titers were the same in both asymptomatic, immunosuppressed, and in dying nonimmunosuppressed LCM virus-infected mice. Clinching evidence that acute LCM disease in mice was immune-mediated was provided by the ability of LCM virus-sensitized spleen cells transferred to syngeneic recipient mice to cause these asymptomatic cyclophosphamide-induced LCM virus carriers to die of acute choriomeningitis (Gilden et al. 1972). Incubation of these LCM virus-sensitized immunocytes (spleen cells) with anti-theta serum (which depletes the population of thymus-derived T lymphocytes), prior to adoptive transfer of these cells to LCM-infected syngeneic recipient mice, inhibited the ability of the transferred cells to produce acute LCM. This indicates that the immunopathology of acute LCM disease is in part T-cell dependent (Cole and Nathanson 1974).

Acute hemorrhagic fever

In Argentina and Bolivia, this disease is caused by Junin and Machupo viruses, respectively. The disease starts with fever of several days' duration and is followed by viremia accompanied by hemorrhages and shock. Between 10–50% of patients die. Pathologic changes are found mostly in the kidneys, but often also in the CNS. After one–three weeks from onset of the disease, antibodies appear in the blood.

Lassa virus

This virus causes human infections in Africa (Nigeria, Liberia, and Sierra Leone). The virus came to the attention of the medical world in 1969 when three U.S. missionary nurses working in Lassa, Northern Nigeria, contracted the disease (Buckley and Casals 1978). This disease, which is highly fatal, is manifested by a fever lasting two–three weeks, pleural and peritoneal effusions, acute arthritis, and encephalitis. During the disease period, IgM and IgG antibodies appear in the blood. The virus is transmitted by nose and mouth excretions.

BIBLIOGRAPHY

Buchheimer, M.J.; Elder, J.H.; and Oldstone, M.A. Protein structure of lymphocytic choriomeningitis virus: identification of the virus structural and cell associated polypeptides. *Virology 89*:133–145, 1978.

Buckley, S.M., and Casals, J. Pathobiology of Lassa fever. *Int. Rev. Exp. Pathol. 18*:97–136, 1978.

Cole, G.A., and Nathanson, W. Lymphocytic choriomeningitis. *Prog. Med. Virol. 18*:94–110, 1974.

Gard, G.P.; Vezza, A.C.; Bishop, D.H.L.; and Compans, R.W. Structural proteins of Tacaribe and Tamiami virions. *Virology 83*:84–96, 1977.

Gilden, D.H.; Cole, G.A.; and Nathanson, N. Immunopathogenesis of acute central nervous system disease produced by lymphocytic choriomeningitis virus. II. Adoptive immunization of virus carriers. *J. Exp. Med. 135*:874–889, 1972.

Hirsch, M.S.; Moellering, R.C.; Pope, H.G.; and Poskanzer, D.C. Lymphocytic-choriomeningitis virus infection traced to a pet hamster. *N. Engl. J. Med. 291*:610–612, 1974.

Hotchin, J. Persistent and slow virus infections. *Monogr. Virol. 3*:2–71, 1971.

Lehmann-Grube, F. Lymphocytic choriomeningitis virus. *Virol. Monogr.* Vol. 10 (Springer Verlag, Vienna), 1971.

Oldstone, M.B.A., and Dixon, F.J. Pathogenesis of chronic disease associated with persistent LCM viral infection. II. Relationship of acute LCM viral response to tissue injury in chronic disease. *J. Exp. Med. 131*:1–19, 1970.

Vezza, A.C., and Bishop, D.H.L. Recombination between temperature-sensitive mutants of the arenavirus Pichinde. *J. Virol. 24*:712–715, 1977.

Vezza, A.C.; Clewley, J.P.; Gard, G.P.; Abraham, N.Z.; Compans, R.W.; and Bishop, D.H.L. Virion RNA species of the arenaviruses Pichinde, Tacaribe and Tamiami. *J. Virol. 26*:485–497, 1978.

RECOMMENDED READING

Bro-Jørgensen, K. The interplay between lymphocytic choriomeningitis virus, immune function and hemopoiesis in mice. *Adv. Virus Res. 22*:327–369, 1978.

Howard, C.R., and Simpson, D.I.H. The biology of the arenaviruses. *J. Gen. Virol. 51*:1–14, 1980.

Pedersen, I.R. Structural components and replication of arenaviruses. *Adv. Virus Res. 24*:277–330, 1979.

Rawls, W.E., and Leung, W.-C. Arenaviruses. *In: Comprehensive Virology* (H. Fraenkel-Conrat and R.R. Wagner, eds.), Vol. 14, 1979, Plenum Press, New York, pp. 157–192.

17. SINGLE-STRANDED RNA MINUS VIRUSES WITH A FRAGMENTED CIRCULAR GENOME: BUNYAVIRUSES

This is a taxonomic subgroup of arboviruses since these viruses have circular fragmented RNA minus genomes.

FAMILY BUNYAVIRIDAE

These are round virions with a lipid envelope and a diameter of 90–100 nm. The lipid envelope contains at least one viral glycoprotein and covers an elongated ribonucleoprotein. The viral RNA^- genome is made up of three circular ssRNA molecules of 3, 2, and 0.5×10^6 daltons. The virions are formed by budding through smooth membranes in the Golgi region of the infected cell.

Genus Bunyamwera supergroup (Bunyavirus)

Bunyamwera virus (114 antigenically related virus strains, arranged into 13 subgroups, mostly mosquito transmitted). Other members include the Anopheles A group, Bunyamwera group, Bwamba group, C group, California group, Capim group, Guama group, Koongol group, Simbu group, and several other groups.

Unnamed genera

There are at least 95 viruses in 11 serological groups. Members are transmitted by ticks or mosquitoes. These genera include: Uukuniemi group, Anopheles B group, Bakau group, Crimean hemorrhagic fever, Congo group, and several other groups.

Table 9. Bunyaviruses: viral RNA and proteins

	RNA molecular weight × 10^6			Protein molecular weight × 10^3			
Virus	L	M	S	L	G1	G2	N
Uukuniemi	2.3–4.1	1.0–1.9	0.4–0.88	170	75	65	25
La Crosse	2.9	1.6	0.4	180	110–120	35–39	25
Lumbo	2.9	2.0	0.5	—	115	35–38	22–25
Snowshoe hare	2.8–3.0	1.6–1.9	0.41–0.45	—	115	38	21
Main drain	3.1	2.0	0.4	—	115	38	21
Bunyamwera	3.0	1.9	0.34	—	115	30–38	19–23
California encephalitis (BFS-283)	6.7	2.2	0.4	—	82–125	30–39	17.5–25

(After Obijeski and Murphy, 1977.)

The name *Bunyamwera* comes from a place in Uganda where the type species was isolated.

Viral RNA and proteins

The viral genome is divided into three RNA minus circular molecules (large, medium, and small: L, M, S) that code for four polypeptides. The relationship between the size of the viral RNA and the polypeptides coded from them is presented in table 9. Protein L is found in small amounts in the viral nucleocapsid; N is a major protein of the nucleocapsid; G1 and G2 are external membrane glycoproteins of the virions.

The ribonucleoprotein (RNP) complexes (nucleocapsids) were isolated from Uukuniemi virions and by centrifugation in sucrose gradients were resolved into three components with sedimentation coefficients of 150S (L), 120S (M) and 90S (S) (figure 54). Electron microscopy revealed that the RNP complexes have a circular conformation. Each nucleocapsid size class contains its corresponding size class of RNA—namely, 29S (L), 22S (M), and 19S (S). The S RNA codes for the N protein and the M RNA codes for the G1 and G2 polypeptides (Gentsch and Bishop 1978, 1979).

RNA-dependent RNA polymerase in the virions

This enzyme was isolated from purified virions of Uukuniemi and Lumbo viruses (Ranki and Pettersson 1975; Bouloy and Hannoun 1976). The RNA molecules synthesized by the enzyme are complementary to the viral RNA minus genome. The enzymatic activity was inhibited by pancreatic RNase.

Virus replication in infected cells

The viral mRNA species that are translated by the cellular polyribosomes have sedimentation coefficients of 32S, 24S, and 14S. Viral nucleocapsids are formed in the cell cytoplasm and they have sedimentation coefficients of 105S, 85S, and 45S. The four viral proteins of La Crosse virus (L, G1, G2, and N) could be identified in the cytoplasm, the site of virus replication (see Obijeski and Murphy 1977). Defective interfering particles are produced in Bunyamwera virus-infected cells (Kascsak and Lyons 1978).

Virus recombinants

Recombinants between two Bunyaviruses are obtained when two viruses infect the same cell. The viral recombinants contain RNA molecules of both viruses (Gentsch et al. 1977).

DISEASES IN MAN AND ANIMALS

Rift Valley fever (RVF)

This is a viral disease of ruminants, rodents, and man. RVF virus is transmitted by the bite of mosquitoes. The disease was first described in the Rift Valley in Kenya in 1912 and since then it has been noted in Uganda, Rhodesia, South Africa, Zambia, Mozambique, Somalia, Sudan, and more recently, in Egypt (Ellis et al. 1979).

RVF primarily affects domestic animals, causing abortion in pregnant sheep, cows, and goats, and has an exceptionally high mortality in young lambs and calves. Sheep are the animals most sensitive to RVF infection whereas cattle are less susceptible. Buffalo and camels develop a slightly different disease, and goats can also be infected, as can rats and other rodents, for which the virus is fatal. Rabbits, horses, donkeys, and pigs are resistant and do not develop the disease. The acute stage of the disease in young animals is accompanied by vomiting, bloody diarrhea, and weakness. Virus can be found in the blood, liver, and spleen, but the damage is predominantly in the liver.

Transmission of virus

Mosquitoes such as *Aedes* or *Culex* are the transmitters of RVF virus in Africa. In addition to insect transmission, humans are infected by direct contact with blood or tissues of diseased animals since in viremic animals a high virus concentration is present in the blood.

The symptoms of the disease in man resemble those of influenza, starting with fever, tremors, vomiting, muscle ache, and headache. Complications can occur in the retina, followed by partial blindness.

Virus stability

The virus is stable under hot climatic conditions. In a blood sample, the virus can be kept for one year at 4°C, three months at room temperature, and 24 hr at 37°C or in the sun. Heat inactivation at 56°C requires 40 min to destroy the infectivity of the virus. The virus can also be inactivated by a 1:1000 dilution of formalin or by pasteurization.

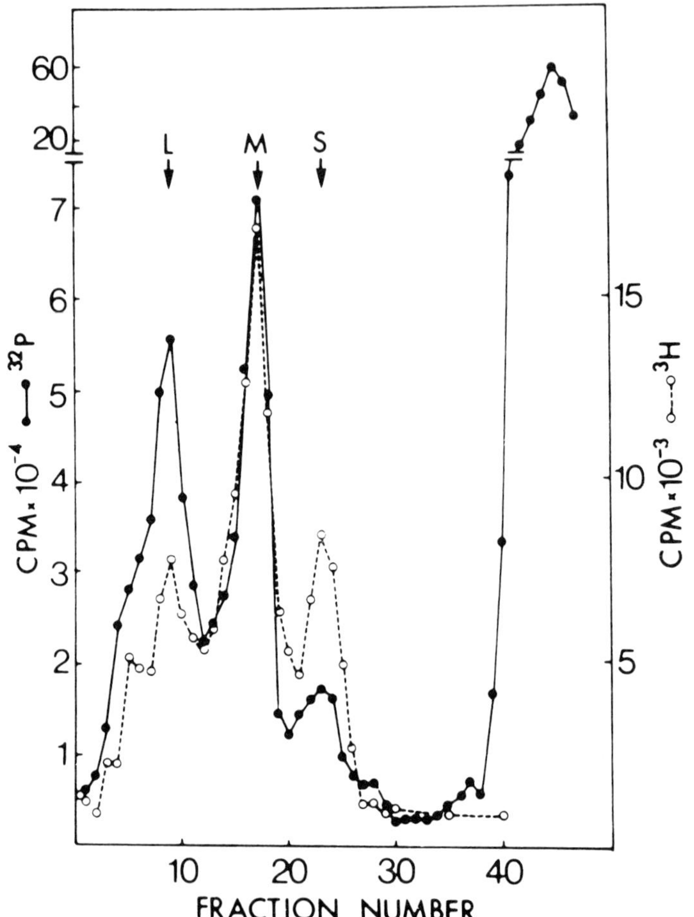

Figure 54. Separation of Uukuneimi virus ribonucleoprotein complexes by centrifugation in a sucrose gradient. The virus labeled with ^{32}P was disrupted with Triton X-100 and the released ribonucleoproteins separated on a linear sucrose gradient. A standard assay for polymerase activity was performed and the incorporation of ^{3}H-UTP was determined.
(Ranki and Petterson 1978. Reprinted by permission from *Negative Strand Viruses and the Host Cell*. B.W.J. Mahy and R.D. Barry, eds. Copyright by Academic Press Inc. [London] Ltd. 1978, Fig. 3, p. 362.)

Diagnosis

Isolation of the virus in laboratory animals is achieved by injecting a liver or brain suspension from diseased animals into mice, which develop hepatitis as a result.

Serological tests like agar diffusion, complement fixation, hemagglutination inhibition, or neutralization are used to identify the isolated virus. The antigen used is inactivated virus. Serum antibodies to the virus obtained from infected animals that survived the infection are used to identify RVF.

Vaccines and vaccination

A killed virus vaccine similar to the Salk polio vaccine is available for the immunization of humans in those regions where RVF virus occurs. A live attenuated virus vaccine was developed in the Veterinary Research Laboratories in Kenya to immunize sheep. It is inadvisable to use these vaccines in virus-free countries so as not to mask a naturally occurring epidemic.

Prevention of virus spread

RVF, which is a life-threatening disease in humans and sheep, is spread by mosquitoes, making eradication of the virus virtually impossible. Prevention of virus spread, as recommended by the United Nations Food and Agriculture Organization, involves:

1. Spraying all airplanes coming from RVF enzootic countries with insecticides.
2. Transport of live sensitive animals from enzootic countries is prohibited.
3. If the disease is diagnosed in animals, the whole area should be sprayed with insecticides. Animals must be kept under conditions that repel insects.
4. All dead animals must be buried. Consumption of meat from infected animals is prohibited.
5. Migration of herds from an infected area to an uninfected area is prohibited.
6. In an enzootic area, humans and animals must be vaccinated.

Sandfly fever virus

This virus is prevalent in the Mediterranean basin and is transmitted by sandflies. The natural hosts are birds.

Crimean hemorrhagic fever

This fever occurs in Central Asia and Africa and is transmitted by ticks. Mammals are the natural hosts.

California encephalitis virus

This virus is prevalent in North America and is transmitted by a mosquito. The virus reservoir is in rabbits and rodents.

REFERENCES

Bouloy, M., and Hannoun, C. Studies on Lumbo virus replication. 1. RNA-dependent RNA polymerase associated with virions. *Virology 69*:258–264, 1976.

Ellis, D.S.; Simpson, D.I.H.; Stamford, S.; and Abdel Wahab, K.S.E. Rift Valley fever virus: some ultrastructural observations on material from the outbreak in Egypt 1977. *J. Gen. Virol. 42*:329–337, 1979.

Gentsch, J.R., and Bishop, D.H.L. M viral RNA segment of Bunyaviruses codes for two glycoproteins, G1 and G2. *J. Virol. 30*:767–770, 1979.

Gentsch, J.R., and Bishop, D.H.L. Small viral RNA segment of bunyaviruses codes for viral nucleocapsid protein. *J. Virol. 28*:417–419, 1978.

Gentsch, J.; Wynne, R.; Clewley, J.P.; Shope, R.E.; and Bishop, D.H.L. Formation of recombinants between snowshoe hare and La Crosse Bunyaviruses. *J. Virol.* *24*:893–902, 1977.

Kascsak, R., and Lyons, M.J. Bunyamwera virus: II. The generation and nature of defective interfering particles. *Virology* *89*:539–546, 1978.

Obijeski, J.F., and Murphy, F.A. Bunyaviridae: recent biochemical developments. *J. Gen. Virol.* *37*:1–14, 1977.

Ranki, M., and Pettersson, R. Uukuniemi virus contains an RNA polymerase. *J. Virol.* *16*:1420–1425, 1975.

Ranki, M., and Pettersson, R. Transcription in vitro of the segmented RNA genome of Uukuniemi virus. *In: Negative Strand Viruses and the Host Cell* (B.W.J. Mahy and R.D. Barry, eds.), Academic Press, London, New York, San Francisco, 1978, pp. 357–365.

D. VIRUSES WITH SINGLE-STRANDED RNA PLUS GENOMES

18. RNA PLUS GENOME THAT SERVES AS MESSENGER RNA: PICORNAVIRUSES

FAMILY PICORNAVIRIDAE

The virions are spherical particles with no envelope or core and a diameter of 20–30 nm. The icosahedral capsid consists of 60 copies of coat proteins VP1 and VP3, 58–59 copies of VP2 and VP4, and 1–2 copies of VPC, the precursor to VP2 and VP4. The virions contain a single-stranded RNA$^+$ genome of 2.6 $\times 10^6$ daltons. The unique property of this virus is the ability of the plus strand viral genome to serve as messenger RNA in the infected cell. After release from the virions in the cytoplasm, the viral RNA interacts with the cellular ribosomes and serves as templates for the synthesis of the viral proteins.

Genus Enterovirus

These viruses are resistant to acid conditions (as low as pH 3) and are therefore able to replicate in the alimentary tract after passing through the stomach.

Human viruses

Human polioviruses 1–3.
Human coxsackieviruses A1–22, 24 (A23 = echovirus 9).
Human coxsackieviruses B1–6.
Human echoviruses 1–9, 11–27, 29–34.
Human enteroviruses 68–71.
Hepatitis A is a possible member.

Animal viruses
Bovine enteroviruses 1–7
Porcine enterovirus 1–18
Simian enteroviruses 1–18
Murine poliovirus

Genus Cardiovirus (EMC virus group)

Encephalomyocarditis (EMC) virus
Murine encephalomyelitis
Mengoviruses (probably same as EMC virus)

Genus Rhinovirus

Human rhinoviruses 1A, 2–113
Bovine rhinoviruses 1–2

These viruses are unstable below pH 5.

Genus Aphthovirus

Foot and mouth disease virus strains O, A, C, SAT 1, 2, 3, Asia 1 (Figure 55)

These viruses are unstable below pH 7.

Equine rhinoviruses 1–2 (not yet assigned to genus)

There are also a number of unclassified picornaviruses of invertebrates. The name *picorna* is derived from pico (micromicro) and the suffix RNA.

Properties of poliomyelitis virus

The diameter of the virion is 30 nm with a molecular weight of 8.6×10^6. The viral single-stranded RNA genome has a molecular weight of 2.6×10^6; a polyadenosine residue is attached at the 3′ end of the molecule (Yogo and Wimmer 1972). A small virus-coded protein (designated VPg) of >7,000 daltons (Lee et al 1977) is covalently linked to the genome through a tyrosine residue and a phosphate attached to the first nucleotide at the 5′ end of the RNA molecule:

VPg-tyr-O-P-O-UUAAAACAG . . .

An enzyme that cleaves the bond between tyrosine and phosphate was found to be present in animal cells.

ADSORPTION AND UNCOATING OF THE VIRAL RNA. The virions adsorb to a specific receptor on the cell membrane and are incorporated by pinocytosis into the cytoplasm where they appear in phagocytic vacuoles. The coat proteins are digested by the cellular proteolytic enzymes which enter the vacuoles, and the viral RNA genomes are uncoated and released into the cytoplasm. Of

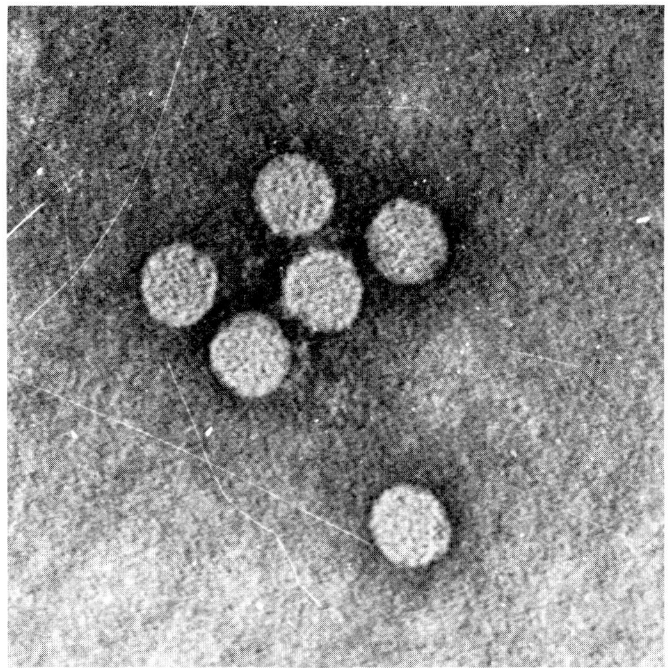

Figure 55. Electron micrograph (negative staining) of virions of foot and mouth disease virus (FMDV).
(By courtesy of Dr. Daniel Dekegel, Institut Pasteur du Brabant, Brussels, Belgium.)

these genomes, 94% are digested by the cytoplasmic nucleases. The remaining viral RNA genomes attach to cellular ribosomes and serve as mRNA in the translation process. Because of the degradation of the viral RNA by the cellular nucleases, 100 virions are required for the production of 1 plaque (i.e., 1 pfu).

GENE IN CHROMOSOME 19 DETERMINES THE CELL SENSITIVITY TO POLIOVIRUS. The use of human-mouse cell hybrids made it possible to demonstrate that a gene in human chromosome 19 codes for the protein that is the virus receptor on the outer surface of the cell membrane (see chapter 4). The properties of this cellular gene and its protein product are still to be studied.

Synthesis of viral RNA

Two poliovirus-specific RNA polymerase activities were identified: a poly(U) polymerase and a replicase (Flanegan and Baltimore 1979; Dasgupta et al. 1980). The virus-coded RNA-dependent RNA polymerase (replicase) and poly(U) polymerase both have the same polypeptide component of 63,000 daltons designated p63. Synthesis of poliovirions and mRNA takes place in a replication complex that consists of the replicase and a replicative intermediate (RI), which is composed of one complete strand of complementary (negative-

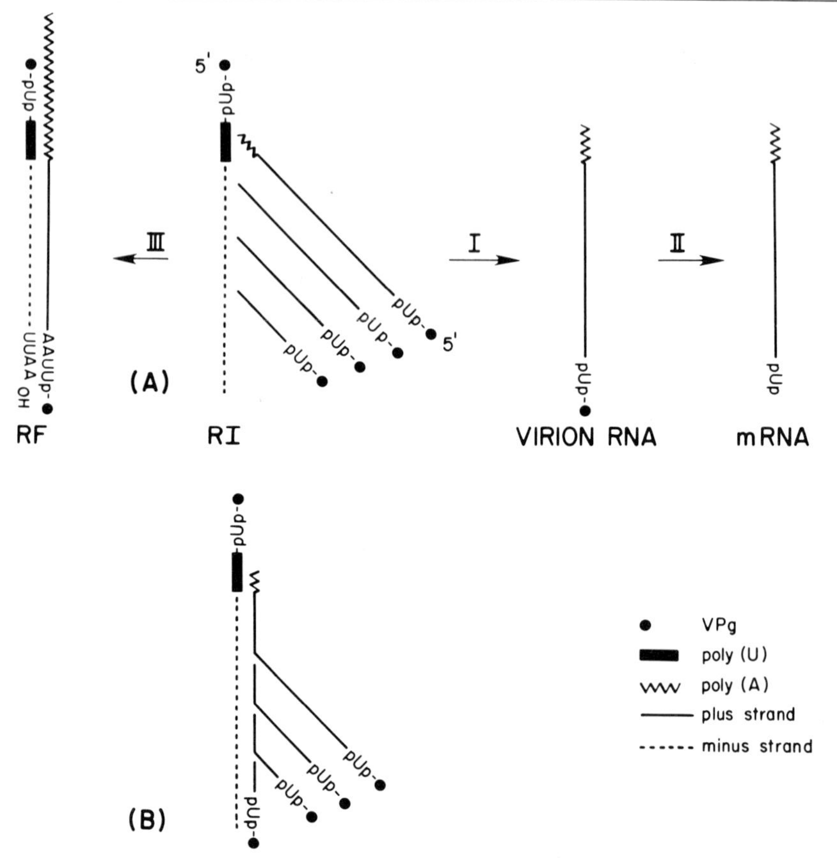

Figure 56. Schematic representation of terminal structures of poliovirus RNAs. The arrows (I, II, and III) suggest the relationship between different RNA structures. Whether structure (A) or (B) of polio RI functions in vivo is unknown, but structure (B) certainly arises upon deproteinization of structure (A). The 3′-end of minus-strand RNA of RF terminates in . . . $GUUUUAA_{OH}$, which we assume is true for all minus strands.
(Kitamura et al. 1980. Reprinted by permission from *Ann. NY Acad. Sci.* 354, Fig. 3, p. 188.)

strand) RNA that has several nascent chains of plus-strand RNA attached (Flanegan and Baltimore 1979). The VPg protein is covalently linked at the 5′ end to all nascent strands of the RI, to the minus-strand RNA of the RI, as well as to the replicative form (RF) (Pettersson et al. 1978; Kitamura et al. 1980). While the RI of poliovirus is involved in the synthesis of single-stranded progeny RNA, the polio RF has been shown to be a by-product of replication: the RF is formed as a result of a nascent strand annealing to template RNA, thus generating double-stranded RNA (Nomoto et al. 1977).

Since all newly synthesized viral RNA species are protein-bound (except mRNA) (figure 56), it has been suggested that VPg acts as a primer in the initiation of RNA synthesis (Flanegan et al. 1977; Kitamura et al. 1980). It has also been proposed that VPg undergoes proteolytic cleavage at the moment of initiation of RNA synthesis. Using rapid sequencing techniques as illustrated in figure 57, the primary nucleotide sequence of the entire single-stranded genome of poliovirus type 1 was determined (figure 58) (Kitamura et al. 1980, 1981; Semler et al. 1981).

Messenger RNA

The mRNAs of a number of animal viruses contain a capping group of methylated nucleotides with the general structure $m^7G(5')ppp(5')N^{m}p$ at the 5′ terminus. This capping group, which is important for viral protein synthesis, has not been identified in poliovirus mRNA, and it appears that pUp is the 5′-terminal structure (Nomoto et al. 1976; Hewlett et al. 1976). Viral mRNA differs from virion RNA in that it lacks VPg; it is formed by removal of VPg from the phosphate residue at the 5′-terminal nucleotide without loss of infectivity.

Inhibition of cellular protein synthesis

Infection with poliovirus leads to inhibition of cellular protein synthesis at an early stage of infection. The site of inhibition of host protein synthesis is located on the host polyribosomes (Kaufmann et al. 1976). With eukaryotic mRNA, an initiation factor (eIF-4B) interacts with the capped 5′ end of the molecule for translation to occur and the capped 5′ end is important for ribosome recognition of mRNA. Since poliovirus mRNA lacks a 5′ end, the 5′-cap-dependent recognition mechanism has to be bypassed; inactivation of an initiation factor involved in 5′-cap recognition would then favor poliovirus mRNA translation. Thus it appears that inactivation of the eIF-4B initiation factor is the mechanism by which poliovirus infection selectively inhibits the translation of host cell proteins (Rose et al. 1978).

Synthesis of viral proteins and virions

For translation of mRNA, the ribosomes must attach at a binding sequence at or near the 5′ end and move along the molecule from the initiation site to the termination codon at the 3′ end. The nucleotide sequence is translated, and the corresponding viral propeptide is synthesized. The viral propeptide is then cleaved (processed) by proteolytic enzymes, as shown in figure 59 (Kitamura et al. 1981). The polypeptide map has been divided into three regions: P_1 corresponds to the region that produces the four capsid polypeptides VP1, VP2, VP3, and VP4; P2 corresponds to the central portion of the genome; and P3 to a group of nonstructural polypeptides that include the putative proteinase P3-7c, VPg, and the RNA-dependent RNA polymerase P3-4b.

Assembly of the poliovirus shell is a three-stage process (figure 60): Protomers form pentamers (five uncleaved protomers form a 13S pentamer); pentamers are cleaved; and a dodecahedron is formed. Virions are formed in two

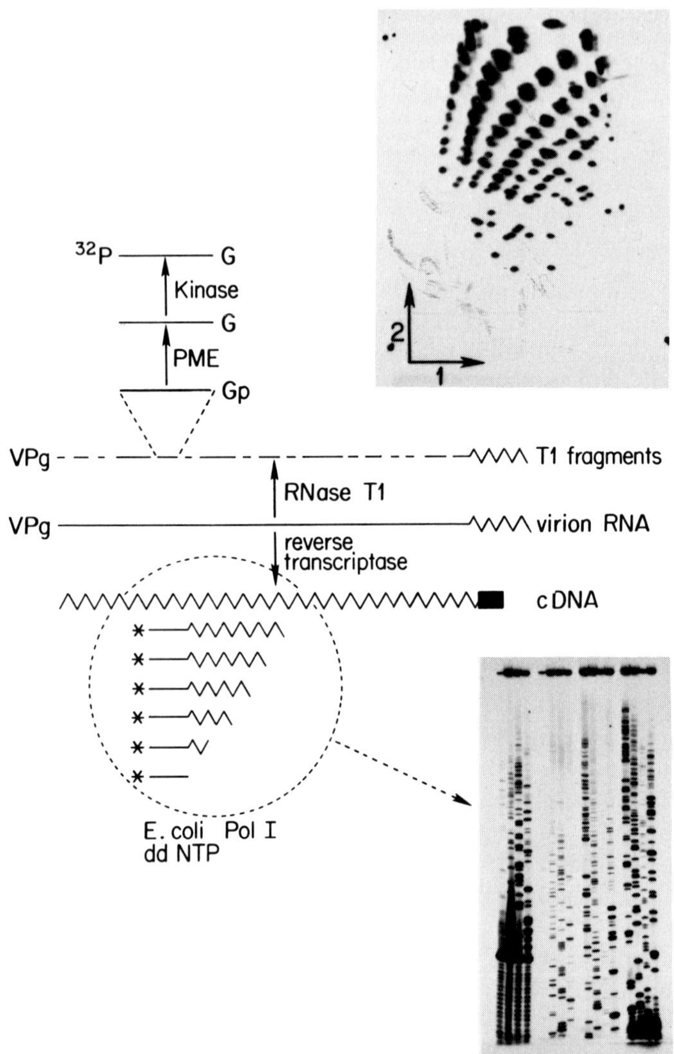

Figure 57. Strategy to determine the primary structure of poliovirus RNA. Separation of RNase T1-resistant oligonucleotides of poliovirus is achieved by two-dimensional gel electrophoresis (autoradiogram in the upper right). An oligonucleotide is recovered from the two-dimensional gel, dephosphorylated with phosphomonoesterase (PME), phosphorylated with polynucleotide kinase, and used as a primer for cDNA-dependent DNA synthesis catalyzed by *E. coli* DNA polymerase I. DNA synthesis is terminated randomly by dideoxynucleoside triphosphates (ddNTP) and the fragments are separated by gel electrophoresis. This method can be applied to the sequence determination of any high molecular weight RNA.
(Kitamura et al. 1980. Reprinted by permission from *Ann. NY Acad. Sci.* 354, Fig. 4, p. 189.)

additional steps: The RNA-VPg genome is encapsidated to form a provirion of 12 pentamers which then undergoes maturation cleavage to finally yield infective virions. During the final stage, the four capsid proteins (VP1 to VP4) are formed (Rueckert 1978).

Phenotypic mixing of picornaviruses

Infection of one cell with two viruses (e.g., polio and Coxsackie viruses) leads to the replication of both viruses at one cytoplasmic site. No genetic recombination between the two viral RNA genomes takes place, but the newly formed virions contain one of the two RNA genomes in a capsid made up of capsomeres donated by both viruses. Phenotypically mixed virions are not neutralized by antiserum to one of the viruses, and antibodies to both viruses are needed to neutralize this phenotypically mixed virus.

POLIOMYELITIS VIRUS AS A HUMAN PATHOGEN

Studies by A. Lwoff revealed that poliovirus strains differ in their ability to replicate at low (37°C) and high (41°C) temperatures. This property of poliovirus was further investigated by H. Koprowski and A. Sabin and led to the development of avirulent mutants of poliovirus. The poliovirus plaques that developed at 41°C were found to be avirulent, as compared to the plaques formed at 37°C, which are highly virulent. Sabin selected the strains of the three poliovirus types (1, 2, and 3) which were able to grow at 41°C and used them to develop the avirulent live virus vaccine. The Sabin attenuated poliovirus vaccine which came into use throughout the world in 1957 (see chapter 24) was preceded by the formalin-inactivated virus vaccine developed by Jonas Salk. Immunization of children with poliovirus vaccine stopped the spread of the disease.

Poliovirus, which is found in sewage, affects children and adults, leading in most cases to an inapparent infection that results in lifelong immunity. In some infected individuals, the disease is accompanied by fever, sore throat, headache, and vomiting. In severe cases, symptoms include muscular aches, followed by stiffness of the neck and finally paralysis of the legs and/or arms, as well as of the diaphragm. Patients afflicted with paralysis of the diaphragm were kept in mechanical respirators (iron lungs) to assist them in breathing. The paralysis is due to the destruction of CNS neurones. The virus replicates in the intestine from where it enters the blood stream and penetrates into certain defined regions of the CNS.

The extent of poliomyelitis past and present

Until 1955, the poliomyelitis virus caused epidemics all over the world. In 1955, 76,000 poliomyelitis patients were reported in the United States, Canada, Australia, New Zealand, USSR, and 23 European countries. After the introduction of the inactivated Salk vaccine and the Sabin attenuated vaccine for large-scale immunization, a sharp decline occurred in the extent of the

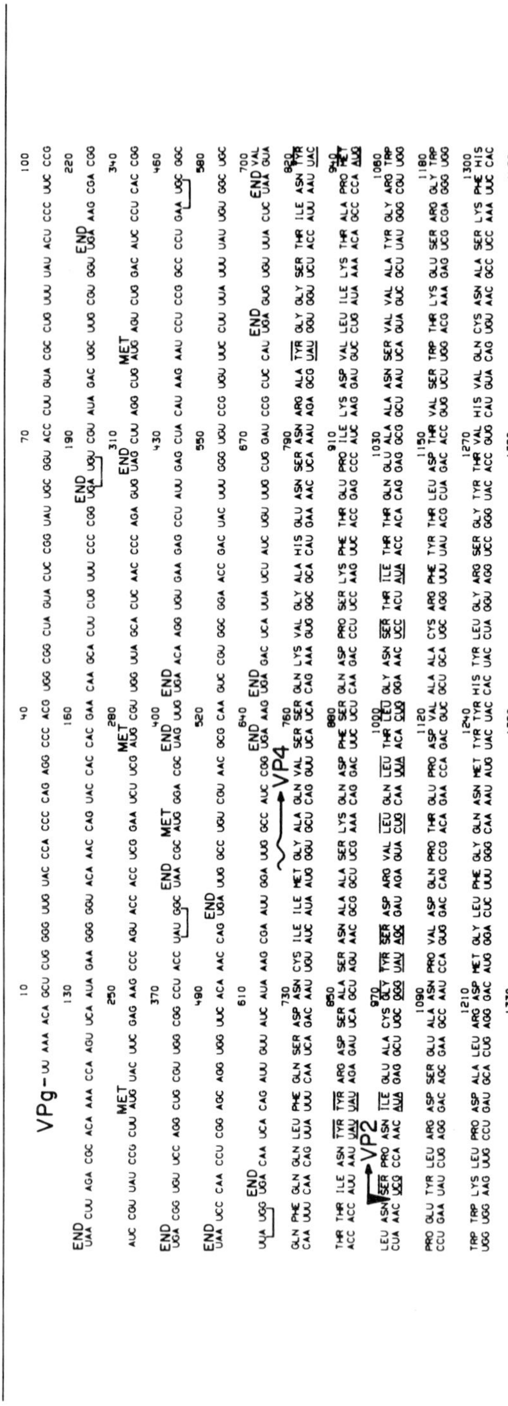

```
                10                                 40                                 70                                 100
VPg-UU AAA ACA GCU CUG GGG UUG UAC CCA CCC CAG AGG CCC ACG UGG CGG CUA GUA CUC CGG UAU UGC GGU ACC CUU GUA CGC CUG UUU UAU ACU CCC UUC CCG

END                                 130                                160                           END     190                     END        220
UAA CUU AGA CGC ACA AAA CCA AGU UCA AUA GAA GGG GGU ACA AAC CAG UAC CAC CAC GAA CAA GCA CUU CUG UUU CCC CGG UGA UGU CGU AUA GAC UGC UUG CGU GGU UGA AAG CGA CGG

                        MET         250                                MET 280                         END 310         MET                   340
AUC CGU UAU CCG CUU AUG UAC UUC GAG AAG CCC AGU ACC ACC UCG GAA UCU UCG AUG CGU UGG UUA GCA CUC AAC CCC AGA GUG UAG CUU AGG CUG AUG AGU CUG GAC AUC CCU CAC CGG

END                                 370             END     MET         END 400 END                                    430                                460
UGA CGG UGU UCC AGG CUG CGU UGG CGG CCU ACC UAU GGC UAA CGC AUG GGA CGC UAG UUG UGA ACA AGG UGU GAA GAG CCU AUU GAG CUA CAU AAG AAU CCU CCG GCC CCU GAA UGC GGC

END                                 490             END                 520                                550                                580
UAA UCC CAA CCU CGG AGC AGG UGG UUC ACA AAC CAG UGA UUG GCC UGU CGU AAC GCG CAA GUC CGU GGC GGA ACC GAC UAC UUU GGG UGU CCG UGU UUC CUU UUA UUU UAU UGU GGC UGC

        END                         610                            END 640 END                             670             END                 END 700 VAL
UUA UGG UGA CAA UCA CAG AUU GUU AUC AUA AAG CGA AUU GGA UUG GCC AUC CGG UGA AAG UGA GAC UCA UUA UCU AUC UGU UUG CUG GAU CCG CUC CAU UGA GUG UGU UUA CUC UAA GUA

                                    730                 ~~~~~~> VP4 760                                    790                                820
GLN PHE GLN GLN LEU PHE GLN SER ASP ASN CYS ILE ILE MET GLY ALA GLN VAL SER SER GLN LYS VAL GLY ALA HIS GLU ASN SER ASN ARG ALA TYR GLY GLY SER THR ILE ASN TYR
CAA UUU CAA CAG UUA UUU CAA UCA GAC AAU UGU AUC AUA AUG GGU GCU CAG GUU UCA UCA CAG AAA GUG GGC GCA CAU GAA AAC UCA AAU AGA GCG UAU GGU GGU UCU ACC AUU AAU UAC

                                    850                                 880                                910                                940
THR THR ILE ASN TYR TYR ARG ASP SER ALA SER ASN ALA ALA SER LYS GLN ASP PHE SER GLN ASP PRO SER LYS PHE THR GLU PRO ILE LYS ASP VAL LEU ILE LYS THR ALA PRO MET
ACC ACC AUU AAU UAU UAU AGA GAU UCA GCU AGU AAC GCG GCU UCG AAA CAG GAC UUC UCU CAA GAC CCU UCC AAG UUC ACC GAG CCC AUC AAG GAU GUC CUG AUA AAA ACA GCC CCA AUG

       --> VP2                      970                                1000                               1030                               1060
LEU ASN SER PRO ASN ILE GLU ALA CYS GLY TYR SER ASP ARG VAL LEU GLN LEU THR LEU GLY ASN SER THR ILE THR THR GLN GLU ALA ALA ASN SER VAL VAL ALA TYR GLY ARG TRP
CUA AAC UCG CCA AAC AUA GAG GCU UGC GGG UAU AGC GAU AGA GUA CUG CAA UUA ACA CUG GGA AAC UCC ACU AUA ACC ACA CAG GAG GCG GCU AAU UCA GUA GUC GCU UAU GGG CGU UGG

                                    1090                                1120                               1150                               1180
PRO GLU TYR LEU ARG ASP SER GLU ALA ASN PRO VAL ASP GLN PRO THR GLU PRO ASP VAL ALA ALA CYS ARG PHE TYR THR LEU ASP THR VAL SER TRP THR LYS GLU SER ARG GLY TRP
CCU GAA UAU CUG AGG GAC AGC GAA GCC AAU CCA GUG GAC CAG CCG ACA GAA CCA GAC GUC GCU GCA UGC AGG UUU UAU ACG CUA GAC ACC GUG UCU UGG ACG AAA GAG UCG CGA GGG UGG

                                    1210                                1240                               1270                               1300
TRP TRP LYS LEU PRO ASP ALA LEU ARG ASP MET GLY LEU PHE GLY GLN ASN MET TYR TYR HIS TYR LEU GLY ARG SER GLY TYR THR VAL HIS VAL GLN CYS ASN ALA SER LYS PHE HIS
UGG UGG AAG UUG CCU GAU GCA CUG AGG GAC AUG GGA CUC UUU GGG CAA AAU AUG UAC UAC CAC UAC CUA GGU AGG UCC GGG UAC ACC GUG CAU GUA CAG UGU AAC GCC UCC AAA UUC CAC

                                    1330                                1360                               1390                               1420
GLN GLY ALA LEU GLY VAL PHE ALA VAL PRO GLU MET CYS LEU ALA GLY ASP SER ASN THR THR THR MET HIS THR SER TYR GLN ASN ALA ASN PRO GLY GLU LYS GLY GLY THR PHE THR
CAG GGG GCA CUA GGG GUA UUC GCC GUA CCA GAG AUG UGU CUG GCC GGG GAU AGC AAC ACC ACU ACC AUG CAC ACC AGC UAU CAA AAU GCC AAU CCU GGC GAG AAA GGA GGC ACU UUC ACG

                                    1450                                1480                               1510                               1540
GLY THR PHE THR PRO ASP ASN ASN GLN THR SER PRO ALA ARG SER SER ALA ARG TRP ILE THR SER LEU GLU MET ALA ARG CYS TRP GLY MET PRO LEU CYS SER ALA GLN ILE ILE ASN
GGU ACG UUC ACU CCU GAC AAC AAC CAG ACA UCA CCU GCC CGC AGU UCU GCC CGG UGG AUU ACC UCC UUG GAA AUG GCA CGU UGU UGG GGA AUG CCU UUG UGU UCC GCA CAG AUA AUA AAC

                                    1570                                1600                               1630                               1660
LEU ARG THR ASN ASN CYS ALA THR LEU VAL LEU PRO TYR VAL ASN SER LEU SER LEU ASP SER MET VAL LYS HIS ASN ASN TRP GLY ILE ALA ILE LEU PRO LEU ALA PRO LEU ASN PHE
UUA CGG ACC AAC AAC UGU GCU ACA CUG GUA CUA CCU UAC GUG AAC UCC CUC UCG CUA GAU AGU AUG GUA AAG CAC AAU AAU UGG GGA AUU GCA AUA UUA CCA UUG GCC CCA UUA AAU UUU

                                    1690                                1720                               1750            --> VP3            1780
VAL SER GLU SER SER PRO GLU ILE PRO ILE THR LEU THR ILE ALA PRO MET CYS CYS GLU PHE ASN GLY LEU ARG ASN ILE THR LEU PRO ARG LEU GLN GLY LEU PRO VAL MET ASN THR
GUU AGU GAG UCC UCC CCA GAG AUU CCA AUC ACC UUG ACC AUA GCC CCU AUG UGC UGU GAG UUC AAU GGA UUA AGA AAC AUC ACC CUG CCA CGC UUA CAG GGC CUG CCG GUC AUG AAC ACC

                                    1810                                1840                               1870                               1900
PRO GLY SER ASN GLN TYR LEU THR ALA ASP ASN PHE GLN SER PRO CYS ALA LEU PRO GLU PHE ASP VAL THR PRO PRO ILE ASP ILE PRO GLY GLU VAL LYS ASN MET MET GLU LEU ALA
CCU GGU AGC AAU CAA UAU CUU ACU GCA GAC AAC UUC CAG UCA CCG UGU GCG CUG CCU GAA UUU GAU GUG ACC CCA CCU AUU GAC AUA CCC GGU GAA GUA AAG AAC AUG AUG GAA UUG GCA

                                    1930                                1960                               1990                               2020
GLU ILE ASP THR MET ILE PRO PHE ASP LEU SER ALA THR LYS LYS ASN THR MET GLU MET TYR ARG VAL ARG LEU SER ASP LYS PRO HIS THR ALA ALA SER ILE LEU CYS LEU SER LEU
GAA AUC GAC ACC AUG AUU CCC UUU GAC UUA AGU GCC ACA AAA AAG AAC ACC AUG GAA AUG UAU AGG GUU CGG UUA AGU GAC AAA CCA CAU ACA GCC GCC UCC AUA CUC UGC CUG UCA CUC

                                    2050                                2080                               2110                               2140
SER PRO ALA SER ASP PRO ARG LEU SER HIS THR MET LEU GLY GLU ILE LEU ASN TYR TYR THR HIS TRP ALA GLY SER LEU LYS PHE THR PHE LEU PHE CYS GLY SER MET MET ALA THR
UCU CCA GCC UCA GAU CCU AGG UUG UCA CAU ACU AUG CUU GGA GAA AUC CUA AAU UAC UAC ACA CAC UGG GCA GGA UCC CUG AAG UUC ACG UUU CUG UUC UGU GGA UCC AUG AUG GCA ACU

                                    2170                                2200                               2230                               2260
GLY LYS LEU LEU VAL SER TYR ALA PRO PRO GLY ALA ASP PRO PRO LYS LYS ARG LYS GLU ALA MET LEU GLY THR HIS VAL ILE TRP ASP ILE GLY LEU GLN SER SER CYS THR MET VAL
GGC AAA CUG UUG GUG UCA UAC GCG CCU CCU GGA GCC GAC CCA CCA AAG AAG CGU AAG GAG GCG AUG UUG GGA ACA CAU GUG AUC UGG GAC AUA GGA CUG CAG UCC UCA UGU ACU AUG GUA

                                    2290                                2320                               2350                               2380
VAL PRO TRP ILE SER ASN SER THR TYR ARG GLN THR ILE ASP ASP SER PHE THR GLU GLY GLY TYR ILE SER VAL PHE TYR GLN THR ARG ILE VAL VAL PRO LEU SER THR PRO ARG GLU
GUG CCA UGG AUU AGC AAC AGC ACG UAU CGG CAA ACC AUA GAU GAU AGU UUC ACC GAA GGC GGA UAC AUC AGC GUC UUC UAC CAA ACU AGA AUA GUC GUC CCU CUU UCG ACA CCC AGA GAG

                                    2410                                2440                               2470    --> VP1                    2500
MET ASP ILE LEU GLY PHE VAL SER ALA CYS ASN ASP PHE SER VAL ARG LEU LEU ARG ASP THR THR HIS ILE GLU GLN LYS ALA LEU ALA GLN GLY LEU GLY GLN MET LEU GLU SER MET
AUG GAC AUC CUU GGU UUU GUG UCA GCG UGU AAU GAC UUC AGC GUG CGC UUG UUG CGA GAU ACC ACA CAU AUA GAG CAA AAA GCG CUA GCA CAG GGG UUA GGU CAG AUG CUU GAA AGC AUG
```

```
2530 2560 2590 2620
ILE ASP ASN THR VAL ARG GLU THR VAL GLY ALA ALA THR SER ARG ASP ALA LEU PRO ASN THR GLU ALA SER GLY PRO THR HIS SER LYS GLU ILE PRO ALA LEU THR ALA VAL GLU THR
AUU GAC AAC ACA GUC CGU GAA ACG GUG GGG GCG GCA ACA UCU AGA GAC GCU CUC CCA AAC ACU GAA GCC AGU GGA CCA ACA CAC UCC AAG GAA AUU CCG GCA CUC ACC GCA GUG GAA ACU

2650 2680 2710 2740
GLY ALA THR ASN PRO LEU VAL PRO SER ASP THR VAL GLN THR ARG HIS VAL VAL GLN HIS ARG SER ARG SER GLU SER SER ILE GLU SER PHE PHE ALA ARG GLY ALA CYS VAL THR ILE
GGG GCC ACA AAU CCA CUA GUC CCU UCU GAU ACA GUG CAA ACC AGA CAU GUU GUA CAA CAU AGG UCA AGG UCA GAG UCU AGC AUA GAG UCU UUC UUC GCG CGG GGU GCA UGC GUG ACC AUU

2770 2800 2830 2860
MET THR VAL ASP ASN PRO ALA SER THR THR ASN LYS ASP LYS LEU PHE ALA VAL TRP LYS ILE THR TYR LYS ASP THR VAL GLN LEU ARG ARG LYS LEU GLU PHE PHE THR TYR SER ARG
AUG ACC GUG GAU AAC CCA GCU UCC ACC ACG AAU AAG GAU AAG CUA UUU GCA GUG UGG AAG AUC ACU UAU AAA GAU ACU GUC CAG UUA CGG AGG AAA UUG GAG UUC UUC ACC UAU UCU AGA

2890 2920 2950 2980
PHE ASP MET GLU LEU THR PHE VAL VAL THR ALA ASN PHE THR GLU THR ASN ASN GLY HIS ALA LEU ASN GLN VAL TYR GLN ILE MET TYR VAL PRO PRO GLY ALA PRO VAL PRO GLU LYS
UUU GAU AUG GAA CUU ACC UUU GUG GUU ACU GCA AAU UUC ACU GAG ACU AAC AAU GGG CAU GCC UUA AAU CAA GUG UAC CAA AUU AUG UAC GUA CCA CCA GGC GCU CCA GUG CCC GAA AAA

3010 3040 3070 3100
TRP ASP ASP TYR THR TRP GLN THR SER SER ASN PRO SER ILE PHE TYR THR TYR GLY THR ALA PRO ALA ARG ILE SER VAL PRO TYR VAL GLY ILE SER ASN ALA TYR SER HIS PHE TYR
UGG GAC GAC UAC ACA UGG CAA ACC UCA UCA AAU CCA UCA AUC UUU UAC ACC UAC GGG ACA GCU CCA GCC CGG AUC UCG GUA CCG UAU GUU GGU AUU UCG AAC GCC UAU UCA CAC UUU UAC

3130 3160 3190 3220
ASP GLY PHE SER LYS VAL PRO LEU LYS ASP GLN SER ALA ALA LEU GLY ASP SER LEU TYR GLY ALA ALA SER LEU ASN ASP PHE GLY ILE LEU ALA VAL ARG VAL VAL ASN ASP HIS ASN
GAC GGU UUU UCC AAA GUA CCA CUG AAG GAC CAG UCG GCA GCA CUA GGU GAC UCC CUU UAU GGU GCA GCA UCU CUA AAU GAC UUC GGU AUU UUG GCU GUU AGA GUA GUC AAU GAU CAC AAC

3250 3280 3310 3340
PRO THR LYS VAL THR SER LYS ILE ARG VAL TYR LEU LYS PRO LYS HIS ILE ARG VAL TRP CYS PRO ARG PRO PRO ARG GLN LEU ALA TYR TYR GLY PRO GLY VAL ASP TYR LYS ASP GLY
CCG ACC AAG GUC ACC UCC AAA AUC AGA GUG UAU CUA AAA CCC AAA CAC AUC AGA GUC UGG UGC CCG CGU CCA CCG AGG CAG CUG GCG UAC UAC GGC CCU GGA GUG GAU UAC AAG GAU GGU

                                                  →3b
3370 3400 3430 3460
THR LEU THR PRO LEU SER THR LYS ASP LEU THR THR TYR GLY PHE GLY HIS GLN ASN LYS ALA VAL TYR THR ALA GLY TYR LYS ILE CYS ASN TYR HIS LEU ALA THR GLN ASP ASP LEU
ACG CUU ACA CCC CUC UCC ACC AAG GAU CUG ACC ACA UAU GGA UUC GGA CAC CAA AAC AAA GCG GUG UAC ACU GCA GGU UAC AAA AUU UGC AAC UAC CAC UUG GCC ACU CAG GAU GAU UUG

3490 3520 3550 3580
GLN ASN ALA VAL ASN VAL MET TRP SER ARG ASP LEU LEU VAL THR GLU SER ARG ALA GLN GLY THR ASP SER ILE ALA ARG CYS ASN CYS ASN ALA GLY VAL TYR TYR CYS GLU SER ARG
CAA AAC GCA GUG AAC GUC AUG UGG AGU AGA GAC CUC UUA GUC ACA GAA UCA AGA GCC CAG GGC ACC GAU UCA AUC GCA AGG UGC AAU UGC AAC GCA GGG GUG UAC UAC UGC GAG UCU AGA

3610 3640 3670 3700
ARG LYS TYR TYR PRO VAL SER PHE VAL GLY PRO THR PHE GLN TYR MET GLU ALA ASN ASN TYR TYR PRO VAL ARG TYR GLN SER HIS MET LEU ILE GLY HIS GLY PHE GLU SER PRO GLY
AGG AAA UAC UAC CCA GUA UCC UUC GUU GGC CCA ACG UUC CAG UAC AUG GAG GCU AAU AAC UAU UAC CCA GUU AGG UAC CAG UCC CAU AUG CUC AUU GGC CAU GGA UUC GAA UCU CCA GGG

3730 3760 3790 3820
ASP CYS GLY GLY ILE LEU ARG CYS HIS HIS GLY VAL ILE GLY ILE ILE THR ALA GLY GLY GLU GLY LEU VAL ALA PHE SER ASP ILE ARG ASP LEU TYR ALA TYR GLU GLU GLU ALA MET
GAU UGU GGU GGC AUA CUC AGA UGU CAC CAC GGG GUG AUA GGG AUC AUU ACU GCU GGU GGA GAA GGG UUG GUU GCA UUU UCA GAC AUU AGA GAC UUG UAU GCC UAC GAA GAA GAA GCC AUG

        →5b
3850 3880 3910 3940
GLU GLN GLY ILE THR ASN TYR ILE GLU SER LEU GLY ALA ALA PHE GLY SER GLY PHE THR GLN GLN ILE SER ASP LYS ILE THR GLU LEU THR ASN MET VAL THR SER THR ILE THR GLU
GAA CAA GGC AUC ACC AAU UAC AUA GAG UCA CUU GGG GCC GCA UUU GGA AGU GGA UUU ACU CAG CAG AUU AGC GAC AAA AUA ACA GAG UUG ACC AAU AUG GUG ACC AGU ACC AUC ACU GAA

3970 4000 4030 4060
LYS LEU LEU LYS ASN LEU ILE LYS ILE ILE SER SER LEU VAL ILE ILE THR ARG ASN TYR GLU ASP THR THR THR VAL LEU ALA THR LEU ALA LEU LEU GLY CYS ASP ALA SER PRO TRP
AAG CUA CUU AAG AAC UUG AUC AAG AUC AUA UCC UCA CUA GUU AUU AUA ACU AGG AAC UAU GAA GAC ACC ACA ACA GUG CUC GCU ACC CUG GCC CUU CUU GGG UGU GAU GCU UCA CCA UGG

                                                                  →X                      *
4090 4120 4150 4180
GLN TRP LEU ARG LYS LYS ALA CYS ASP VAL LEU GLU ILE PRO TYR VAL ILE LYS GLN GLY ASP SER TRP LEU LYS LYS PHE THR GLU ALA CYS GLN ARG ALA LYS GLY LEU GLU TRP VAL
CAG UGG CUU AGA AAG AAA GCA UGC GAU GUU CUG GAG AUA CCU UAU GUC AUC AAG CAA GGU GAC AGU UGG UUG AAG AAG UUU ACU GAA GCA UGC CAA CGA GCU AAG GGC CUG GAG UGG GUG

4210 4240 4270 4300
SER ASN LYS ILE SER LYS PHE ILE ASP TRP LEU LYS GLU LYS ILE ILE PRO GLN ALA ARG ASP LYS LEU GLU PHE VAL THR LYS LEU ARG GLN LEU GLU MET LEU GLU ASN GLN ILE SER
UCA AAC AAA AUC UCA AAA UUC AUU GAU UGG CUC AAG GAG AAA AUU AUC CCA CAA GCU AGA GAU AAG UUG GAA UUU GUA ACA AAA CUU AGA CAA CUA GAA AUG CUG GAA AAC CAA AUC UCA

4330 4360 4390 4420
THR ILE HIS GLN SER CYS PRO SER GLN GLU HIS GLN GLU ILE LEU PHE ASN ASN VAL ARG TRP LEU SER ILE GLN SER LYS ARG PHE ALA PRO LEU TYR ALA VAL GLU ALA LYS ARG ILE
ACU AUA CAC CAA UCA UGC CCU AGU CAG GAA CAC CAG GAA AUU CUA UUC AAU AAU GUC AGA UGG UUA UCC AUC CAG UCU AAG AGG UUU GCC CCU CUU UAC GCA GUG GAA GCC AAA AGA AUA

4450 4480 4510 4540
GLN LYS LEU GLU HIS THR ILE ASN ASN TYR ILE GLN PHE LYS SER LYS HIS ARG ILE GLU PRO VAL CYS LEU LEU VAL HIS GLY SER PRO GLY THR GLY LYS SER VAL ALA THR ASN LEU
CAG AAA CUA GAG CAU ACU AUU AAC AAC UAC AUA CAG UUC AAG AGC AAA CAC CGU AUU GAA CCA GUA UGU UUG CUA GUA CAU GGC AGC CCC GGA ACA GGU AAA UCU GUA GCA ACC AAC CUG

4570 4600 4630 4660
ILE ALA ARG ALA ILE ALA GLU ARG GLU ASN THR SER THR TYR SER LEU PRO PRO ASP PRO SER HIS PHE ASP GLY TYR LYS GLN GLN GLY VAL VAL ILE MET ASP ASP LEU ASN GLN ASN
AUU GCU AGA GCC AUA GCU GAA AGA GAA AAC ACG UCC ACG UAC UCG CUA CCC CCG GAU CCA UCA CAC UUC GAC GGA UAC AAA CAA CAG GGA GUG GUG AUU AUG GAC GAC CUG AAU CAA AAC

4690 4720 4750 4780
PRO ASP GLY ALA ASP MET LYS LEU PHE CYS GLN MET VAL SER THR VAL GLU PHE ILE PRO PRO MET ALA SER LEU GLU GLU LYS GLY ILE LEU PHE THR SER ASN TYR VAL LEU ALA SER
CCA GAU GGU GCG GAC AUG AAG CUG UUC UGU CAG AUG GUA UCA ACA GUG GAG UUU AUA CCA CCC AUG GCA UCC CUG GAG GAG AAA GGA AUC CUG UUU ACU UCA AAU UAC GUU CUA GCA UCC

4810 4840 4870 4900
THR ASN SER SER ARG ILE SER PRO PRO THR VAL ALA HIS SER ASP ALA LEU ALA ARG ARG PHE ALA PHE ASP MET ASP ILE GLN VAL MET ASN GLU TYR SER ARG ASP GLY LYS LEU ASN
ACA AAC UCA AGC AGA AUU UCC CCC CCC ACU GUG GCA CAC AGU GAU GCA UUA GCC AGG CGC UUU GCG UUC GAC AUG GAC AUU CAG GUC AUG AAU GAG UAU UCU AGA GAU GGG AAA UUG AAC

4930 4960 4990 5020
MET ALA MET ALA THR GLU MET CYS LYS ASN CYS HIS GLN PRO ALA ASN PHE LYS ARG CYS CYS PRO LEU VAL CYS GLY LYS ALA ILE GLN LEU MET ASP LYS SER SER ARG VAL ARG TYR
AUG GCC AUG GCU ACU GAA AUG UGU AAG AAC UGU CAC CAA CCA GCA AAC UUU AAG AGA UGC UGU CCU UUA GUG UGU GGU AAG GCA AUU CAA UUA AUG GAC AAA UCU UCC AGA GUU AGA UAC
```

```
                                    5050                                    5080                        →1b                                    5140
SER ILE ASP GLN ILE THR THR MET ILE ILE ASN GLU ARG ASN ARG ARG SER ASN ILE GLY ASN CYS MET GLU ALA LEU PHE GLN GLY PRO LEU GLN TYR LYS ASP LEU LYS ILE ASP ILE
AGU AUU GAC CAG AUC ACU ACA AUG AUU AUC AAU GAG AGA AAC AGA AGA UCC AAC AUU GGC AAU UGU AUG GAG GCU UUG UUU CAA GGC CCA CUC CAG UAU AAA GAC UUG AAA AUU GAC AUC

                                    5170                                    5200                                    5230                                    5260
LYS THR SER PRO PRO PRO GLU CYS ILE ASN ASP LEU LEU GLN ALA VAL ASP SER GLN GLU VAL ARG ASP TYR CYS GLU LYS LYS GLY TRP ILE VAL ASN ILE THR SER GLN VAL GLN THR
AAG ACG AGU CCC CCU CCU GAA UGU AUC AAU GAC UUG CUC CAA GCA GUU GAC UCC CAG GAG GUG AGA GAU UAC UGU GAG AAG AAG GGU UGG AUA GUC AAC AUC ACC AGC CAG GUU CAA ACA

                                    5290                                    5320                                    5350                  →VPg          5380
GLU ARG ASN ILE ASN ARG ALA MET THR ILE LEU GLN ALA VAL THR THR PHE ALA ALA VAL ALA GLY VAL VAL TYR VAL MET TYR LYS LEU PHE ALA GLY HIS GLN GLY ALA TYR THR GLY
GAA AGG AAC AUC AAC AGG GCA AUG ACA AUU CUA CAA GCG GUG ACA ACC UUC GCC GCA GUG GCU GGA GUU GUC UAU GUC AUG UAU AAA CUG UUU GCU GGA CAC CAG GGA GCA UAC ACU GGU

                                    5410                        →2;7c                                           5470                                    5500
LEU PRO ASN LYS LYS PRO ASN VAL PRO THR ILE ARG THR ALA LYS VAL GLN GLY PRO GLY PHE ASP TYR ALA VAL ALA MET ALA LYS ARG ASN ILE VAL THR ALA THR THR SER LYS GLY
UUA CCA AAC AAA AAA CCC AAC GUG CCC ACC AUU CGG ACA GCA AAG GUA CAA GGA CCA GGG UUC GAU UAC GCA GUG GCU AUG GCU AAA AGA AAC AUU GUU ACA GCA ACU ACU AGC AAG GGA

                                    5530                                    5560                                    5590                                    5620
GLU PHE THR MET LEU GLY VAL HIS ASP ASN VAL ALA ILE LEU PRO THR HIS ALA SER PRO GLY GLU SER ILE VAL ILE ASP GLY LYS GLU ALA GLU ILE LEU ASP ALA LYS VAL PHE GLU
GAG UUC ACU AUG UUA GGA GUC CAC GAC AAC GUG GCU AUU UUA CCA ACC CAC GCU UCA CCU GGU GAA AGC AUU GUG AUC GAU GGC AAA GAA GCG GAG AUC UUG GAU GCC AAA GUG UUU GAA

                                    5650                                    5680                                    5710                                    5740
ASP GLN ALA GLY THR ASN PHE GLU ILE THR ILE ILE THR LEU LYS ARG ASN GLU LYS PHE ARG ASP ILE ARG PRO HIS ILE PRO THR GLN ILE THR GLU THR ASN ASP GLY VAL LEU ILE
GAU CAA GCA GGA ACC AAU UUU GAA AUC ACU AUA AUC ACU CUA AAG AGA AAU GAA AAG UUC AGA GAC AUU AGA CCA CAU AUA CCU ACU CAA AUC ACU GAG ACA AAU GAU GGA GUC UUG AUC

                                    5770                                    5800                                    5830                                    5860
VAL ASN THR SER LYS TYR PRO ASN MET TYR VAL PRO VAL ARG ALA VAL THR GLU GLN GLY TYR LEU ASN LEU GLY GLY ARG GLN THR ALA ARG THR LEU MET TYR ASN PHE PRO THR ARG
GUG AAC ACU AGC AAG UAC CCC AAU AUG UAU GUU CCU GUC CGU GCU GUG ACU GAA CAG GGA UAU CUA AAU CUC GGU GGG CGC CAA ACU GCU CGU ACU CUA AUG UAC AAC UUU CCA ACC AGA

                                    5890                                    5920                                    5950                                        →
ALA GLY GLN CYS GLY GLY VAL ILE THR CYS THR GLY LYS SER SER GLY CYS MET LEU VAL ASP GLY SER HIS GLY PHE ALA ALA ALA LEU LYS ARG SER LEU PHE THR GLN SER GLN GLY
GCA GGA CAG UGU GGU GGA GUC AUC ACA UGU ACU GGG AAG UCA UCG GGA UGC AUG UUG GUG GAC GGU UCA CAC GGG UUU GCA GCG GCC CUG AAG CGA UCA UUA UUC ACU CAG AGU CAA GGU
4b
                                    6010                                    6040                                    6070                                    6100
GLU ILE PRO TRP MET ARG PRO SER LYS ASP ALA GLY TYR PRO ILE ILE ASN ALA PRO SER LYS THR LYS LEU GLU PRO SER ALA PHE HIS TYR VAL PHE GLU GLY VAL LYS GLU PRO ALA
GAA AUC CCG UGG AUG AGA CCU UCG AAG GAC GCG GGA UAU CCA AUC AUA AAU GCC CCG UCC AAA ACC AAG CUU GAA CCC AGU GCU UUC CAC UAU GUG UUU GAA GGG GUG AAG GAA CCA GCA

                                    6130                                    6160                                    6190                                    6220
VAL LEU THR LYS ASN ASP PRO ARG LEU LYS THR ASP PHE GLU GLU ALA ILE PHE SER LYS TYR VAL GLY ASN LYS ILE THR GLU VAL ASP GLU TYR MET LYS GLU ALA VAL ASP HIS TYR
GUC CUC ACU AAA AAC GAU CCC AGG CUU AAG ACA GAC UUU GAG GAG GCA AUU UUC UCC AAG UAC GUG GGU AAC AAA AUU ACU GAA GUG GAU GAG UAC AUG AAA GAG GCA GUA GAC CAC UAU

                                    6250                                    6280                                    6310                                    6340
ALA GLY GLN LEU MET SER LEU ASP ILE ASN ILE GLU GLN MET CYS LEU GLU ASP ALA MET TYR GLY THR ASP GLY LEU GLU ALA LEU ASP LEU SER THR SER ALA GLY TYR PRO TYR VAL
GCU GGC CAG CUC AUG UCA CUA GAC AUC AAC AUA GAA CAA AUG UGC UUG GAG GAU GCC AUG UAU GGC ACU GAU GGU CUA GAA GCA CUU GAU UUG UCC ACC AGU GCU GGC UAC CCU UAU GUA

                                    6370                                    6400                                    6430                                    6460
ALA MET GLY LYS LYS LYS ARG ASP ILE LEU ASN LYS GLN THR ARG ASP THR LYS GLU MET GLN LYS LEU LEU ASP THR TYR GLY ILE ASN LEU PRO LEU VAL THR TYR VAL LYS ASP GLU
GCA AUG GGA AAG AAG AAG AGA GAC AUC UUG AAC AAA CAA ACC AGA GAC ACU AAG GAA AUG CAA AAA CUG CUC GAC ACA UAU GGA AUC AAC CUC CCA CUG GUG ACU UAU GUA AAG GAU GAA

                                    6490                                    6520                                    6550                                    6580
LEU ARG SER LYS THR LYS VAL GLU GLN GLY LYS SER ARG LEU ILE GLU ALA SER SER LEU ASN ASP SER VAL ALA MET ARG MET ALA PHE GLY ASN LEU TYR ALA ALA PHE HIS LYS ASN
CUU AGA UCC AAA ACA AAG GUU GAG CAG GGG AAA UCC AGA UUA AUU GAA GCU UCU AGU UUG AAU GAC UCA GUG GCA AUG AGA AUG GCU UUU GGG AAC CUA UAU GCU GCU UUU CAC AAA AAC

                                    6610                                    6640                                    6670                                    6700
PRO GLY VAL ILE THR GLY SER ALA VAL GLY CYS ASP PRO ASP LEU PHE TRP SER LYS ILE PRO VAL LEU MET GLU GLU LYS LEU PHE ALA PHE ASP TYR THR GLY TYR ASP ALA SER LEU
CCA GGA GUG AUA ACA GGU UCA GCA GUG GGG UGC GAU CCA GAU UUG UUU UGG AGC AAA AUU CCG GUA UUG AUG GAA GAG AAG CUG UUU GCU UUU GAC UAC ACA GGG UAU GAU GCA UCU CUC

                                    6730                                    6760                                    6790                                    6820
SER PRO ALA TRP PHE GLU ALA LEU LYS MET VAL LEU GLU LYS ILE GLY PHE GLY ASP ARG VAL ASP TYR ILE ASP TYR LEU ASN HIS SER HIS HIS LEU TYR LYS ASN LYS THR TYR CYS
AGC CCU GCU UGG UUC GAG GCA CUA AAG AUG GUG CUU GAG AAA AUC GGA UUC GGA GAC AGA GUU GAC UAC AUC GAC UAC CUA AAC CAC UCA CAC CAC CUG UAC AAG AAU AAA ACA UAC UGU

                                    6850                                    6880                                    6910                                    6940
VAL LYS GLY GLY MET PRO SER GLY CYS SER GLY THR SER ILE PHE ASN SER MET ILE ASN ASN LEU ILE ILE ARG THR LEU LEU LEU LYS THR TYR LYS GLY ILE ASP LEU ASP HIS LEU
GUC AAG GGC GGU AUG CCA UCU GGC UGC UCA GGC ACU UCA AUU UUU AAC UCA AUG AUU AAC AAC UUG AUU AUC AGG ACA CUC UUA CUG AAA ACC UAC AAG GGC AUA GAU UUA GAC CAC CUA

                                    6970                                    7000                                    7030                                    7060
LYS MET ILE ALA TYR GLY ASP ASP VAL ILE ALA SER TYR PRO HIS GLU VAL ASP ALA SER LEU LEU ALA GLN SER GLY LYS ASP TYR GLY LEU THR MET THR PRO ALA ASP LYS SER ALA
AAA AUG AUU GCC UAU GGU GAU GAU GUA AUU GCU UCC UAC CCC CAU GAA GUU GAC GCU AGU CUC CUA GCC CAA UCA GGA AAA GAC UAU GGA CUA ACU AUG ACU CCA GCU GAC AAA UCA GCU

                                    7090                                    7120                                    7150                                    7180
THR PHE GLU THR VAL THR TRP GLU ASN VAL THR PHE LEU LYS ARG PHE PHE ARG ALA ASP GLU LYS TYR PRO PHE LEU ILE HIS PRO VAL MET PRO MET LYS GLU ILE HIS GLU SER ILE
ACA UUU GAA ACA GUC ACA UGG GAG AAU GUA ACA UUC UUG AAG AGA UUC UUC AGG GCA GAC GAG AAA UAC CCA UUU CUU AUU CAU CCA GUA AUG CCA AUG AAG GAA AUU CAU GAA UCA AUU

                                    7210                                    7240                                    7270                                    7300
ARG TRP THR LYS ASP PRO ARG ASN THR GLN ASP HIS VAL ARG SER LEU CYS LEU LEU ALA TRP HIS ASN GLY GLU GLU GLU TYR ASN LYS PHE LEU ALA LYS ILE ARG SER VAL PRO ILE
AGA UGG ACU AAA GAU CCU AGG AAC ACU CAG GAU CAC GUU CGC UCU CUG UGC CUU UUA GCU UGG CAC AAU GGC GAA GAA GAA UAU AAC AAA UUC CUA GCU AAA AUC AGG AGU GUG CCA AUU

                                    7330                                    7360                                    7390                                    7420
GLY ARG ALA LEU LEU LEU PRO GLU TYR SER THR LEU TYR ARG ARG TRP LEU ASP SER PHE END END
GGA AGA GCU UUA UUG CUC CCA GAG UAC UCA ACA UUG UAC CGC CGU UGG CUU GAC UCA UUU UAG UAA CCC UAC CUC AGU CGA AUU GGA UUG GGU CAU ACU GCU GUA GGG GUA AAU UUU UCU

UUA AUU CGG AGG - poly (A)
```

disease, and in 1967 only 1,013 poliomyelitis patients were identified, a reduction of 98.7%.

In spite of the vast decrease in poliomyelitis in the world, cases still occur, especially in countries where immunization programs are not carried out systematically. Out of four million children immunized with the Sabin vaccine, one child developed paralytic polio. In addition, two unimmunized adults who came into contact with children immunized with the Sabin vaccine also contracted the disease. Nonetheless, Sabin's vaccine is the best and most effective vaccine for immunization of children and adults against polio. Identification of individuals sensitive to the vaccine will prevent any ill effects resulting from immunization. The prospects of, and problems found in, vaccination against poliomyelitis are discussed in chapter 24.

Laboratory diagnosis of poliovirus types 1, 2, and 3

The virus is isolated from the feces of the patient by inoculation into cultured cells. The typing of the virus is done by neutralization of the isolated virus with antibodies to the three poliovirus types.

Since attenuated live poliovirus vaccine is widespread and the virus is excreted in the feces, it is necessary to differentiate between a vaccine strain and the wild type when poliovirus is isolated. The best method is to determine the neurovirulence of the virus by injection into the brain of a monkey, but this is highly expensive and impractical. Laboratory diagnosis is based on the genetic properties of the wild type and the attenuated viruses (Nakano et al. 1978). The vaccine strains are able to replicate in infected cells incubated at 35.5°C, 39.5°C, and 40.1°C, whereas the wild type virus replicates only at 35.5°C. It is also possible to differentiate between the attenuated and wild type viruses by analyzing the structural proteins, using electrophoresis in polyacrylamide gels. This is based on a change in the properties of the structural proteins in the attenuated virus strains.

Figure 58. The complete nucleotide sequence and encoded information of virion RNA of poliovirus type 1 (Mahoney). As judged by fingerprint analyses this isolate has not undergone detectable genetic variation in our laboratory during multiple passages in HeLa cells. A reading frame of 2,207 consecutive coding triplets within the nucleotide sequence was determined by computer analysis and confirmed by radiochemical microsequence analysis of 12 virus-specific polypeptides. The cleavage sites involved in processing viral polypeptides and the N-termini of viral proteins are indicated by a solid triangle and arrow; amino acids that were identified by stepwise Edman degradation of radiochemically pure polypeptides are indicated by bars. The asterisk at N4159 indicates an amino acid ambiguity. The coding sequence of capsid protein VP4 has been determined by analysis of tryptic peptides. The N-terminal amino acid of VP4 is blocked as are those of P1-1a and VPO. Preliminary evidence suggests that translation may be initiated with the methionine at N741. Termination codons in the non-coding regions are designated as 'END'.
Potential initiation codons of translation in the 5′-terminal, non-coding sequence preceding N741 are indicated as MET if in phase with the major reading frame. MET codons out of phase with the major reading frame in this region are indicated with brackets.
(Kitamura et al. 1981. Reprinted by permission from *Nature* 291, Fig. 2, p. 549. Copyright © 1981 Macmillan Journals Limited.)

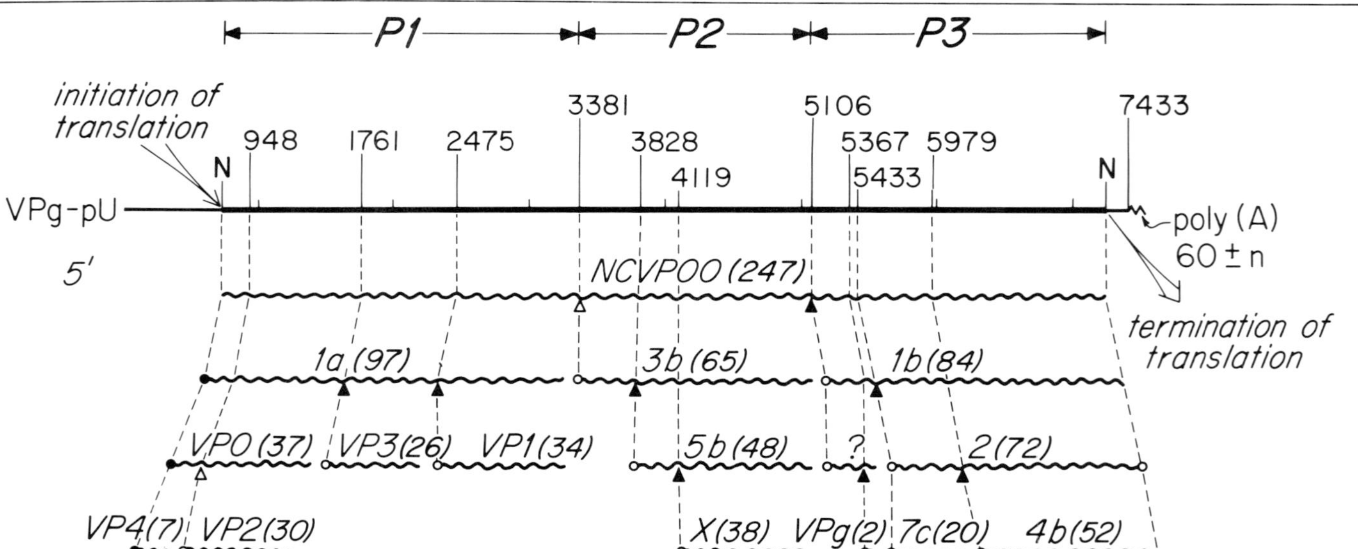

Figure 59. Gene organization of poliovirus RNA. Virion RNA, terminated at the 5′ end with the genome-linked protein VPg and at the 3′ end with poly(A), is shown as a solid line, the translated region being more pronounced than the non-coding regions. N indicates the proposed sites at which initiation (N741) and termination (N7361) of translation occurs. The numbers above the virion RNA line refer to the first nucleotide of the codon specifying the N-terminal amino acid for the viral specific proteins. The coding region has been divided into three regions (P1, P2, P3), corresponding to the primary cleavage products of polyprotein NCVPOO. Accordingly, polypeptides (waved line, designated in italics) are referred to in the text as P1-1a, P3-1b, etc. No 'leader peptide' that would map in a region preceding P1 has been found for poliovirus RNA translation. Numbers in parentheses are molecular weights calculated from the amino acid sequences shown in Fig. 59, assuming that C-terminal 'trimming' does not occur. Open circles indicate that the exact position of the terminal acid is known. The N-termini are glycine in all cases except for VP2 where it is serine. The C-terminal amino acid of P3-2 is phenylalanine. Closed circles indicate that the N-termini are known to be blocked. Closed triangles: Gln-Gly pairs that are cleaved during proteolytic processing of a polypeptide. Open triangles: exceptional cleavage sites in polypeptide processing (Asn-Ser at N948; Tyr-Gly at N3381). This figure is not meant to imply that the sequence of processing occurs via all possible intermediate polypeptides. For example, it is not known whether P2-3b is always processed to P2-X via P2-5b, or occasionally directly to P2-X.

(Kitamura et al. 1981. Reprinted by permission from *Nature* 291, Fig. 4, p. 552. Copyright © 1981 Macmillan Journals Limited.)

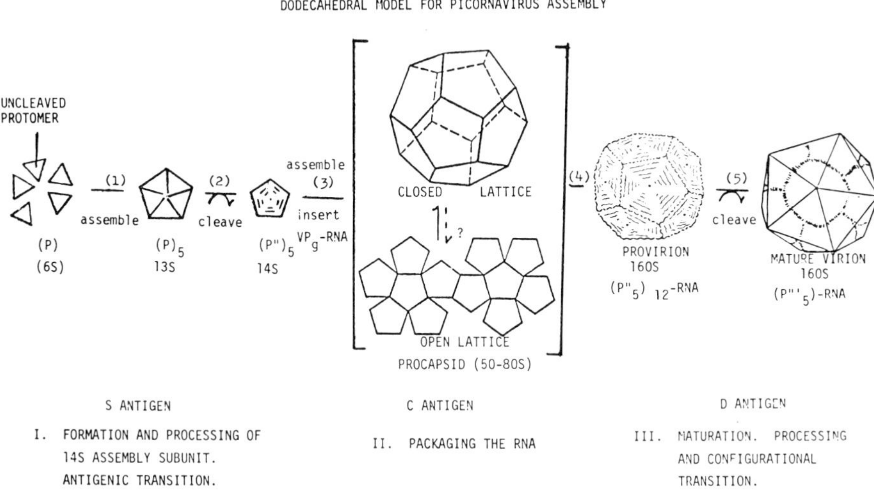

Figure 60. Dodecahedral model for picornavirus assembly.
(Rueckert 1978. Reprinted by permission from *International Virology* 4, Fig. 3, p. 45. Centre for Agricultural Publishing and Documentation, Wageningen.)

COXSACKIE VIRUS INFECTIONS

These members of the genus Enterovirus are divided into two groups: Human Coxsackie viruses A (24 virus strains) and B (6 virus strains).

Diseases caused by Coxsackie viruses

1. Herpangina infection in the pharynx of children is caused by Coxsackie viruses A 1–6, 8, and 10.
2. An infection that causes a skin rash is caused by Coxsackie viruses A 4, 6, 9, and 16.
3. Bornholm's disease, named after a disease first discovered on a Danish island in the Baltic Sea, is caused by Coxsackie B. The disease is manifested by fever, headache, and muscular aches, followed by chest pain.
4. Pericarditis is caused by Coxsackie B virus, due to infection of the pericardium, and can result in the death of the patient.
5. Acute pancreatitis is also caused by Coxsackie B; the virus infects the acinar cells of the pancreas and has been associated with diabetes mellitus.

Laboratory diagnosis of Coxsackie viruses

Diagnosis of the virus isolated from the feces is based on its ability to infect cultured cells and newborn mice. Coxsackie A viruses cause paralysis in the newborn mice, while Coxsackie B viruses do not infect newborn mice but replicate in cultured cells. The isolated virus stains are typed with specific antisera.

ECHOVIRUSES AND ENTEROVIRUSES AS HUMAN PATHOGENS

Diseases caused by these viruses are similar to those caused by Coxsackie viruses but are less pathogenic.

1. Upper respiratory tract infections. These infections of the upper respiratory tract in children are caused by echoviruses 11 and 20.
2. Pneumonitis and bronchitis are caused by enterovirus 68.
3. Encephalitis is caused by enterovirus 71.
4. Meningitis is caused by echoviruses 4, 6, 9, 11, 14, 16, or 30.

GENUS RHINOVIRUS: HUMAN COMMON COLD VIRUS

Human rhinovirus 1A and strains 2 to 113 cause the common cold in humans, and all the virus strains differ in their antigenicity. As opposed to the enteroviruses, these viruses are sensitive to low pH and are destroyed by solutions of pH 3–5. The virus can be transmitted to chimpanzees. D. Tyrrell and his colleagues succeeded in growing these viruses in cell cultures at 33°C, which is the temperature in the nose, at a pH around 7.0 by lowering the concentration of $NaHCO_3$ in the medium. These cells need good oxygenation for growth.

The virus infects the nasal mucosa and replicates in the epithelial cells. As a result of virus infection, there is a watery discharge from the nose, accompanied by headache, coughing, and sore throat. Because of the high mutation rate of the rhinoviruses and the existence of over 100 virus strains, it is not possible to develop a vaccine for immunization against the common cold.

FOOT AND MOUTH DISEASE VIRUS (FMDV)

Foot and mouth disease was described in 1514 in northern Italy by Hyronimus Proctorius. It was also recorded in animals in Germany in 1751 and in animals and man in 1756. During the 19th century, the disease spread to England. In 1898, Loeffler and Frosch demonstrated that the agent causing foot and mouth disease is a filterable virus.

The virus is present in vesicles in the mouth and on the feet of infected animals, as well as in the blood and in the milk. Initially, two antigenic types of FMDV (named O and A) were found in 1922 in France and Germany, respectively. A third type of virus was found four years later in Germany. In 1948, three additional virus types were found in Southern Africa; these were named South African Territories (SAT) 1, 2, and 3 viruses. In 1954, a new antigenic type of FMDV was found in Pakistan and other Asian countries and was named Asia 1. With the aid of the complement fixation test, subtypes of FMDV were recognized: type O has ten subtypes; type A fifteen subtypes; SAT1 six subtypes; SAT2 three subtypes; and SAT3 three subtypes. There is no cross protection between the types, but partial immunity exists between the subtypes.

The disease in cattle, sheep, and pigs

In cattle, FMDV causes an acute infection characterized in dairy cattle by a marked decrease in milk production, accompanied by dullness and a rise in temperature to 41°C. In a matter of hours from the onset of the disease (after an incubation period of three–six days) vesicles appear on the tongue, lips, muzzle, and foot pads. The vesicles coalesce, and the epithelium becomes necrotic and susceptible to bacterial infection accompanied by salivation. When the vesicles start to heal, the animal is more willing to eat and drink. Vesicular lesions may develop on the udders and teats of cows and heifers; sometimes bacterial mastitis develops.

Mortality, which is less than 2%, is mainly due to myocardial collapse on the day five or six of the sickness. In young animals, mortality is 50%; animals die without symptoms of infection.

Infected sheep and pigs develop a sudden onset of acute lameness in all four extremities. People coming into contact with infected animals can contract the disease.

Vaccination

Inactivated FMDV vaccines are currently used; acetylethyleneimine is the inactivating agent. The virus is produced in the baby hamster kidney (BHK) cell line grown in suspension in large quantities (2,000 liters or more) or in surviving tongue epithelial cells collected from animals slaughtered for meat. Vaccination is complicated since 7 types and over 60 subtypes of the virus exist. As a result, a polyvalent vaccine is required containing 3 or 4 virus types, and 2 or 3 annual vaccinations of animals are necessary. New virus strains may suddenly appear that render the vaccines impotent. Live attenuated virus vaccine is not in use; attenuation of the virus for cattle does not guarantee attenuation for other animals.

Recently, Küpper and associates (1981) succeeded in cloning a DNA copy of the RNA sequence coding for the major antigen (VP1) of FMDV in a plasmid vector. They were able to demonstrate synthesis of the VP1 antigen in the *E. coli* host. VP1 is the coat protein responsible for the antibody response to FMDV. Boothroyd and associates (1981) also constructed and analyzed recombinant plasmids containing complementary DNA copies of FMDV RNA. This breakthrough in genetic engineering holds promise for the production of a new, effective vaccine against FMDV.

BIBLIOGRAPHY

Boothroyd, J.C.; Highfield, P.E.; Cross, G.A.M.; Rowlands, D.J.; Lowe, P.A.; Brown, F.; and Harris, T.J.R. Molecular cloning of foot and mouth disease virus genome and nucleotide sequences in the structural protein genes. *Nature 290*:800–802, 1981.

Dasgupta, A.; Zabel, P.; and Baltimore, D. Dependence of the activity of the poliovirus replicase on a host cell protein. *Cell 19*:423–429, 1980.

Flanegan, J.B., and Baltimore, D. Poliovirus polyuridylic acid polymerase and RNA replicase have the same viral polypeptide. *J. Virol. 29*:352–360, 1979.

Flanegan, J.B.; Petterson, R.F.; Ambros, V.; Hewlett, M.J.; and Baltimore, D. Covalent linkage of a protein to a defined nucleotide sequence at the 5′-terminus of virion and replicative intermediate RNAs of poliovirus. *Proc. Natl. Acad. Sci. USA 74*:961–965, 1977.

Hewlett, M.J.; Rose, J.K.; and Baltimore, D. 5′-Terminal structure of poliovirus polyribosomal RNA is pUp. *Proc. Natl. Acad. Sci. USA 73*:327–330, 1976.

Kaufman, Y.; Goldstein, E.; and Penman, S. Poliovirus-induced inhibition of polypeptide initiation in vitro on native polyribosomes. *Proc. Natl. Acad. Sci. USA 73*:1834–1838, 1976.

Kitamura, N.; Adler, C.; and Wimmer, E. Structure and expression of the picornavirus genome. *Ann. NY Acad. Sci. 354*:183–201, 1980.

Kitamura, N.; Semler, B.L.; Rothberg, P.G.; Larsen, G.R.; Adler, C.J.; Dorner, A.J.; Emini, E.A.; Hanecak, R.; Lee, J.J.; van der Werf, S.; Anderson, C.W.; and Wimmer, E. Primary structure, gene organization and polypeptide expression of poliovirus RNA. *Nature 291*:547–553, 1981.

Küpper, H.; Keller, W.; Kurz, C.; Forss, S.; Schaller, H.; Franze, R.; Strohmaier, K.; Marquandt, O.; Zaslavsky, F.G.; and Hofschneider, P.H. Cloning of cDNA of major antigen of foot and mouth disease virus and expression in *E. coli*. *Nature 289*:555–559, 1981.

Lee, A.F.; Nomoto, A.; Detjen, B.M.; and Wimmer, E. A protein covalently linked to poliovirus genome RNA. *Proc. Natl. Acad. Sci. USA 74*:59–63, 1977.

Nakano, J.H.; Hatch, M.H.; Thieme, M.L.; and Nottay, B. Parameters for differentiating vaccine-derived and wild poliovirus strains. *Prog. Med. Virol. 24*:178–206, 1978.

Nomoto, A.; Detjen, B.; Pozzatti, R.; and Wimmer, E. The location of the polio genome protein in viral RNAs and its implication for RNA synthesis. *Nature 268*:208–213, 1977.

Nomoto, A.; Lee, Y.F.; and Wimmer, E. The 5′ end of poliovirus mRNA is not capped with $m^7G(5')ppp(5')Np$. *Proc. Natl. Acad. Sci. USA 73*:375–380, 1976.

Pettersson, R.F.; Ambros, V.; and Baltimore, D. Identification of a protein linked to nascent poliovirus RNA and to the polyuridylic acid of negative-strand RNA. *J. Virol. 27*:357–365, 1978.

Rose, J.K.; Trachsel, H.; Leong, K.; and Baltimore, D. Inhibition of translation by poliovirus: inactivation of a specific initiation factor. *Proc. Natl. Acad. Sci. USA 75*:2732–2736, 1978.

Rueckert, R. *Structure and Assembly of Picornaviruses*. International Virology IV. Centre for Agricultural Publishing and Documentation, Wageningen, 1978, pp. 44–45.

Semler, B.L.; Anderson, C.W.; Kitamura, N.; Rothberg, P.G.; Wishart, W.L.; and Wimmer, E. Poliovirus replication proteins: RNA sequence encoding P3-1b and the sites of proteolytic processing. *Proc. Natl. Acad. Sci. USA 78*:3464–3468, 1981.

Yogo, Y., and Wimmer, E. Polyadenylic acid at the 3′-terminus of poliovirus RNA. *Proc. Natl. Acad. Sci. USA 69*:1877–1882, 1972.

RECOMMENDED READING

Becker, Y. Do toxic tetrapeptides play a role in fatal viral diseases? *JAMA*:258–260, 1981.

Cole, C.N. Defective interfering (DI) particles of poliovirus. *Prog. Med. Virol. 20*:180–207, 1975.

Cooper, P.D. Genetics of picornaviruses. *In: Comprehensive Virology* (H. Fraenkel-Conrat and R.R. Wagner, eds.) Plenum Press, New York, 1977, Vol. 9, pp. 138–208.

Craighead, J.E. The role of viruses in the pathogenesis of pancreatic disease and diabetes mellitus. *Prog. Med. Virol. 19*:161–214, 1975.

Fenner, F., and White, D.O. Picornaviridae. *In: Medical Virology*, 2nd ed. Academic Press, New York, San Francisco, London, 1976, pp. 332–354.

Levintow, L. Reproduction of picornaviruses. *In: Comprehensive Virology* (H. Fraenkel-Conrat and R.R. Wagner, eds.), Plenum Press, New York, 1974, Vol. 2, pp. 109–170.

Longworth, J.F. Some isometric viruses of invertebrates. *Adv. Virus Res. 23*:103–157, 1978.

Sangar, D.V. Review article: The replication of picornaviruses. *J. Gen. Virol. 45*:1–13, 1979.

Stott, E.J., and Killington, R.A. Rhinoviruses. *Annu. Rev. Microbiol. 26*:503–524, 1972.

Wimmer, E. Structure and replication of picornaviruses. *Curr. Top. Microbiol. Immunol.*, 1981.

19. RNA PLUS GENOME THAT SERVES AS MESSENGER RNA: TOGAVIRUSES

FAMILY TOGAVIRIDAE

The virions are spherical and enveloped, with a diameter of 40–70 nm. The genome is a single RNA plus molecule with a molecular weight of 4×10^6. The RNA genome is contained in an icosahedral nucleocapsid that is assembled in the cytoplasm and obtains its lipoprotein envelope during budding through the plasma membrane of the host cell in the final stage of virus maturation.

Genus Alphavirus (Arbovirus group A)

The virus species include: Aura, Bebaru, Chikungunya, Eastern equine encephalomyelitis (EEE), Everglades, Getah, Mayaro, Middleburg, Mucambo, Ndumu, O'nyong-nyong, Pixuma, Ross river, Semliki Forest, Sindbis, Venezuela equine encephalomyelitis (VEE), Western equine encephalomyelitis (WEE) (figure 61), and Whataroa. These viruses multiply in arthropods as well as in vertebrates.

Genus Flavivirus (Arbovirus group B)

Species that are mosquito-borne include: Yellow fever; dengue types 1, 2, 3, and 4; Japanese encephalitis; Spondweni; St. Louis; Uganda S; Wesselsbron; West Nile; and Zika.

Tick-borne species include: Kyasanur Forest disease; Langat; Louping ill;

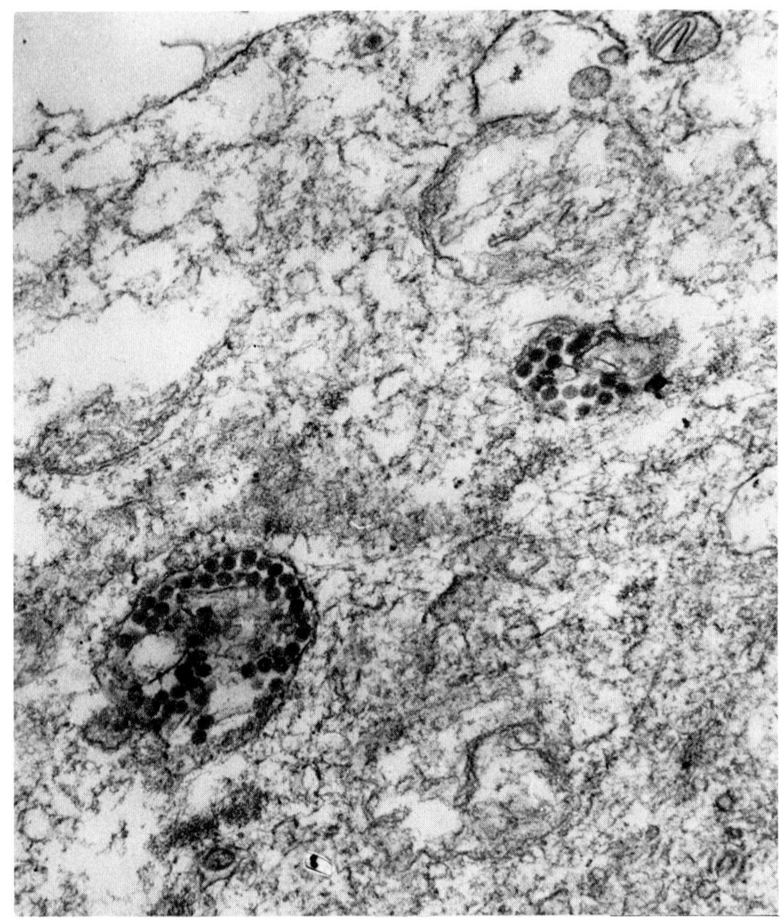

Figure 61. Electron micrograph of Western equine encephalitis (WEE) virus in the cytoplasm of infected cells.
(Courtesy of Dr. Daniel Dekegel, Institut Pasteur du Brabant, Brussels, Belgium.)

Omsk hemorrhagic fever; Royal Farm; Saumarez Reef; and tick-borne encephalitis (European and Far Eastern). There are also a number of species with unknown vectors.

Genus Rubivirus

Rubella virus occurs only in man.

Genus Pestivirus

Mucosal disease virus, border disease of sheep, hog cholera (European swine fever) and other possible members of the Togaviridae are not arthropod-borne.

Toga comes from the Latin *toga* (gown, cloak), alpha from the Greek letter *A*; and flavi, rubi, and pesti from the Latin *flavus* (yellow), *rubeus* (reddish), and *pestis* (plague), respectively.

ORGANIZATION OF THE VIRIONS

The virions have a diameter of 60 nm (alpha- and rubiviruses) and a sedimentation coefficient of 240–300 S. Flavi- and pestiviruses have a diameter of 45 nm and a sedimentation coefficient of 170–220S. Virions have a nucleocapsid complex of 240 capsid proteins (C protein, m.w. 30,000) containing the RNA genome (42 S), enveloped by a lipid bilayer in which the virus-coded glycoproteins are present as spikes. There are about 240 spikes in the envelope of each virion. The spike is made up of three polypeptides: E1 (m.w. 49,000), E2 (m.w. 52,000), and E3 (m.w. 10,000). E1 and E3 have one attached oligosaccharide and E2 contains two attached oligosaccharides. Polypeptides E1 and E2 of the spike are attached to the lipid bilayer by their hydrophobic tails (COOH-terminal ends), while E3 is present on the outer side of the envelope bound to E1 and/or E2 (Garoff and Söderlund 1978). The E2 polypeptide spans the membrane: The terminus of the E2 polypeptide passes through the lipid bilayer; the amino terminal end is present on the outer surface, and the carboxy terminal end on the inner surface of the lipid bilayer envelope. It is possible that each spike is attached to the capsid protein C beneath the lipid bilayer. The viral protein C present in the nucleocapsid is rich in lysine.

The 42 S virion RNA contains 12,000 nucleotides with a poly(A) sequence attached to the 3′ end and a cap (m^7GpppAp) present in the 5′ end. The virion RNA serves as messenger RNA after uncoating in the cytoplasm and can be found in the polyribosomes of the infected cell.

Alphaviruses contain neutral lipids arranged in a bilayer structure in the virus membrane: mainly cholesterol, as well as phosphatidylethanolamine, phosphatidylserine, sphingomyelin, and phosphatidylcholine. Oleic acid, palmitic acid, and stearic acid are the major fatty acids in the lipid bilayer. The virion envelope has a lipid composition similar to that of the host cell membrane, but no host cell proteins are found in the viral membrane. The membrane is most likely derived from a segment of the plasma membrane that does not contain host cell proteins.

ANTIGENIC STRUCTURE

The classification of togaviruses is based on immunological cross reactivity, as determined by hemagglutination inhibition, neutralization tests, and agar diffusion, as well as by radioimmunoassay, using specific antibodies against each virus isolate. Members of a genus are serologically related to each other, but not to other members of the family. The protein E1 (a glycoprotein) of Sindbis virus has the ability to hemagglutinate red blood cells, whereas protein E2 is responsible for the ability of the virus to infect cells. Antibodies to the E2 protein, therefore, have the ability to neutralize virus infectivity. The glycoprotein E1 of Sindbis virus is antigenically related to that of WEE, while E2 is specific for both viruses. Relatedness of alphaviruses determined by RNA-RNA hybridization showed that RNA from Chikungunya and O'nyong-nyong viruses had 13% base sequence homology whereas sequence homologies of 1% or less were detected between these two viruses and Semliki Forest and Sindbis viruses (Wengler et al. 1977). Thus the rather low sequence homologies between the nucleic acids of these viruses do not affect their antigenic relationships. This may be explained by the assumption that the antigenic sites are composed of only a small number of amino acids.

MOLECULAR EVENTS IN ALPHAVIRUS REPLICATION

Alphaviruses replicate in the cytoplasm of both vertebrate and invertebrate cell cultures. In chick embryo fibroblasts, the virus reaches maximal yields within five hr. In actinomycin D-treated infected cells, the virus yield is higher than in untreated cultures, possibly due to inhibition of interferon synthesis. Actinomycin D inhibits DNA-dependent RNA synthesis, which shows that this virus does not require host cell nuclear functions for replication.

The virus enters the cell by absorptive endocytosis (Helenius et al. 1980). Inside the lysosomes of the cell, the low pH probably causes the viral membrane to fuse with the lysosomal membrane (White and Helenius 1980). This allows the nucleocapsid to enter the cytoplasm, where the viral genome is uncoated. The parental viral RNA acts as mRNA for the synthesis of the viral RNA-dependent RNA polymerase that binds to the smooth cytoplasmic membranes where replication of viral RNA takes place. The replicative intermediates are RNA molecules that are partially double-stranded and include one complete RNA molecule that serves as template and a number of progeny RNA molecules that are hydrogen-bonded to the RNA template. The RNA polymerase is responsible for the transcription of the viral RNA^+ genomes from the RNA^- strands.

Synthesis of the viral RNA and proteins

In the alphaviruses, two species of single-stranded RNA are synthesized, 42S RNA (m.w. 4.2×10^6) and 26S RNA (m.w. 1.6×10^6). Both RNA molecules contain poly(A) sequences of heterogeneous length, and the 26S molecules represent subgenomic RNA species, having the same polarity as the virion RNA. The 26S subgenomic RNA is homologous to the 3′ end of the viral genome (Kennedy 1976) and functions as mRNA for the structural proteins of the virus. This is an efficient mechanism of reiteration of a portion (about one-third) of the viral genome and allows excess synthesis of the viral structural proteins independently of the synthesis of the nonstructural proteins by the 42S RNA molecules.

The nucleotide sequence of the gene coding for the capsid (C) protein is now known, as is the nucleotide sequence of cloned cDNA made from the 26S

RNA of Semliki Forest virus (Garoff et al. 1980*a,b*). The amino acid sequences of the different membrane proteins deduced from the nucleotide sequences are shown in figure 62. The coding region of the 26S RNA starts with the aminoterminus of the capsid protein C and terminates 3759 nucleotides later when the first stop codon UAA is reached. The regions coding for the E3, E2 and E1 polypeptides were localized in this sequence.

Translation of the 26S RNA

The structural proteins consisting of the four polypeptides E1, E2, and E3, and the capsid protein are translated from the 26S RNA using a single initiation site in the following order: capsid, p62, and E1. The p62 polypeptide is the precursor of the E3 and E2 proteins. Figure 63 shows that the C protein appears after the first cleavage of the precursor polypeptide p130. The C protein molecules subsequently bind to the 42S genomic RNA to form the nucleocapsid. The first cleavage product p97 is further cleaved to yield polypeptide p62 and the glycoprotein E1. P62 is cleaved to yield the glycoproteins E3 and E2 (Kääriäinen and Söderlund 1978).

A scheme for the assembly of the Semliki Forest virus spike glycoprotein E1 in the membrane of the endoplasmic reticulum as suggested by Garoff and associates (1980*b*) is depicted in figure 64.

Translation of the 42S RNA

The viral 42S RNA acts as polycistronic mRNA that is translated by the cellular ribosomes to yield the nonstructural viral proteins. This is demonstrated in figure 65, which shows that the viral RNA is not completely translated. In Semliki Forest virus, the precursor polypeptide is cleaved at three sites to yield four stable proteins of m.w. 70,000, 86,000, 72,000, and 60,000 (ns 70, ns 86, ns 72, and ns 60, respectively); ns 70 is the N-terminal protein. Two large, short-lived, intermediate polypeptides of 155,000 and 135,000 daltons have been identified that are probably precursors to the four stable proteins. Another polypeptide of 220,000 daltons (containing the N-terminus) has also been found.

Biosynthesis of virions

In alphaviruses, the viral nucleocapsid aligns below the plasma membrane and the virus is formed by budding through the plasma membrane. During the budding, the virus obtains the lipids from the host cell plasma membrane (Kääriäinen and Söderlund 1978). The formation of virions is explained schematically in figure 66.

VIRUS MUTANTS

Ts mutants of Sindbis virus can be divided into five complementation groups. Viruses in groups A and B are incapable of virus-specific RNA synthesis (RNA^-) at the restrictive temperature, and the parental RNA fails to enter the

```
GAA GAG TGG TCC GCC CCG CTG ATT ACT GCC ATG TGT GTC CTT GCC AAT GCT ACC TTC CCG TGC TTC CAG CCC CCG
GLU GLU TRP|SER ALA PRO LEU ILE THR ALA MET CYS VAL LEU ALA ASN ALA THR PHE PRO CYS PHE GLN PRO PRO
          C | E3                                        ●

TGT GTA CCT TGC TGC TAT GAA AAC AAC GCA GAG GCC ACA CTA CGG ATG CTC GAG GAT AAC GTG GAT AGG CCA GGG
CYS VAL PRO CYS CYS TYR GLU ASN ASN ALA GLU ALA THR LEU ARG MET LEU GLU ASP ASN VAL ASP ARG PRO GLY   E3(47)

TAC TAC GAC CTC CTT CAG GCA GCC TTG ACG TGC CGA AAC GGA ACA AGA CAC CGG CGC AGC GTG TCG CAA CAC TTC
TYR TYR ASP LEU LEU GLN ALA ALA LEU THR CYS ARG ASN GLY THR ARG HIS ARG ARG SER VAL SER GLN HIS PHE
                                                ●
                                                             E3  E2
AAC GTG TAT AAG GCT ACA CGC CCT TAC ATC GCG TAC TGC GCC GAC TGC GGA GCA GGG CAC TCG TGT CAT AGC CCC
ASN VAL TYR LYS ALA THR ARG PRO TYR ILE ALA TYR CYS ALA ASP CYS GLY ALA GLY HIS SER CYS HIS SER PRO   E2(31)

GTA GCA ATT GAA GCG GTC AGG TCC GAA GCT ACC GAC GGG ATG CTG AAG ATT CAG TTC TCG GCA CAA ATT GGC ATA
VAL ALA ILE GLU ALA VAL ARG SER GLU ALA THR ASP GLY MET LEU LYS ILE GLN PHE SER ALA GLN ILE GLY ILE   E2(56)

GAT AAG AGT GAC AAT CAT GAC TAC ACG AAG ATA AGG TAC GCA GAC GGG CAC GCC ATT GAG AAT GCC GTC CGG TCA
ASP LYS SER ASP ASN HIS ASP TYR THR LYS ILE ARG TYR ALA ASP GLY HIS ALA ILE GLU ASN ALA VAL ARG SER   E2(81)

TCT TTG AAG GTA GCC ACC TCC GGA GAC TGT TTC GTC CAT GGC ACA ATG GGA CAT TTC ATA CTG GCA AAG TGC CCA
SER LEU LYS VAL ALA THR SER GLY ASP CYS PHE VAL HIS GLY THR MET GLY HIS PHE ILE LEU ALA LYS CYS PRO   E2(106)

CCG GGT GAA TTC CTG CAG GTC TCG ATC CAG GAC ACC AGA AAC GCG GTC CGT GCC TGC AGA ATA CAA TAT CAT CAT
PRO GLY GLU PHE LEU GLN VAL SER ILE GLN ASP THR ARG ASN ALA VAL ARG ALA CYS ARG ILE GLN TYR HIS HIS   E2(131)

GAC CCT CAA CCG GTG GGT AGA GAA AAA TTT ACA ATT AGA CCA CAC TAT GGA AAA GAG ATC CCT TGC ACC ACT TAT
ASP PRO GLN PRO VAL GLY ARG GLU LYS PHE THR ILE ARG PRO HIS TYR GLY LYS GLU ILE PRO CYS THR THR TYR   E2(156)

CAA CAG ACC ACA GCG GAG ACC GTG GAG GAA ATC GAC ATG CAT ATG CCG CCA GAT ACG CCG GAC AGG ACG TTG CTA
GLN GLN THR THR ALA GLU THR VAL GLU GLU ILE ASP MET HIS MET PRO PRO ASP THR PRO ASP ARG THR LEU LEU   E2(181)

TCA CAG CAA TCT GGC AAT GTA AAG ATC ACA GTC GGA GGA AAG AAG GTG AAA TAC AAC TGC ACC TGT GGA ACC GGA
SER GLN GLN SER GLY ASN VAL LYS ILE THR VAL GLY GLY LYS LYS VAL LYS TYR ASN CYS THR CYS GLY THR GLY   E2(206)
                                                                        ●

AAC GTT GGC ACT ACT AAT TCG GAC ATG ACG ATC AAC ACG TGT CTA ATA GAG CAG TGC CAC GTC TCA GTG ACG GAC
ASN VAL GLY THR THR ASN SER ASP MET THR ILE ASN THR CYS LEU ILE GLU GLN CYS HIS VAL SER VAL THR ASP   E2(231)

CAT AAG AAA TGG CAG TTC AAC TCA CCT TTC GTC CCG AGA GCC GAC GAA CCG GCT AGA AAA GGC AAA GTC CAT ATC
HIS LYS LYS TRP GLN PHE ASN SER PRO PHE VAL PRO ARG ALA ASP GLU PRO ALA ARG LYS GLY LYS VAL HIS ILE   E2(256)

CCA TTC CCG TTG GAC AAC ATC ACA TGC AGA GTT CCA ATG GCG CGC GAA CCA ACC GTC ATC CAC GGC AAA AGA GAA
PRO PHE PRO LEU ASP ASN ILE THR CYS ARG VAL PRO MET ALA ARG GLU PRO THR VAL ILE HIS GLY LYS ARG GLU   E2(281)
                     ●

GTG ACA CTG CAC CTT CAC CCA GAT CAT CCC ACG CTC TTT TCC TAC CGC ACA CTG GGT GAG GAC CCG CAG TAT CAC
VAL THR LEU HIS LEU HIS PRO ASP HIS PRO THR LEU PHE SER TYR ARG THR LEU GLY GLU ASP PRO GLN TYR HIS   E2(306)

GAG GAA TGG GTG ACA GCG GCG GTG GAA CGG ACC ATA CCC GTA CCA GTG GAC GGG ATG GAG TAC CAC TGG GGA AAC
GLU GLU TRP VAL THR ALA ALA VAL GLU ARG THR ILE PRO VAL PRO VAL ASP GLY MET GLU TYR HIS TRP GLY ASN   E2(331)

AAC GAC CCA GTG AGG CTT TGG TCT CAA CTC ACC ACT GAA GGG AAA CCG CAC GGC TGG CCG CAT CAG ATC GTA CAG
ASN ASP PRO VAL ARG LEU TRP SER GLN LEU THR THR GLU GLY LYS PRO HIS GLY TRP PRO HIS GLN ILE VAL GLN   E2(356)

TAC TAC TAT GGG CTT TAC CCG GCC GCT ACA GTA TCC GCG GTC GTC GGG ATG AGC TTA CTG GCC TTG ATA TCG ATC
TYR TYR TYR GLY LEU TYR PRO ALA ALA THR VAL SER ALA VAL VAL GLY MET SER LEU LEU ALA LEU ILE SER ILE   E2(381)

TTC GCG TCG TGC TAC ATG CTG GTT GCG GCC CGC AGT AAG TGC TTC ACC CCT TAT GCT TTA ACA CCA GGA GCT GCA
PHE ALA SER CYS TYR MET LEU VAL ALA ALA ARG SER LYS CYS LEU THR PRO TYR ALA LEU THR PRO GLY ALA ALA   E2(406)

GTT CCG TGG ACG CTG GGG ATA CTC TGC TGC GCC CCG CGG GCG CAC GCA GCT AGT GTG GCA GAG ACT ATG GCC TAC
VAL PRO TRP THR LEU GLY ILE LEU CYS CYS ALA PRO ARG ALA HIS ALA|ALA SER VAL ALA GLU THR MET ALA TYR
                                                         E2 | 6K
TTG TGG GAC CAA AAC CAA GCG TTG TTC TGG TTG GAG TTT GCG GCC CCT GTT GCC TGC ATC CTC ATC ATC ACG TAT
LEU TRP ASP GLN ASN GLN ALA LEU PHE TRP LEU GLU PHE ALA ALA PRO VAL ALA CYS ILE LEU ILE ILE THR TYR   6K(34)

TGC CTC AGA AAC GTG CTG TGT TGC TGT AAG AGC CTT TCT TTT TTA GTG CTA CTG AGC CTC GGG GCA ACC GCC AGA
CYS LEU ARG ASN VAL LEU CYS CYS CYS LYS SER LEU SER PHE LEU VAL LEU LEU SER LEU GLY ALA THR ALA ARG   6K(59)
```

Figure 62. The nucleotide sequence of the membrane protein genes. The deduced amino acid sequences of the membrane proteins are shown. The amino acid residues are numbered from the amino terminus of each polypeptide. The positions are shown in the parentheses to the right. Membrane-spanning segments are underlined and potential glycosylation sites are marked (●).

(Garoff et al. 1980b. Reprinted by permission from *Nature* 288, Fig. 3, p. 239. Copyright © 1980 Macmillan Journals Limited.)

```
GCT TAC GAA CAT TCG ACA GTA ATG CCG AAC GTG GTG GGC TTC CCG TAT AAC GCT CAC ATT GAA AGG CCA GGA TAT
ALA|TYR GLU HIS SER THR VAL MET PRO ASN VAL VAL GLY PHE PRO TYR LYS ALA HIS ILE GLU ARG PRO GLY TYR
6K |E1

AGC CCC CTC ACT TTG CAG ATG CAG GTT GTT GAA ACC AGC CTC GAA CCA ACC CTT AAT TTG GAA TAC ATA ACC TGT
SER PRO LEU THR LEU GLN MET GLN VAL VAL GLU THR SER LEU GLU PRO THR LEU ASN LEU GLU TYR ILE THR CYS   E1(49)

GAG TAC AAG ACG GTC GTC CCG TCG CCG TAC GTG AAG TGC TGC GGC GCC TCA GAG TGC TCC ACT AAA GAG AAG CCT
GLU TYR LYS THR VAL VAL PRO SER PRO TYR VAL LYS CYS CYS GLY ALA SER GLU CYS SER THR LYS GLU LYS PRO   E1(74)

GAC TAC CAA TGC AAG GTT TAC ACA GGC GTG TAC CCG TTC ATG TGG GGA GGG GCA TAT TGC TTC TGC GAC TCA GAA
ASP TYR GLN CYS LYS VAL TYR THR GLY VAL TYR PRO PHE MET TRP GLY GLY ALA TYR CYS PHE CYS ASP SER GLU   E1(99)

AAC ACG CAA CTC AGC GAG GCG TAC GTC GAT CGA TCG GAC GTA TGC AGG CAT GAT CAC GCA TCT GCT TAC AAA GCC
ASN THR GLN LEU SER GLU ALA TYR VAL ASP ARG SER ASP VAL CYS ARG HIS ASP HIS ALA SER ALA TYR LYS ALA   E1(124)

CAT ACA GCA TCG CTG AAG GCC AAA GTG AGG GTT ATG TAC GGC AAC GTA AAC CAG ACT GTG GAT GTT TAC GTG AAC
HIS THR ALA SER LEU LYS ALA LYS VAL ARG VAL MET TYR GLY ASN VAL ASN GLN THR VAL ASP VAL TYR VAL ASN   E1(149)
                                                                ●
GGA CAC CAT GCC GTC ACG ATA GGC GGT ACT CAG TTC ATA TTC GGG CCC CTG TCA TCG GCC TGG ACC CCG TTC GAC
GLY ASP HIS ALA VAL THR ILE GLY GLY THR GLN PHE ILE PHE GLY PRO LEU SER SER ALA TRP THR PRO PHE ASP   E1(174)

AAC AAG ATA GTC GTG TAC AAA GAC GAA GTG TTC AAT CAG GAC TTC CCG CCG TAC GGA TCT GGG CAA CCA GGG CGC
ASN LYS ILE VAL VAL TYR LYS ASP GLU VAL PHE ASN GLN ASP PHE PRO PRO TYR GLY SER GLY GLN PRO GLY ARG   E1(199)

TTC GGC GAC ATC CAA AGC AGA ACA GTG GAG AGT AAC GAC CTG TAC GCG AAC ACG GCA CTG AAG CTG GCA CGC CCT
PHE GLY ASP ILE GLN SER ARG THR VAL GLU SER ASN ASP LEU TYR ALA ASN THR ALA LEU LYS LEU ALA ARG PRO   E1(224)

TCA CCC GGC ATG GTC CAT GTA CCG TAC ACA CAG ACA CCT TCA GGG TTC AAA TAT TGG CTA AAG GAA AAA GGG ACA
SER PRO GLY MET VAL HIS VAL PRO TYR THR GLN THR PRO SER GLY PHE LYS TYR TRP LEU LYS GLU LYS GLY THR   E1(249)

GCC CTA AAT ACG AAG GCT CCT TTT GGC TGC CAA ATC AAA ACG AAC CCT GTC AGG GCC ATG AAC TGC GCC GTG GGA
ALA LEU ASN THR LYS ALA PRO PHE GLY CYS GLN ILE LYS THR ASN PRO VAL ARG ALA MET ASN CYS ALA VAL GLY   E1(274)

AAC ATC CCT GTC TCC ATG AAT TTG CCT GAC AGC GCC TTT ACC CGC ATT GTC GAG GCG CCG ACC ATC ATT GAC CTG
ASN ILE PRO VAL SER MET ASN LEU PRO ASP SER ALA PHE THR ARG ILE VAL GLU ALA PRO THR ILE ILE ASP LEU   E1(299)

ACT TGC ACA GTG GCT ACC TGT ACG CAC TCC TCG GAT TTC GGC GGC GTC TTG ACA CTG ACG TAC AAG ACC AAC AAG
THR CYS THR VAL ALA THR CYS THR HIS SER SER ASP PHE GLY GLY VAL LEU THR LEU THR TYR LYS THR ASN LYS   E1(324)

AAC GGG GAC TGC TCT GTA CAC TCG CAC TCT AAC GTA GCT ACT CTA CAG GAG GCC ACA GCA AAA GTG AAG ACA GCA
ASN GLY ASP CYS SER VAL HIS SER HIS SER ASN VAL ALA THR LEU GLN GLU ALA THR ALA LYS VAL LYS THR ALA   E1(349)

GGT AAG GTG ACC TTA CAC TTC TCC ACG GCA AGC GCA TCA CCT TCT TTT GTG GTG TCG CTA TGC ACT GCT AGG GCC
GLY LYS VAL THR LEU HIS PHE SER THR ALA SER ALA SER PRO SER PHE VAL VAL SER LEU CYS SER ALA ARG ALA   E1(374)

ACC TGT TCA GCG TCG TGT GAG CCC CCG AAA GAC CAC ATA GTC CCA TAT GCG GCT AGC CAC AGT AAC GTA GTG TTT
THR CYS SER ALA SER CYS GLU PRO PRO LYS ASP HIS ILE VAL PRO TYR ALA ALA SER HIS SER ASN VAL VAL PHE   E1(399)

CCA GAC ATG TCG GGC ACC GCA CTA TCA TGG GTG CAG AAA ATC TCG GGT GGT CTG GGG GCC TTC GCA ATC GGC GCT
PRO ASP MET SER GLY THR ALA LEU SER TRP VAL GLN LYS ILE SER GLY GLY LEU GLY ALA PHE ALA ILE GLY ALA   E1(424)
                                                    ------------------------------------------------
ATC CTG GTG CTG GTT GTG GTC ACT TGC ATT GGG CTC CGC AGA TAA GTT AGG GTA GGC AAT GGC ATT GAT ATA GCA
ILE LEU VAL LEU VAL VAL VAL THR CYS ILE GLY LEU ARG ARG
------------------------------------------------
AGA AAA TTG AAA ACA GAA AAA GTT AGG GTA AGC AAT GGC ATA TAA CCA TAA CTG TAT AAC TTG TAA CAA ACC GCA

ACA AGA CCT GCG CAA TTG GCC CCG TGG TCC GCC TCA CGG AAA CTC GGG GCA ACT CAT ATT GAC ACA TTA ATT GGC

AAT AAT TGG AAG CTT ACA TAA GCT TAA TTC GAC GAA TAA TTG GAT TTT TAT TTT ATT TTG CAA TTG GTT TTT AAT

ATT TCC
```

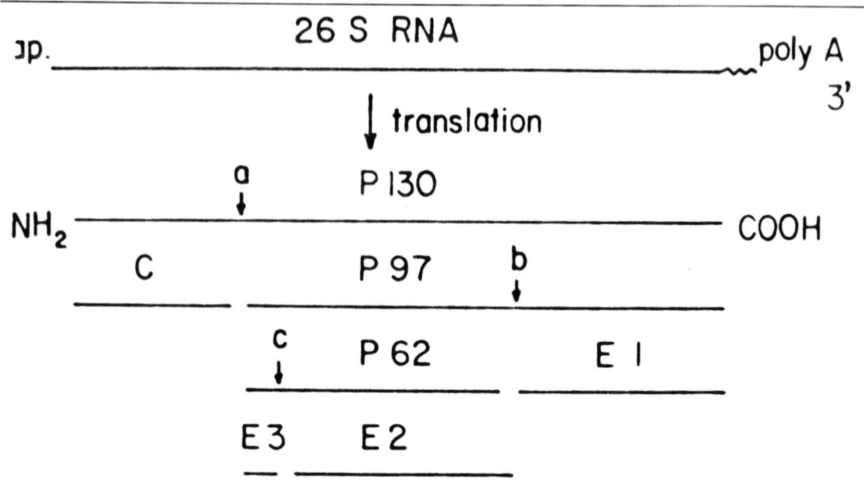

Figure 63. Translation of structural protein as a polyprotein from the 26 S RNA. The cleavage of capsid protein is nascent *(a)*, cleavage between p62 and E1 takes place rapidly after translation *(b)*, whereas p62 is cleaved during the maturation of the virus *(c)*.
(Kääriäinen and Söderlund 1978. Reprinted by permission from *Curr. Top. Microbiol. Immunol.* 82, Fig. 6, p. 35. Copyright Springer-Verlag, 1978, Heidelberg.)

replicative form. Mutants in groups C, D, and E have no apparent lesion in viral RNA synthesis (RNA^+), although in group C nucleocapsids are not assembled; in group D hemadsorption is absent, and in group E infectious virions are not produced, even though hemadsorption and nucleocapsid assembly are normal (Pfefferkorn and Shapiro 1974).

Effect of virus infection on the host cell

Alphavirus infection leads to a decay in the synthesis of cellular proteins and phospholipids. The effect of flaviviruses on the host cell metabolism is milder and leads to a mild cytopathogenic effect.

Defective interfering particles

The defective interfering particles formed during serial passage of high concentrations of virus contain RNA molecules that are smaller than the 42S genome. These particles stimulate the synthesis of abnormal single-stranded RNA molecules that is accompanied by a reduction in the synthesis of 42S and 26S RNA. Double-stranded RNAs of 12–16S have also been found.

DISEASES CAUSED BY TOGAVIRUSES

Alphaviruses and flaviviruses cause encephalitis and are transmitted by mosquitoes and ticks.

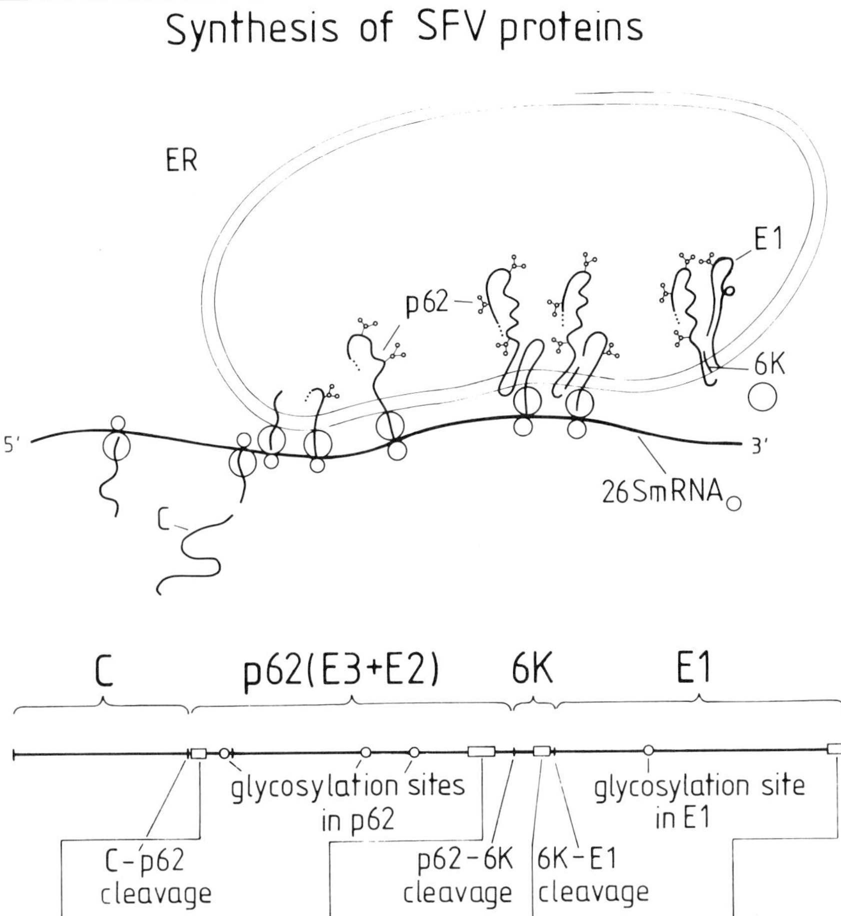

Figure 64. Scheme for the synthesis of Semliki Forest virus (SFV) spike glycoprotein E1 in the membrane of the cytoplasmic reticulum. The signal peptide of the p62 polypeptide mediates insertion into the membrane after the capsid protein C is cleaved. The polypeptide chain continues to grow, cleavage of the 6K peptide occurs, and the E1 polypeptide is synthesized when the signal peptide for E1 emerges.

(Garoff et al. 1980b. Reprinted by permission from *Nature* 288, Fig. 5, p. 240. Copyright © 1980, Macmillan Journals Limited.)

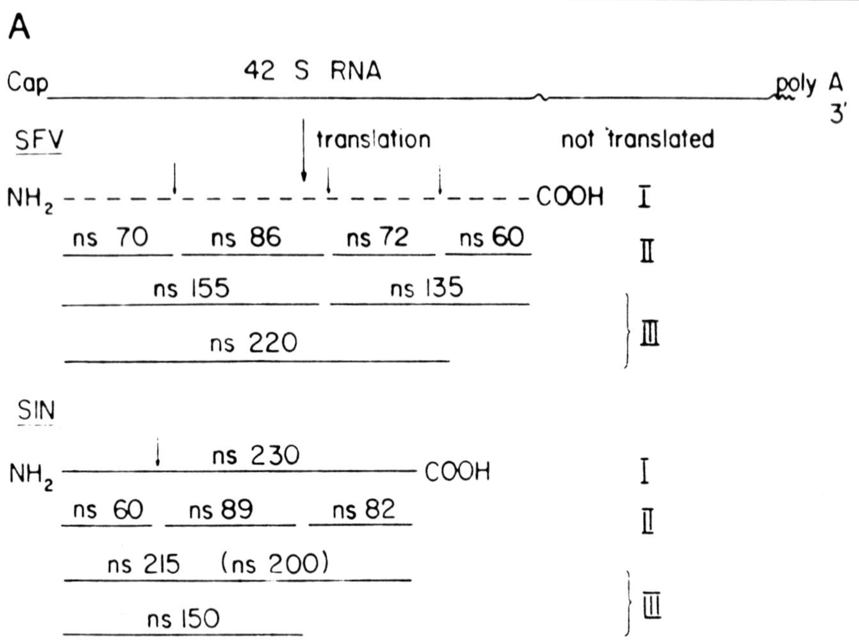

Figure 65. Translation of SFV and Sindbis virus (SIN) nonstructural proteins from 42 S RNA. I indicates the postulated primary translational product; II the stable cleavage products; and III the detected intermediates. Arrows indicate the cleavage sites. Dotted line refers to a protein that has not been found.

(Kääriäinen and Söderlund 1978. Reprinted by permission from *Curr. Top. Microbiol. Immunol.* 82, Fig. 7A, p. 40. Copyright Springer-Verlag 1978, Heidelberg.)

Equine encephalitis

Three members of the genus alphavirus—EEE, WEE, and VEE—cause acute encephalitis in horses. The horses are infected by mosquitoes carrying the virus, and people attending the sick horses can also be infected. In an epidemic in Texas, children and adults who were in contact with infected horses fell ill, and fatal cases were reported among the children.

Yellow fever

Yellow fever virus, a member of the genus flavivirus, has a life cycle in the mosquito and the monkey. In areas in central Africa endemic for yellow fever, people were infected by the bite of the *Aedes aegypti* mosquito carrying the virus. In some infected persons, a fatal disease occurred, but in others the infection was mild, with an incubation period of three–six days, fever, and headache. In severe cases, symptoms included jaundice, haemorrhages in the intestine, vomiting, and a drop in blood pressure followed by coma.

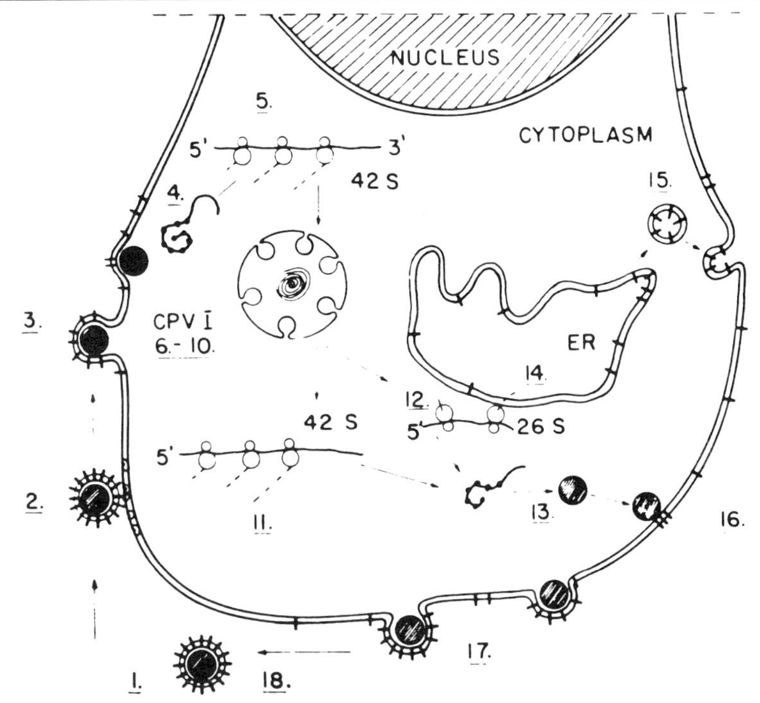

Figure 66. Simplified scheme of α-virus replication. The virus (1) adsorbs to specific receptors (2) at the plasma membrane, fusing (?) with it (3). The nucleocapsid is released to the cytoplasm and uncoated (4). Ribosomes associate with the 42 *S* RNA genome, translating nonstructural proteins which are components of the RNA polymerase (*primary translation,* 5). When enough RNA polymerase has been assembled, the translation of the parental RNA is replaced by *primary transcription,* first producing 42 *S* RNA-negative strands (6) and by their transcription a first progeny of 42 *S* RNA-positive strands (7). These are used as messengers in the production of more nonstructural proteins during the *secondary translation* (8). Part of the positive strands are used parallely as templates for synthesis of more negative 42 *S* RNAs, which in turn are used as templates for synthesis of positive strands (*secondary transcription,* 9). When the concentration of the interconversion protein is high enough, the synthesis of 26 *S* RNA begins (10). The RNA synthesis probably takes place in cytoplasmic vacuoles (CPV I). The translation of 42 *S* RNA takes place in free polysomes (11), and the structural proteins are translated into membrane-bound polysomes (12). The progeny 42 *S* RNA and the capsid protein assemble into nucleocapsids (13). The envelope proteins are protruded into the cisternal side (14) of the endoplasmic reticulum membrane (ER), become glycosylated, first in ER and finally in the Golgi apparatus from which they are transported (15) to the plasma membrane. The nucleocapsid recognizes the spanning part of the envelope protein dimer (p62-E1), preventing its free lateral mobility (16). Increasing interactions between nucleocapsid and envelope proteins lead to protrusion of nucleocapsid into the plasma membrane (17). At this stage, the host cell proteins are excluded from the forming virion and p62 is cleaved to E2 and E3 (?). Finally, the mature virion is released into the medium (18).

(Kääriäinen and Söderlund 1978. Reprinted by permission from *Curr. Top. Microbiol. Immunol.* 82, Fig. 11, p. 50. Copyright Springer-Verlag, 1978, Heidelberg.)

St. Louis encephalitis

During the 1930s, this virus caused epidemics in the central and southern regions of the United States. The disease is accompanied by elevated body temperature, severe headache for several days, followed by rapid recovery. In some individuals, mainly older people, the disease is severe, accompanied by vomiting, loss of coordination, and a lengthy convalescence.

West Nile fever

West Nile virus, which is prevalent in Africa and Southern Asia, has a life cycle in mosquitoes and rodents. Infection causes a rise in temperature after an incubation period of three days, accompanied by severe headache and a skin rash. Fatal cases have been reported.

Rubella

Rubella is usually a childhood infection, but young adults who escaped the disease in childhood can also be infected. The incubation period of rubella is 12–23 days, and a typical skin rash appears about 13 days after exposure to the virus. The disease starts with malaise, fever, headache, and irritation of the conjunctiva, as well as sore throat. The rash starts on the face and progresses downward on the body; the lymph nodes swell. At a certain stage of viremia, the virus spreads throughout the body and into the capillaries of the skin, where it replicates and causes the rash.

During early pregnancy, the virus is transmitted to the fetus. Infection of the fetus during the first trimester of pregnancy leads to damage to the internal ear and the heart due to virus replication in the stem cells. Damage to the eyes, teeth, and CNS can also occur but is less common than damage to the ear and heart. Women infected with rubella virus during the first trimester of pregnancy are advised to have an abortion; infection of pregnant women during the fourth month of pregnancy can also be harmful to the fetus.

A live attenuated rubella virus vaccine is available and is currently being used for the immunization of children. It is advisable that women of childbearing age be tested for immunity to rubella virus by determining the antibody level in the blood. Those women found to lack antibodies should be immunized with the attenuated live rubella virus vaccine at least three months before becoming pregnant to allow the development of the immune response (see chapter 24).

DISEASES CAUSED BY PESTIVIRUSES

These viruses cause diseases in domestic animals.

Hog cholera (European swine fever)

This disease is prevalent in Europe and also in South America. Infection of pigs with hog cholera virus causes fever and rash. In an epidemic form, the disease causes severe financial losses to the farmers, since the pigs must be eliminated in an infected area in order to block the spread of the virus. Both a live attenuated virus vaccine (Chinese strain) and a killed virus vaccine are available. Countries differ in their policies regarding immunization. In some countries immunization of pigs is compulsory, whereas others, where the disease is under control, immunization is not required so as to allow naturally infected pigs to be identified at the onset of an epidemic. In these countries, the infected animals are eradicated.

The virions contain two antigens, one of which cross reacts with another, unrelated virus—bovine diarrhea virus (BDV)—which infects mainly cattle, but also pigs. A specific diagnostic procedure for the detection of antibodies to the hog cholera virus-specific antigen that does not cross react with BDV antibodies is still needed.

Border disease in sheep seems to be caused by a virus related to hog cholera, but its properties are not yet known.

BIBLIOGRAPHY

Garoff, H.; Frischauf, A.-M.; Simons, K.; Lehrach, H.; and Delius, H. The capsid protein of Semliki Forest virus has clusters of basic amino acids and prolines in its amino-terminal region. *Proc. Natl. Acad. Sci. USA* 77:6376–6380, 1980*a*.

Garoff, H.; Frischauf, A.-M.; Simons, K.; Lehrach, H.; and Delius, H. Nucleotide sequence of the cDNA coding for the Semliki Forest virus membrane glycoproteins. *Nature 288*:236–241, 1980*b*.

Garoff, H., and Söderlund, H. The amphiphilic membrane glycoproteins of Semliki Forest virus are attached to the lipid bilayer by their COOH-terminal ends. *J. Mol. Biol. 124*:535–549, 1978.

Helenius, A.; Kartenbeck, J.; Simons, K.; and Fries, E. On the entry of Semliki forest virus into BHK-21 cells. *J. Cell Biol. 84*:404–420, 1980.

Kääriäinen, L., and Söderlund, H. Structure and replication of alphaviruses. *Curr. Top. Microbiol. Immunol. 82*:15–69, 1978.

Kennedy, S.I.T. Sequence relationships between the genome and the intracellular RNA species of standard and defective-interfering Semliki Forest viruses. *J. Mol. Biol. 108*:491–511, 1976.

Pfefferkorn, E.R., and Shapiro, D. Reproduction of togaviruses. *In: Comprehensive Virology* (H. Fraenkel-Conrat and R.R. Wagner, eds.), Plenum Press, New York, 1974, Vol. 2, pp. 171–230.

Wengler, G.; Wengler, W.; and Filipe, A.R. A study of nucleotide sequence homology between the nucleic acids of different alphaviruses. *Virology 78*:124–134, 1977.

White, J., and Helenius, A. pH-dependent fusion between the Semliki Forest virus membrane and liposomes. *Proc. Natl. Acad. Sci. USA* 77:3273–3277, 1980.

RECOMMENDED READING

Hayes, C.G., and Wallis, R.D. Ecology of Western equine encephalomyelitis in the Eastern United States. *Adv. Virus Res. 21*:37–83, 1977.

Mussgay, M.; Enzmann, P.-J.; Horzinek, M.C.; and Weiland, E. Growth cycle of arboviruses in vertebrate and arthropod cells. *Prog. Med. Virol. 19*:257–323, 1975.

Pfefferkorn, E.R. Genetics of togaviruses. *In: Comprehensive Virology* (H. Fraenkel-Conrat and R.R. Wagner, eds.), Plenum Press, New York, 1977, Vol. 9, pp. 208–238.

Schlessinger, R.W. (ed.). *The Togaviridae*. Academic Press, New York, 1981.

20. RNA PLUS GENOME THAT SERVES AS MESSENGER RNA: CORONAVIRUSES

FAMILY CORONAVIRIDAE

These viruses (100 nm in diameter) have chracteristically long (12–24 nm), widely spaced, bulbous surface projections (peplomeres), a lipid-containing envelope, and a single-stranded RNA^+ genome. The virus replicates in the cytoplasm of the infected cell and matures by budding through the intracytoplasmic membranes. Viral inclusion bodies may be seen in the cytoplasm (figure 67): The mechanisms of virus replication are similar to those described for the picorna- and togavirus families.

Actinomycin D, which inhibits DNA-dependent RNA synthesis, markedly inhibits human coronavirus replication. This means that a host cell component is required during the early stages of virus replication (Kennedy and Johnson-Lussenberg 1979).

Genus Coronavirus (from corona, meaning crown)

The virions that have the ability to hemagglutinate red blood cells include:

Avian infectious bronchitis virus (IBV)
Calf neonatal diarrhea coronavirus (figure 67)
Feline infectious peritonitis virus
Human coronavirus
Murine hepatitis virus (MHV), strain JHM

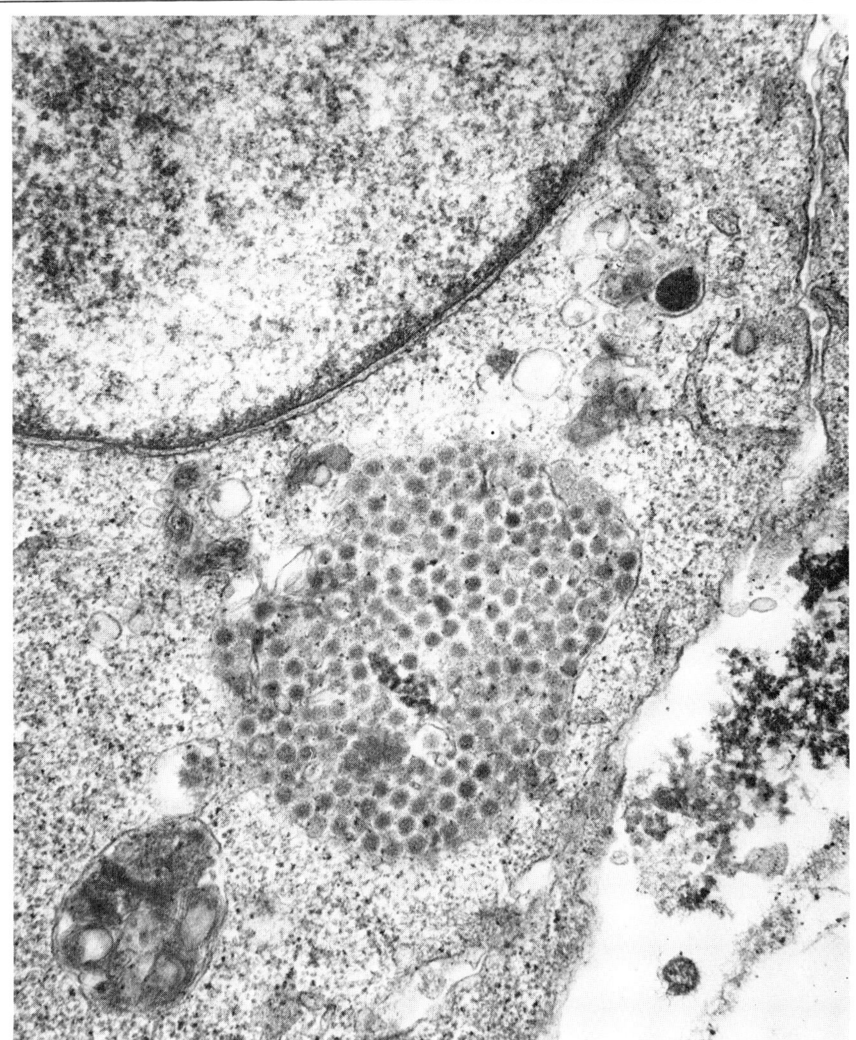

Figure 67. Electron micrograph of a cell infected with a bovine coronavirus (× 24,000). (Courtesy of Dr. Daniel Dekegel, Institut Pasteur du Brabant, Brussels, Belgium.)

Porcine transmissible gastroenteritis virus (TGEV)
Rat coronavirus
Turkey bluecomb disease virus

RNA and proteins

The RNA of several coronaviruses is polyadenylated and infectious and probably acts as mRNA. An infectious RNA of about 8×10^6 daltons, with covalently attached polyadenylic acid sequences, has been extracted from avian coronavirus (Lomniczi and Kennedy 1977). Molecular weights of 5.4×10^6 and 3.8×10^6 were reported for RNA genomes of MHV and bovine coronavirus, respectively (Lai and Stohlman 1978; Guy and Brian 1979).

Purified coronavirus JHM contains six polypeptides: GP1, GP2, GP3, VP4, GP5, and VP6, with apparent molecular weights of 170,000; 125,000; 97,500; 60,800; 24,800; and 22,700, respectively. GP2 and GP3 probably protrude from the lipid envelope and together with GP1 form the spike layer. Protein VP6 and part of GP5 are located within the lipid bilayer (Wege et al. 1979). Nine intracellular virus-specific proteins were identified by Bond and associates (1979), of which four were structural proteins. There is a major nonglycosylated protein of molecular weight 50,000–60,000 located inside the virion that is probably the nucleocapsid protein.

DISEASES CAUSED BY THE VIRUS

Transmissible gastroenteritis (TGE) in pigs

Infection of the small intestine of the pig occurs after ingestion of food containing virus and leads to destruction of the epithelial cells. After the viremic stage, the virus invades the kidneys and lungs. There are virulent and attenuated virus strains. Young pigs exposed to the virulent virus strains usually die within ten days after exposure.

The problem with producing a vaccinal strain of TGE lies in developing a virus that will stimulate an IgA response in the gut of the sow without producing disease in young pigs (Saif and Bohl 1979).

CNS infection

Murine coronavirus strain JHM causes demyelination in the CNS in infected mice and rats. This virus normally produces a rapidly fatal encephalomyelitis. However, under certain conditions (such as low virus dose or selection for ts mutants) nonfatal demyelination can be produced due to a selective destruction of the myelin-synthesizing oligodendrocytes (Haspel et al. 1978; Nagashima et al. 1978).

BIBLIOGRAPHY

Bond, C.W.; Leibowitz, J.L.; and Robb, J.A. Pathogenic murine coronaviruses. II. Characterization of virus-specific proteins of murine coronaviruses JHMV and A59V. *Virology 94*:371–384, 1979.

Guy, J.S., and Brian, D.A. Bovine coronavirus genome. *J. Virol. 29*:293–300, 1979.

Haspel, M.V.; Lampert, P.W.; and Oldstone, M.B.A. Temperature-sensitive mutants of mouse hepatitis virus produce a high incidence of demyelination. *Proc. Natl. Acad. Sci. USA 75*:4033–4036, 1978.

Kennedy, D.A., and Johnson-Lussenburg, C.M. Inhibition of coronavirus 229E replication by actinomycin D. *J. Virol. 29*:401–404, 1979.

Lai, M.M.C., and Stohlman, S.A. RNA of mouse hepatitis virus. *J. Virol. 26*:236–242, 1978.

Lomniczi, B., and Kennedy, I. Genome of infectious bronchitis virus. *J. Virol. 24*:99–107, 1977.

Nagashima, K.; Wege, H.; and ter Meulen, V. Early and late CNS-effects of coronavirus infection in rats. *Adv. Exp. Med. Biol. 100*:395–409, 1978.

Saif, L.J., and Bohl, E.H. Passive immunity in transmissible gastroenteritis of swine: immunoglobulin classes of milk antibodies after oral-intranasal inoculation of sows with a live low cell culture-passaged virus. *Am. J. Vet. Res. 40*:115–117, 1979.

Wege, H.; Wege, H.; Nogashima, K.; and ter Meulen, V. Structural polypeptides of the murine coronavirus JHM. *J. Gen. Virol. 42*:37–47, 1979.

RECOMMENDED READING

Horzinek, M.C., and Osterhaus, A.D.M.E. The virology and pathogenesis of feline infectious peritonitis. Brief Review. *Arch. Virol. 59*:1–15, 1979.

21. SINGLE-STRANDED RNA PLUS GENOMES THAT SYNTHESIZE DNA AS PART OF THEIR LIFE CYCLE: RETROVIRUSES (RNA TUMOR VIRUSES)

FAMILY RETROVIRIDAE

These are RNA viruses that infect animals and replicate through a DNA intermediate called *proviral DNA*. The virions have a diameter of 100 nm with a lipid bilayer enveloping an icosahedral core containing a helical nucleocapsid. The nucleocapsid contains an enzyme, the reverse transcriptase (an RNA-dependent DNA polymerase). The genome is made up of two identical RNA^+ molecules, each with a molecular weight of $7\text{–}10 \times 10^6$. The ultrastructure of the virions is shown in figure 68. Retroviruses resemble other enveloped viruses that are formed by budding through cytoplasmic membranes (figure 69). (The term *retrovirus* comes from the Latin retro, meaning backward; the term also refers to *re*verse *tra*nscriptase.)

Subfamily Oncovirinae (RNA tumor virus group)

Onkos in Greek means tumor.

Genus: Type B oncovirus group
Species: mouse mammary tumor viruses (MTVs)

Genus: Type C oncovirus group that includes mammalian, avian, and reptilian type C oncoviruses.

Mammalian type C oncovirus species include baboon type C oncovirus, bovine leukosis (BLV), gibbon ape leukemia, guinea pig type C oncovirus (guinea pig leukemia), feline sarcoma and feline leukemia viruses (FeLV),

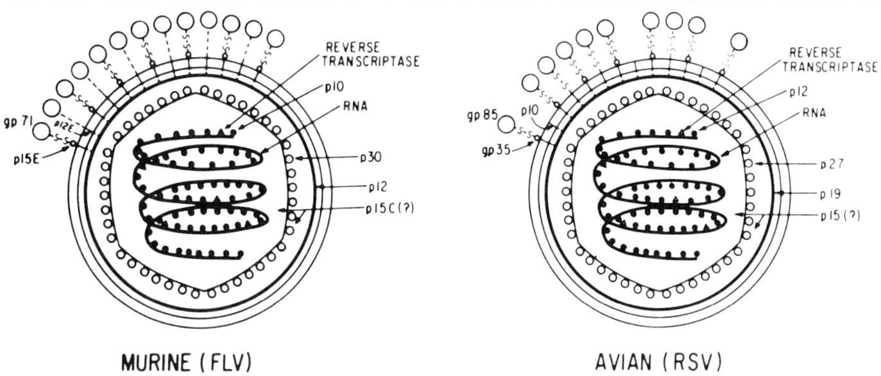

Figure 68. Schematic representation of murine and avian retrovirus particles and the proposed assignment of analogous structural polypeptides. The surface of the virion consists of a lipid bilayer derived from the host cell from which project the loosely attached surface knobs. In avian retroviruses the outer knobs are connected by a spike and thus project further outward. Below the virus envelope is an inner coat (p19 or p17) that surrounds the virus core containing an outer shell and the ribonucleoprotein complex (RNP). The RNP consists of a filamental strand in the form of a spiral.
(Montelaro et al. 1978. Reprinted by permission from *Virology* 84, Fig. 8, p. 29.)

Figure 69. Electron micrograph of a Rous sarcoma virus (RSV)-infected chick embryo fibroblast with a budding virion.
(Department of Molecular Virology, Hebrew University-Hadassah Medical School, Jerusalem, Israel.)

murine sarcoma and murine leukemia viruses (MuLV), porcine type C oncovirus, rat type C oncovirus (rat leukemia), and woolly monkey sarcoma.
Avian type C oncovirus species include avian sarcoma (ASV) and leukemia viruses, avian reticuloendotheliosis, and pheasant type C oncoviruses.
Reptilian type C oncovirus species include viper type C oncoviruses.
The type D retrovirus includes Mason-Pfizer monkey virus.

Subfamily Spumavirinae (foamy virus group)

Spuma, in Latin, means foam. These viruses are symbiotic and evoke little or no response in the host cell.

Species: Bovine syncytial
Feline syncytial
Human foamy
Simian foamy viruses

Subfamily Lentivirinae (Maedi/visna group)

Lenti, in Latin, means slow. These viruses are cytopathic and induce chronic degenerative diseases.

Species: Visna virus

The three major forms of antigenicity of retroviruses are group-specific antigens shared by related viruses derived from a single host species, type-specific antigens that define the various serological subgroups, and interspecies antigens shared by otherwise unrelated viruses derived from different host species.

The different retrovirus strains can be either endogenous or exogenous viruses. The genome of an endogenous virus is encoded in the germ line DNA of a normal animal species and is perpetuated by vertical transmission through the gametes of the species. Exogenous viruses, on the other hand, persist by virtue of horizontal spread among members of susceptible species.

THE VIRAL GENOME

Each virion contains two RNA plus molecules that are hydrogen bonded to each other. The two RNA monomers are joined together close to their 5′ ends in a structure termed the *dimer-linkage structure* (figure 70). The hydrogen-bonded RNA molecules have a sedimentation coefficient in sucrose gradients of 50–70 S. Each genomic monomer has a sedimentation coefficient of 35–39 S.

The viral genome contains four genes and an additional sequence:

1. Group specific antigen gene (*gag*). This gene is responsible for the synthesis of four viral proteins that are common to a group of viruses. These proteins

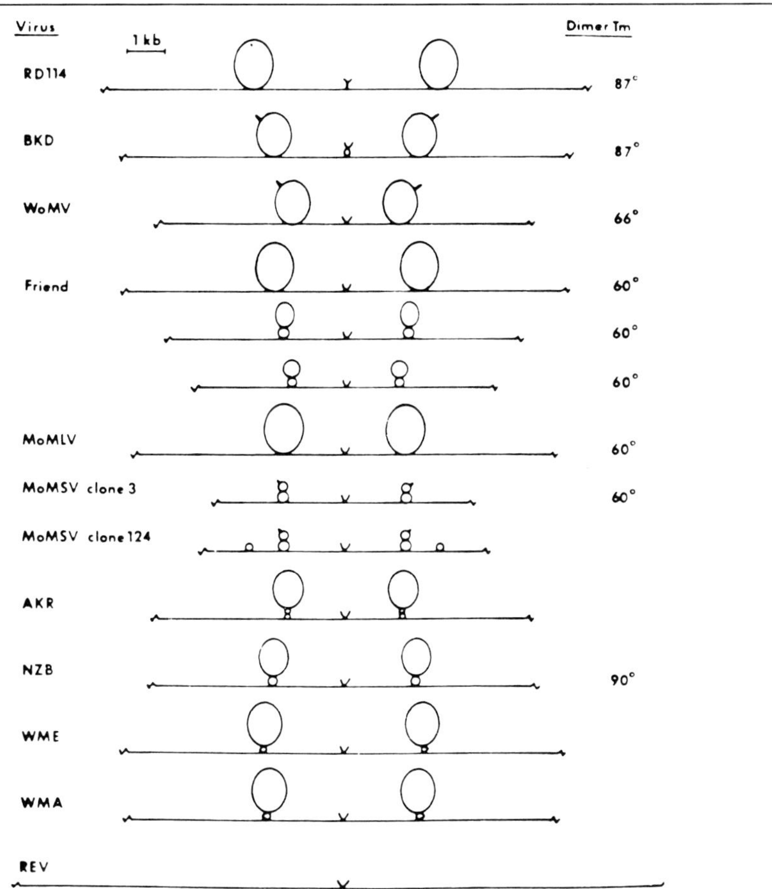

Figure 70. Diagram of the structures of retrovirus RNA dimers constituting the viral genome. The dimer linkage structure is shown in the middle of each molecule; the zig-zag lines at the ends of the molecules symbolize poly(A). Next to each structure is listed the approximate melting temperature of the dimer linkage, where it has been determined. The viruses are: RD-114, the cat endogenous xenotrope; BKD, the baboon endogenous xenotrope; WoMV, the simian sarcoma virus complex; Friend murine virus complex, including the lymphatic leukemia virus and the spleen focus-forming virus (three different viral RNAs were present in the mixture studied); MoMLV, the Moloney MuLV; two isolates, clone 3 and clone 124 of MoMSV, Moloney murine sarcoma virus; AKR murine endogenous ecotrope; NZB murine endogenous xenotrope: WME, a wild mouse ecotrope; WMA, a wild mouse amphotrope; and reticuloendotheliosis virus (REV).

(Bender et al. 1978. Reprinted by permission from *J. Virol.* 25, Fig. 3, p. 891).

Table 10. Structural components of avian and murine oncornaviruses

Virion polypeptide		Structural component	Virion substructure
Avian	Murine		
gp85-S	gp71-S	Knob	Envelope
|	|		
gp35-S	p15E-S	Spike	
p10	p12E	Envelope associated	
p19	p12	Inner coat[a]	Inner coat[a]
p27	p30	Core shell	Core exterior
p15	p15C	Core associated	
p12	p10	Nucleoprotein	Ribonucleoprotein complex
p91(β)	p70(α)	Reverse transcriptase	
p64(α)			

[a]The assignment of viral phosphoproteins (p19 and p12) to the inner coat does not exclude the possibility that a small number of molecules (1%) may bind to viral RNA during maturation. Moreover, the evidence that murine p12 represents the inner coat is based largely on analogy to the avian system. Murine p15C has properties which make it a likely candidate as well. (Bolognesi et al. 1978. Reprinted by permission from *Science* 199, Table 1 p. 184. Copyright 1978 by the American Association for the Advancement of Science)

have molecular weights of 27,000 (p27); 19,000 (p19); 15,000 (p15); and 12,000 (p12) in avian, and 30,000 (p30); 15,000 (p15); 12,000 (p12); and 10,000 (p10) in murine oncoviruses (table 10). The first proteins are translated from near the 5′ end of the RNA molecule as a high molecular weight precursor and subsequently cleaved to finished size. In avian sarcoma-leukosis viruses, a precursor of 76 × 10^3 daltons is synthesized and later cleaved. The intracistronic mapping within the *gag* gene is (5′)-p15-p12-p30-p10(3′) (Reynolds and Stephenson 1977). The major structural protein of the mammalian type C retroviruses is p30.

2. Polymerase gene (*pol*). This gene produces the viral RNA-directed DNA polymerase (reverse transcriptase) responsible for the synthesis of double-stranded DNA on the RNA genome. In avian oncoviruses, two protein subunits of 110 and 70 × 10^3 daltons of overlapping sequence

110
____

____
70

were found, but only one of 70 × 10^3 daltons in murine leukemia viruses (MuLV).

3. Envelope gene (*env*). This gene is responsible for the synthesis of two proteins of 85 and 35 × 10^3 daltons in avian leukemia viruses and 71 and 15 × 10^3 in murine leukemia viruses (the 15 × 10^3 dalton protein is nonglycosylated), and for other glycoproteins present in the envelope of the virions (table 10). The envelope proteins gp71, p15E, and p12E are located on

the surface of murine leukemia viruses and proteins gp85, gp35, and p10 are on the surface of avian retroviruses (table 10 and figure 68) (Montelaro et al. 1978; Bolognesi et al. 1978).

4. The sarcoma gene (*src*) is found only in avian sarcoma viruses, but there are over a dozen different but analogous genes in other retroviruses referred to as the *onc* gene. The *src* gene produces a 60,000 dalton protein responsible for the initiation and maintenance of the transformed state of the infected cell. This protein, called $p60^{src}$, is phosphorylated and displays the properties of a protein kinase. The *onc*-specific, coding RNA sequence is unrelated to virion genes that are essential for virus replication.
5. Common sequence (U_3) (3′ unique). This untranslated sequence of RNA is involved in the replication of the virus, probably by providing various control sequences.

The genes are arranged in the order 5′-cap-gag-pol-env-onc-U_3 for most retroviruses and 5′-cap-gag-pol-env-src-U_3 for avian sarcoma viruses.

The genome contains between 7,000 and 10,000 nucleotides in the two RNA molecules present in each virion.

ORGANIZATION OF THE NUCLEOTIDE SEQUENCE AT THE 5′ END OF THE VIRAL RNA

This region of the genome has an organizational function and is divided into four regions:

1. The terminus of the molecule at the 5′ end is a cap $m^7G^5ppp^{5'}G_m$ nucleotide. This is similar to the caps on mRNA molecules.
2. Immediately after the cap, a sequence of 16–21 nucleotides named *R* constitutes the repeat sequence, which is identical to a sequence present at the 3′ end. For example, the R_5 and R_3 repeat sequences in the RNA of the Prague strain of Rous sarcoma virus (Pr–RSV) are:

5′ m^7G p p p G_m C C A U U U U A C C A U U C A C C A C A . . . . .
3′ G C C A U U U U A C C A U U C A C C A C A . . . . .

3. A unique sequence of eighty nucleotides named U_5. This sequence and others are spliced directly to the *env* and *src* mRNA molecules.
4. A nucleotide sequence (derived from the host cell) which serves as a binding site for the attachment of primer transfer RNA: $tRNA^{trp}$ in ASV, $tRNA^{pro}$ in MLVs and RAV (Rous-associated virus), and $tRNA^{lys}$ in MTVs.

This sequence of 16 nucleotides at the 5′ end is complementary to the sequence of sixteen nucleotides present at the 3′ end of the tRNA. This sequence in the viral genome is named the *primer binding* (PB) site.

Organization of the nucleotide sequence at the 3′ end of the viral RNA

The sequence at the 3′ end of the RNA genome is divided into three regions:

1. Unique sequence (named U_3) that contains about 250 nucleotides in ASV, about 470 in MuLV and about 1,000 in MTV.
2. The R repeat sequence.
3. Poly(A) sequence.

Primer for the synthesis of viral DNA

The tRNA serves as a primer for the synthesis of the viral DNA by forming a hydrogen bond with the 16 nucleotides in the primer binding site at the 5′ end of the viral genome. The 16 nucleotides of the tRNA are complementary to the primer binding sequence. The A-OH at the 3′ terminus of the $tRNA^{trp}$ is covalently linked by the RNA-dependent DNA polymerase to the first deoxyribonucleoside monophosphate at the 5′ end of the DNA.

MOLECULAR EVENTS IN THE REPLICATION PROCESS

The virion envelope adsorbs to the cell membrane and the ribonucleoprotein enters the cytoplasm. Adsorption of virions can also occur at 4°C, but penetration occurs only at 37°C. The entrance of the viral nucleocapsid leads to the initiation of DNA synthesis.

The RNA-dependent DNA polymerase (reverse transcriptase)

This enzyme, which was discovered independently by H. Temin and D. Baltimore in 1970 (Temin and Baltimore 1972) is responsible for the synthesis of double-stranded DNA from the single-stranded viral genomes in cells infected by retroviruses. The reverse transcriptase, which is a core protein, transcribes the viral RNA into DNA that becomes integrated into the chromosomal DNA of the cell. The viral DNA is inserted into the cellular DNA by a recombination process and one or more DNA copies can be integrated. The integrated viral DNA is called *proviral DNA* (Temin 1976). From this proviral DNA, new viral RNA can be transcribed that can act either as mRNA and be translated or as viral genomic RNA and be packed into virion particles. Thus the reverse transcriptase is necessary for the productive infection of cells and for initiating transformation, but not for maintaining the transformed state of the cell. The molecule also has RNase H activity (Collett et al. 1978), which plays a role in viral replication.

The viral RNA-dependent DNA polymerase initiates the polymerization of deoxyribonucleoside triphosphates using the 3′ terminus of the $tRNA^{trp}$ in the direction of the 5′ end of the molecule (about 100 nucleotides away). When the enzyme molecule reaches the 5′ terminus of the viral RNA genome, the enzyme moves by virtue of the complementarity of the copy of R_5 to R_3 to another RNA molecule at the 3′ end and continues DNA synthesis until it

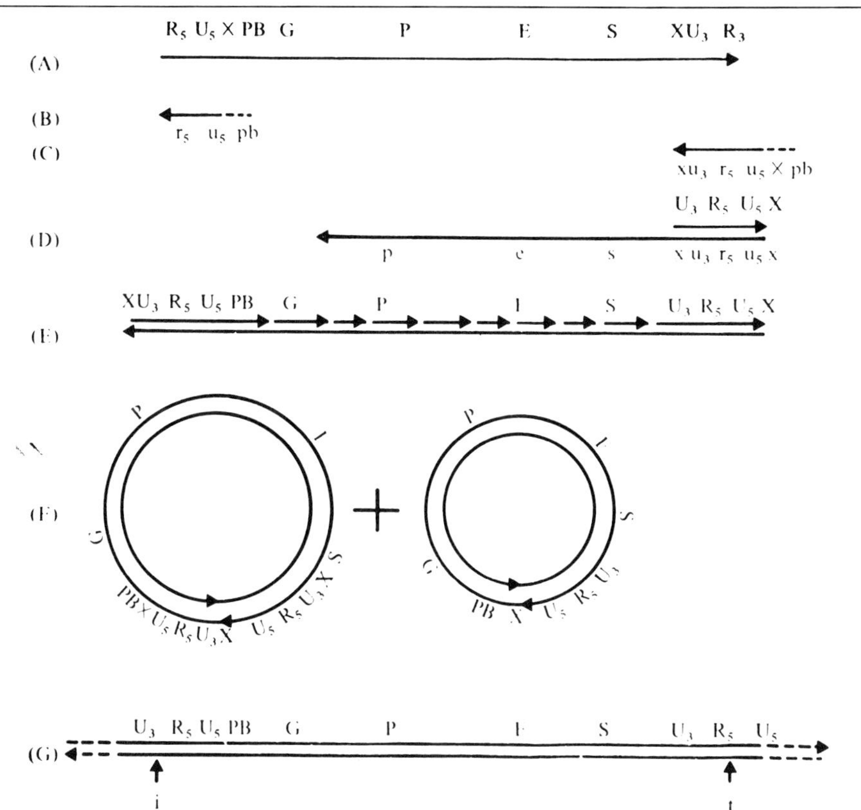

Figure 71. Structures of proviral DNA. (A) The genome of ASV (not to scale) G, P, E, S are the coding regions shown to indicate polarity. X is a possible additional repeat postulated to aid in circle formation; the arrow-head indicates the 3′ end of all molecules. Other symbols are as in figure 73. (B) and (C) In vitro (−) strand forms. The dotted line is the $tRNA^{trp}$ primer. Sequences complementary to the genome are shown in lowercase letters. (B) Strong stop DNA; (C) heterogeneous elongation products of strong stop DNA. (D) and (E) In vivo cytoplasmic forms; (D) a partial (−) strand with a small fragment of (+) strand; (e) complete (−) strand with complete (+) strand in pieces. (F) and (G) nuclear forms: (F) two covalently closed circles differing only in the number of times the U_3-R_5-U_5 sequence appears; (G) integrated provirus. The dashed lines indicate cell DNA. The arrows show initiation (i) and termination (t) sites of transcription if the genome is not transcribed as a larger precursor.

(Coffin 1979. Reprinted by permission from *J. Gen. Virol.* 42, Fig. 2, p. 8. Cambridge University Press, Cambridge.)

reaches the 5′ end of this RNA molecule (Coffin 1979; Gilboa et al. 1979). The process of reverse transcription leading to the production of proviral DNA in avian sarcoma viruses (Coffin 1979) is presented schematically in figure 71.

The viral DNA is present in the cytoplasm in several forms. The different forms of DNA observed in vitro and in vivo are shown aligned with the portion of the genome from which they are believed to be copied, and molecules of the same sense as the virus gene (plus strand) are labeled with capital letters; complementary molecules (minus strand) are designated with small letters. The 3′ end of each strand (and therefore the direction of its synthesis) is shown by an arrowhead.

In in vitro endogenous or reconstructed polymerase reactions, various homogeneous DNA species (B and C in figure 71) can be isolated, each of which is an elongation product of smaller species. The most prominent product (B), called *strong stop DNA*, is a copy of the U_5 and R_5 region of the genome terminating opposite G_m. It is 101 nucleotides long in avian sarcoma and 145 in murine leukemia viruses (Coffin 1979). Strong stop DNA, the structure of which is now known (Van Beveren et al. 1980), represents sequences between the 3′-OH nucleotide of the primer tRNA and the 5′-cap nucleotide of the genomic RNA.

A DNA molecule of variable size, up to the full length of the minus strand, is the first DNA strand to be synthesized on the RNA template. The DNA plus strand is the second DNA strand that is synthesized by the reverse transcriptase using the DNA minus strand as a template. The first of these DNA fragments is synthesized from a unique sequence containing the U_3 and U_5 repeat sequences and the part of $tRNA^{trp}$ complementary to the primer binding sequence. They may be synthesized from the 5′ end of a DNA plus strand prior to synthesis of the DNA minus strand.

Three to four hours after infection, viral double-stranded DNA with a sedimentation coefficient similar to that of complete viral double-stranded DNA appears. One strand (DNA minus) is complete, while the second DNA strand (DNA plus) is composed of a number of DNA fragments. As a consequence of the initiation of the plus and minus strands internally in the molecule, this species has a copy of U_3-R-U_5 (called the *long terminal repeat* [LTR]) (Keshet and Shaul 1981) at either end.

In the nucleus of the infected cell, circular viral double-stranded DNA of approximately genomic size appears. Cellular enzymes like ligase are presumably responsible for sealing the fragments in the DNA plus strand, and the circular DNA molecules are fully double-stranded.

The circular viral DNA molecules appear in two forms: (1) circular DNA, slightly larger than the viral RNA genome (Kung et al. 1980) and containing all the viral genes and sequences, as well as two copies at the LTR, and (2) short circular DNA containing only one copy of the LTR. It is not known whether the linear or one or both of the circular forms is the precursor to the integrated provirus.

Another model for reverse transcription (Gilboa et al. 1979) is given in figure 72 using MuLV as an example. The steps are as follows:

I. Attached to M-MuLV RNA at a distance about 150 nucleotides from its 5′ end is a $tRNA^{pro}$ molecule that serves as the initiator for synthesis of the minus-strand DNA copy of the virion RNA. The primer tRNA is bound to the template by an 18 nucleotide region denoted as (c) on the tRNA and (C) on the genome RNA.

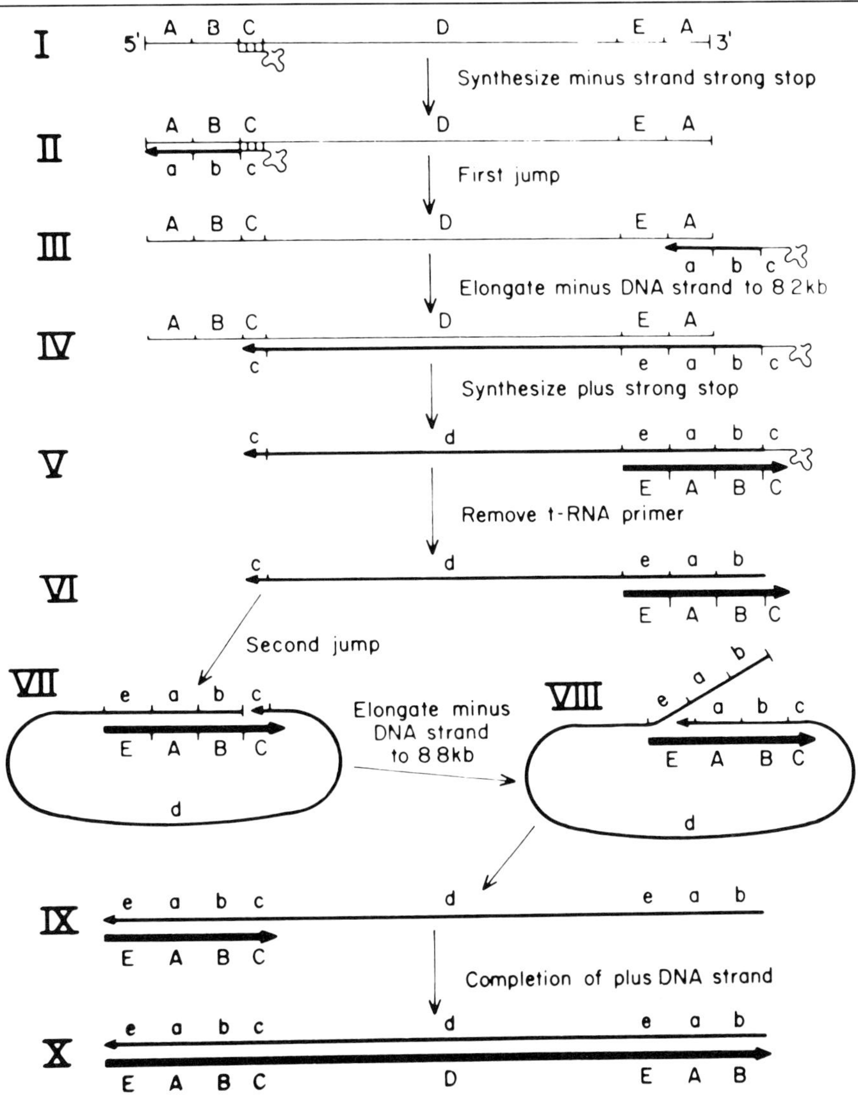

Figure 72. A model for reverse transcription. Structure I represents one molecule of the virion. In II, minus-strand DNA synthesis is initiated, and in V, plus-strand DNA synthesis is initiated. Various regions of the molecules are denoted by letters; their significance is discussed in the text. Uppercase letters refer to sequences in the viral RNA molecule (plus-strand sequences), and lowercase letters refer to minus-strand sequences. The various regions of the molecules are not shown to scale; for instance, region (c/C) is about 18 nucleotides long, while region (d/D) is about 7600 nucleotides long. The 3′ poly(A) of the viral RNA has been omitted.
(Gilboa et al. 1979. Reprinted by permission from *Cell* 18, Fig. 1, p. 94. Copyright © 1979 by MIT).

II. Reverse transcription begins by synthesis of a 150 nucleotide piece of DNA called *minus-strand strong-stop DNA*. This DNA is covalently attached to the $tRNA^{pro}$ primer, and its discrete length is a consequence of the fact that it is initiated at a specific site and continues to the 5′ end of the template.

III. The existence of the direct terminal repeat on the RNA genome allows minus-strong-stop DNA to jump from its original position at the 5′ end of the RNA to a new position at the 3′ end of the RNA. The length of the direct repeat is the distance shown as (A) in figure 72, and its complement is shown as (a). The region of strong-stop DNA not included in the repetition is shown as (b); it is a copy of region (B) of the RNA.

IV. Once minus-strong-stop DNA has jumped to the 3′ end of the template, it is in position to be an initiator for synthesis of a strand of DNA that could in principle continue all the way to the other end of the template. The product that is formed appears to be 8.2 kilobases (kb) long and probably has at its 5′ end the original primer tRNA. The 8.2 kb product has region (c) as a direct repetition; at one end it is DNA and at the other end, it is part of the primer tRNA.

V. Plus-strand DNA synthesis is initiated by the synthesis of a plus-strand strong-stop DNA molecule. The point of initiation is shown as the border of two regions: (d) and (e) of 8.2 kb minus-strand DNA. Plus-strand strong-stop DNA consists of regions (E), (A), (B) and (C). Segment (C) is actually a copy of a portion of the tRNA primer that initiated synthesis of the minus-strand DNA. This is a crucial aspect of the model.

VI. Removal of the original tRNA primer.

VII. The molecule shown in structure VI can easily circularize (or two such molecules could form a dimer). The homology that allows the two ends to pair is regions (C) and (c). Once this partially double-stranded circular DNA is formed, both the plus strand and the minus strand are in position to be finished.

VIII. The minus strand of DNA is finished by copying plus-strong-stop DNA, thus extending its length from 8.2 to 8.8 kb. In this step, the genome acquires its 600 base direct terminal repeat structure.

IX. and X. The extension of plus-strong-stop DNA to copy the minus strand of DNA will ultimately generate a 9 kb, completely double-stranded DNA with two continuous strands.

Transposon-like property of virus DNA

Sequencing of Moloney murine sarcoma virus (M-MSV) (Reddy et al. 1980; Dhar et al. 1980), Moloney murine leukemia virus (M-MLV) (Van Beveren et al. 1980; Sutcliffe et al. 1980), and spleen necrosis virus (Shimotohno et al. 1980) revealed that these viruses have long inverted repeats at the two molecular ends. Integration of these viral genomes into cellular DNA resulted in the duplication of a short host nucleotide sequence at the integration site

(Shimotohno et al. 1980; Shoemaker et al. 1980). The direct repeat at each end of the viral DNA consists of 569 base pairs. The cell-virus junction at each end contains a 5-base pair direct repeat of cell DNA next to a 3-base pair inverted repeat of viral DNA. This structure resembles that of bacterial transposable elements and is consistent with the protovirus hypothesis that retroviruses evolved from the cell genome (Shimotohno et al. 1980.)

RNA transcripts of the integrated viral genomes

The initiation sequence for the transcription of the viral DNA is probably in the U_3 sequence that is close to the R_5 terminus of the viral DNA. The termination sequence for RNA synthesis is present between the U_3 and R_5 sequences on the right-hand side of the viral DNA.

The viral RNA is transcribed from the viral DNA by the cellular RNA polymerase II. The viral RNA is processed from precursor RNA molecules in the nucleus of the infected cell.

The viral mRNA species

Several species of viral mRNA exist (figure 73, B-D) that differ in the amount of the genetic information that they contain. Some mRNA molecules contain all the genetic information, but only the gene at the 5′ end is translated. These consist of the *gag-pol* mRNAs that code for the viral polymerase precursor (pr180) and the *gag* mRNAs that code for the viral group-specific antigen precursor (pr76) (figure 73 D_1 and D_2). Other mRNA species consist of portions of the total genome. The *env* RNA codes only for the envelope proteins but contains the RNA sequence for the *src* protein which is not translated. The mRNA that codes for the src protein also contains the 3′ terminus which is not translated. (This species is only found in avian sarcoma viruses.)

Molecular events in the replication of a retrovirus are schematically summarized in figure 74.

BIOLOGICAL PROPERTIES AND SIGNIFICANCE OF THE INTEGRATED VIRAL DNA GENOMES

The integrated viral DNA is infectious

Hill and Hillova (1972) demonstrated that transfection of chick fibroblasts in vitro with fragmented DNA from a transformed cell that does not produce virus progeny can lead to the production of virions in the chick cells. This study revealed that the viral DNA is present in the DNA of the transformed cell and, when experimentally introduced into sensitive cells, can lead to virus synthesis.

Retrovirus DNA can be incorporated into the germ line

Infection of a mouse fertilized egg with Moloney leukemia virus at the morula stage (4–8 cells), followed by implantation of the egg in the female mouse, led to the development of an embryo containing the viral DNA in the diploid genome of each cell. Some of the newborn mice developed leukemia. Viral

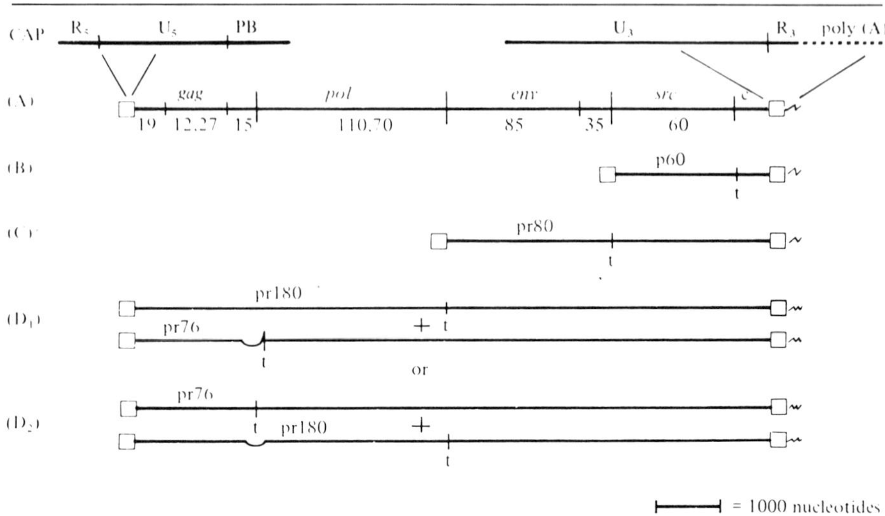

Figure 73. Structure of the ASV genome and mRNAs. (A) The genome of nondefective ASV drawn to approximate scale (bar represents 1,000 nucleotides). The boxes indicate terminal sequences which are shown above the genome enlarged 50-fold. Coding regions are labeled above the genome and the approximate mol. wt. ($\times 10^{-3}$) of the protein products are shown. (B) to (D) Probable mRNA species. The molecular weight ($\times 10^{-3}$) of the primary product is written above each molecule. The translation terminator is shown by t and the box at the 5′ end indicates the region from the 5′ end of the genome added by splicing. (B) *src* mRNA; (C) *env* mRNA; (D_1) and (D_2) two possibilities for *gag-pol* mRNAs.
(Coffin 1979. Reprinted by permission from *J. Gen. Virol.* 42, Fig. 1, p. 5. Cambridge University Press, Cambridge.)

DNA was found in the DNA of the progeny of the in-ovo infected F_1 and F_2 leukemic mouse strains, using molecular hybridization techniques. Infection of newborn mice with MuLV can also lead to the development of leukemia.

Effect of H_2: the histocompatibility locus on mouse leukemia

At least two genes affect the leukomogenesis in the mouse. These are located in the H_2 histocompatibility region of chromosome 17 of the mouse. One gene is associated with the segment K or I of the H_2 region, which affects the ability of the host to develop an immune response to the antigens of virus-transformed leukemic cells. The second gene is associated with segment D of the H_2 region and affects the development of cancer cell clones, which express viral antigens. The Fv-1 cellular locus (see page 241) is not linked to H_2 and obstructs the ability of the virus to replicate by blocking circularization and integration of DNA (Lilly and Pincus 1972).

RETROVIRUSES AND THEIR EVOLUTIONARY PATHWAYS

Laboratory infection of mammalian cells (e.g., rat embryo fibroblasts) with an avian retrovirus (e.g., RSV) causes cell transformation and clones of trans-

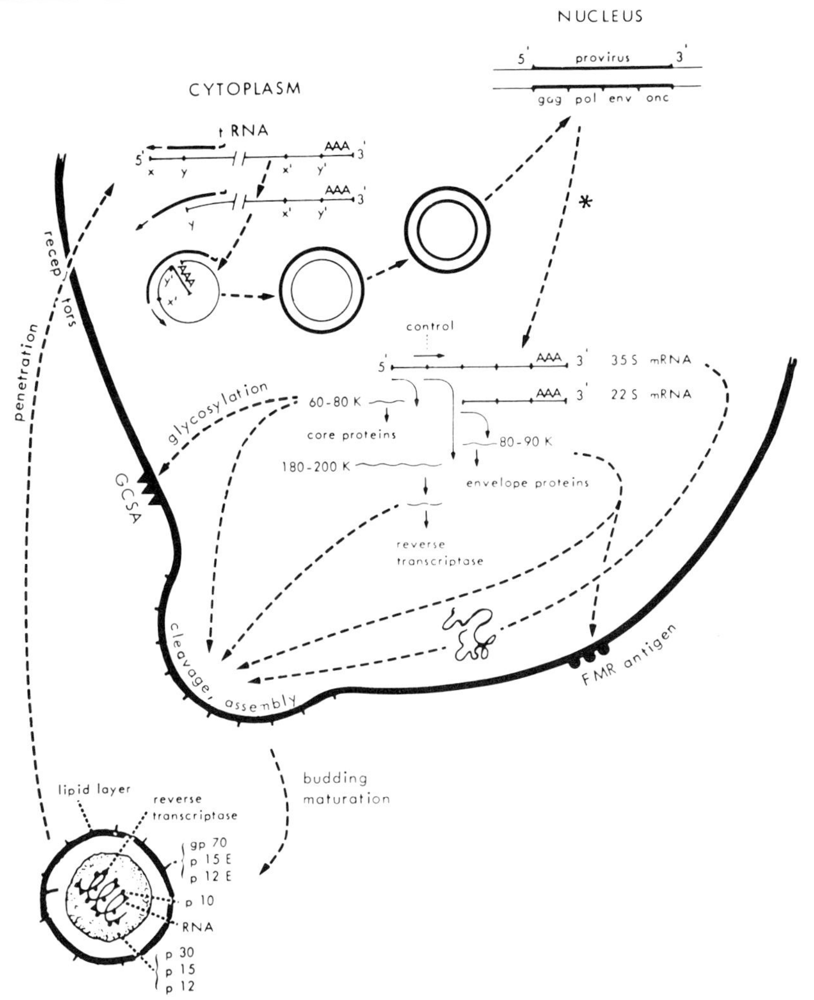

Figure 74. Life cycle of murine leukemia viruses (MuLV). After penetration, the viral RNA is transcribed into DNA (RNA-directed DNA polymerase activity), the RNA template is subsequently removed from the transcription product (RNase H activity), and the DNA molecule is made double-stranded (cDNA-directed DNA polymerase activity). The proviral DNA is then transported to the nucleus where it becomes integrated into the host genome. The integrated proviral DNA is transcribed by the cellular polymerase. The various viral proteins are probably transported to the cell surface, where cleavage of precursor proteins, packaging of polypeptides and RNA, and subsequent budding of virions from the cell membrane take place.

(Reprinted by permission of K. Nooter. Ph.D. thesis, Radiobiological Institute of the Organization for Health Research TNO, Rijswijk, The Netherlands, 1979.)

formed cells can be isolated. These do not produce virions but contain the viral DNA in the chromosomal DNA. Treatment of the cells with iododeoxyuridine or depriving them of arginine leads to the production of virions that contain the avian retrovirus RNA and proteins. Thus infection of a mammalian cell with an avian retrovirus leads to cell transformation without virus synthesis.

Porcine retrovirus was acquired from a rodent

Porcine cells in tissue culture produce a retrovirus that is present in the DNA in all the tissues of the pig. European wild boar and the African bush pig, which are related to domestic swine, contain viral DNA very similar to porcine retrovirus. G. Todaro and his collaborators revealed that the retrovirus of the pig was actually obtained from a small rodent, an ancestor of the mouse. The resemblance between the nucleic acid of the pig retrovirus, which is an endogenous virus, and the nucleotide sequences in the DNA of rodent retrovirus suggested that the rodent retrovirus had been transmitted from the ancestor of the infected pig via the germ line and was retained throughout the porcine evolution.

The monkey retrovirus originated from an ancient Asian rodent

The nucleic acids of retroviruses of monkeys like gibbon ape leukemia virus (GALV) and simian sarcoma virus (SSV) were found to be highly homologous with the type C retrovirus of the mouse, *Mus musculus*. Several structural proteins of the viruses of the mouse and monkey have common antigenic determinants. These results led to the suggestion that GALV and SSV are derived from the Asian mouse (*Mus caroli*) retrovirus by transspecies infection of primates.

Origin of the cat retrovirus

Todaro's studies (Todaro 1975) on the homology between retroviruses of different animal species led to the conclusion that felis, the ancestor of the domestic cat, obtained its retrovirus by infection from old world monkeys three to ten million years ago and from a rat ancestor of the domestic cat (see chapter 25).

Avian retroviruses

V. Ellermann and O. Bang in 1909, and P. Rous in 1911, discovered that tumors in chickens are caused by a filterable agent. This pioneering work, as well as that of J.J. Bittner in the 1940s and the discovery by L. Gross in 1951 that mouse leukemia can be transmitted vertically by a filterable virus (see chapter 1), led to the theories that viruses can cause cancer. Today there are several subgroups of avian retroviruses involved in leukosis and sarcoma. The avian retroviruses are divided into subgroups A–G on the basis of surface receptor specificity of the virus and intracellular restriction phenomena. The susceptibility of permissive avian cells is determined by the presence or absence of receptors that permit attachment and penetration of the virus particle. Subgroups A–F contain both sarcoma and leukosis viruses (namely, exogenous viruses RSV, Rous associated virus, avian myeloblastosis virus, myelocytomatosis virus, avian erythroblastosis virus, and Mill Hill virus 2; the endogenous viruses include the Rous-associated virus types and induced leukosis viruses). Subgroup G includes the endogenous pheasant viruses (Robinson 1978).

THE SRC GENE IS THE TRANSFORMING GENE

Since the identification of *src*, the *onc* gene of avian sarcoma viruses, by Duesberg and Vogt (1970), over a dozen different *onc*-specific sequences have been identified in various oncogenic retroviruses. Seven of these belong to oncogenic viruses of the avian sarcoma leukosis group (figure 75). The sarcoma viruses are divided into four subgroups, while the acute leukemia viruses are divided into three subgroups, based on the *onc*-gene-specific RNA sequences. The RSV genome contains the four viral genes—*gag, pol, env,* and *src*—while the other viruses may lack the *pol, env,* or *src* genes, leading to a marked change in the size of the viral genome. All viruses presented in figure 75 retain the *gag* gene.

There are two types of *onc* genes in avian tumor viruses. The coding sequence of type 1 consists entirely of specific sequences and includes the *src* gene of Rous sarcoma virus which encodes a 60,000 dalton protein (figure 75) responsible for the transformed state of the cell. Another example of type I is the *onc* gene of avian myeloblastosis virus. The coding sequence of type II *onc* genes is a hybrid consisting of a specific sequence and elements of essential virion genes typically including part of the *gag* gene. The *onc* gene of avian myelocytomatosis (MC29) virus consists of both *gag* and a specific sequence termed *mcv* that functions as one genetic unit encoding a 110,000 dalton *gag*-related, probably transforming protein. The hybrid *onc* genes of Fujinami sarcoma virus, other avian sarcoma viruses, avian erythroblastosis virus, and other examples in the avian tumor virus group, are shown in figure 75 (Duesberg 1980; Bister and Duesberg 1982). The discovery of the *onc* genes provides an explanation as to why some retroviruses—namely those with *onc* genes—are acutely oncogenic, while others—like the lymphatic leukemia (leukosis) viruses—are only rarely oncogenic, and if so, only after long latent periods.

ENDOGENOUS VIRUSES AND CANCER

In contrast to exogenous avian sarcoma-leukosis viruses which replicate to high titer in chicken cells, the endogenous viruses replicate poorly, if at all, in most chicken cells. All of the endogenous viruses that have been successfully grown belong to subgroup E. In cells containing endogenous retroviruses integrated into chromosomal DNA, expression of the endogenous virus is

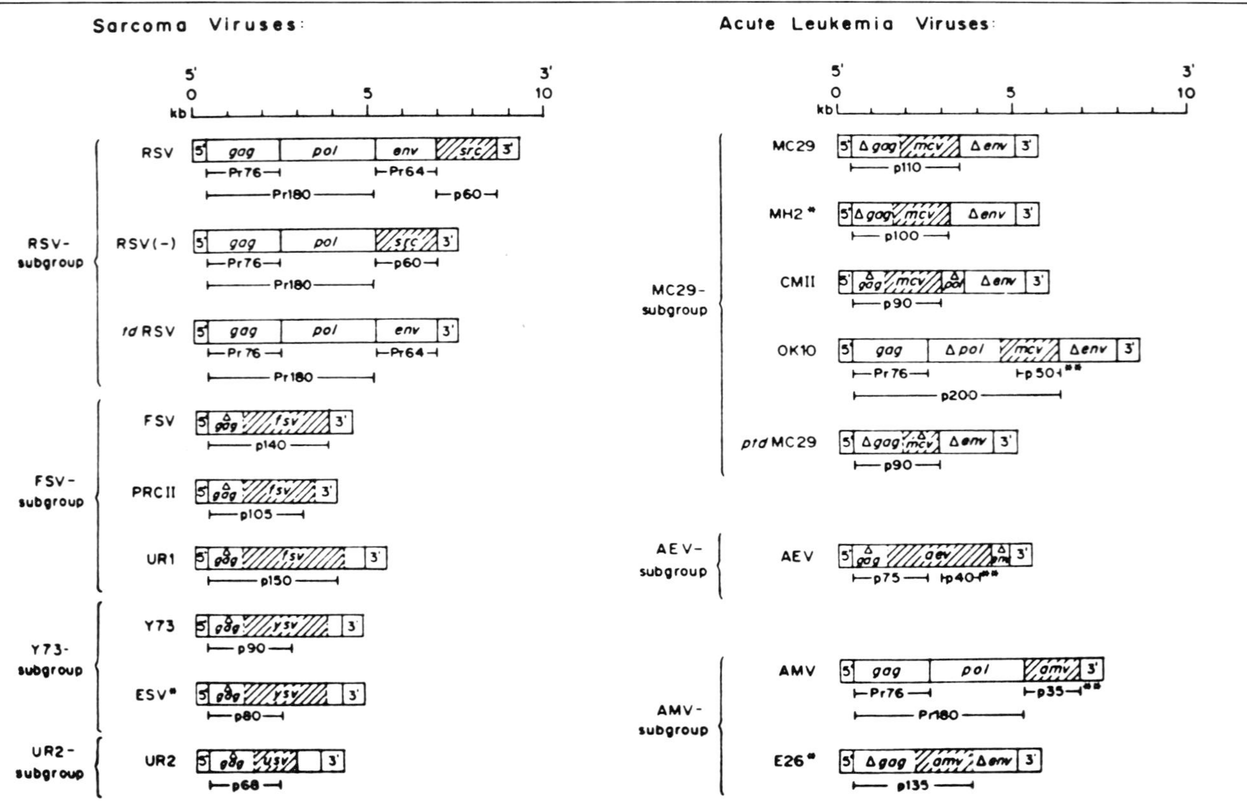

Sarcoma Viruses:
Acute Leukemia Viruses:
5'
3'
0
5
10
kb
RSV-subgroup
RSV
gag
pol
env
src
Pr76
Pr64
Pr180
p60
RSV(-)
td RSV
FSV-subgroup
FSV
fsv
p140
PRCII
p105
UR1
p150
Y73-subgroup
Y73
ysv
p90
ESV*
p80
UR2-subgroup
UR2
p68
MC29-subgroup
MC29
Δgag
mcv
Δenv
p110
MH2*
p100
CMII
p90
OK10
Δpol
p50
p200
ptd MC29
AEV-subgroup
AEV
aev
p75
p40
AMV-subgroup
AMV
amv
p35
Pr180
E26*
p135

determined by cellular genes. Six different phenotypes for the expression of endogenous viruses have been described in chickens (Robinson 1978). The distinctive phenotypes are due to specific germ-like-inherited proviruses (ev loci) that are absent in the negative alleles.

Among the viruses that cause tumors in chickens are viruses with genomes related to the endogenous Rous-associated virus type O (RAV-O). RAV-O is a subgroup E endogenous leukosis virus of chickens that can be spontaneously released but has no oncogenicity in vivo. All chicken cells tested by molecular hybridization techniques contain DNA sequences homologous to RAV-O RNA. Chickens containing the autosomal dominant allele V^+ spontaneously produce infectious RAV-O, while no virus is produced in V^- cells (Linial and Nieman 1976).

It is possible to divide the tumor-causing viruses into two groups: the first group has genes in addition to those of RAV-O (and also deletions), and the second group has the same genes as RAV-O. The *env* gene, however, and therefore the envelope antigen, are different. Both groups also differ from RAV-O by having a distinct U_3 region. Members of the first virus group (myeloblastosis, erythroblastosis, myelocytosis, and sarcoma viruses) cause an acute disease that affects a number of organs within a few days after infection. Members of the second group cause leukosis, a slow-developing disease of transformed B lymphocytes that metastasizes in the liver and spleen and appears four to five months after infection. This lymphoproliferative disease is caused by viruses of all the subgroups.

Viruses belonging to the first group cause cell transformation. The most widely investigated virus is RSV (named after P. Rous) of chickens. RSV can be divided into two subgroups: nondefective (*nd*) viruses that have all the RAV-O genes and defective viruses that lack the *env* gene. Both subgroups contain the *src* gene at the 3′ end of the RNA genome. Transformation-defective (*td*) mutants of RSV can be obtained by spontaneous mutation of

Figure 75. Genetic structures of oncogenic retroviruses of the avian leukosis/sarcoma group: Boxes indicate the size of viral RNAs in kilobases (kb) and segments within boxes indicate map locations in kb of complete or partial (Δ) complements of the three essential virion genes *gag*, *pol*, and *env* of the *onc*-specific sequences and of the noncoding regulatory sequences at the 5′ and 3′ end of viral RNAs. Dotted lines are used to indicate that borders between genetic elements are uncertain. Based on *onc* gene-specific RNA sequences (hatched boxes), four subgroups of sarcoma viruses and three subgroups of acute leukemia viruses can be distinguished. The three-letter code for *onc*-specific RNA sequences extends the one used previously by the authors. *src* represents the *onc*-specific sequence of the RSV-subgroup of avian sarcoma viruses; *fsv* that of the Fujinami-subgroup, *ysv* that of the Y73 subgroup, and *usv* that of the UR2-subgroup of avian sarcoma viruses; *mcv* that of the MC29-subgroup, *aev* that of the avian erythroblastosis virus subgroup and *amv* that of the AMV-subgroup of acute leukemia viruses. Lines and numbers under the boxes symbolize the complexities in kilodaltons (kd) of the precursors (Pr) for viral structural proteins and of the transformation-specific polyproteins (p). For some viruses (*) complete genetic maps are not yet available, and some protein products (**) have only been identified in cell-free translation assays.

(Bister and Duesberg 1982. Reprinted by permission from *Advances in Viral Oncology*. G. Klein, ed. Copyright 1982 by Raven Press, New York).

nondefective RSV. These viruses retain all the normal growth and replicative functions but have lost the ability to induce transformation in infected cultures. The RNA genome of *td* mutants is uniform in size, smaller than that of the transforming parental virus and, in general, very similar to that of the naturally existing avian leukosis viruses, such as RAV-2. Td mutants spontaneously derived from the Schmidt-Ruppin strain of RSV, subgroup A (SR-RSV-A) were found to lack either all or part of the *src* gene (Kawai et al. 1977).

Virus strains that transform cultured chick embryo fibroblasts include: (1) avian sarcoma viruses with the coat gene of an endogenous virus (e.g., members of Prague strain of RSV, subgroup E and Schmidt-Ruppin strain of RSV, subgroup E: PR-RSV-E and SR-RSV-E) and (2) recombinants of avian sarcoma virus and RAV-O (other PR-RSV-E and SR-RSV-E members) (Robinson 1978).

Virus strains that do not cause cell transformation include: (1) strain 7-ILV, an endogenous virus that appeared in cells prepared from a resistant chicken with the phenotype $V^{-}E^{+}$ susceptible to subgroup E virus after treatment of the cells with bromodeoxyuridine; (2) RAV strain O (RAV-O) that is a subgroup E endogenous virus; and (3) RAV strain RAV-60 that is a recombinant with genetic information of an infecting avian leukosis virus and genes coding for the type-specific E coat antigen of the chick cell.

The genomes of these viruses are organized as follows:

1. RAV-O (2.7–2.9 $\times$ 10^6 daltons)

   r-ss-gag-pol-env-U_3-r-poly(A)

2. ASV nondefective (3.3–3.4 $\times$ 10^6 daltons)

   r-ss-gag-pol-env-src-U_3-r-poly(A)

3. BH-RSV (Bryan high titer strain of RSV) (2.7–2.9 $\times$ 10^6 daltons)

   r-ss-gag-pol-src-U_3-r-poly(A)

4. Viruses defective in the ability to transform cells yet that induce leukemia in chickens (2.7–2.9 $\times$ 10^6 daltons)

   r-ss-gag-pol-env-U_3-r-poly(A)

   (in which r = 21 nucleotides, ss = 101 nucleotides, r is part of ss (= U_5 + r), gag = group specific antigen, pol = reverse transcriptase, env = envelope protein, src = sarcoma gene, and U_3 = 3′ untranslated sequence of 250 nucleotides)

The viruses replicate in the infected cells and transcribe RNA from different viral genes. Some of the viral DNA is retained as episomal DNA in the cell nucleus but disappears after a while, and some DNA molecules integrate into the cell DNA. The viruses differ in the mRNA species that they code for. RAV-O virus codes for two species of viral mRNA: (1) mRNA species (1,000–2,000 copies per cell) with a sedimentation coefficient of 21 S containing the *env* gene in the form of ss-env-U_3-poly(A) that is translated to gpr90, the precursor of the envelope protein, and (2) mRNA species with a sedimentation coefficient of 35 S (1,000–2,000 copies per cell) that contain the genes ss-gag-pol-env-U_3-poly(A). The 35S mRNA is translated into pr76 (that is cleaved to the group-specific [gs] antigen) and pr180 (that is cleaved to yield the polymerase). RAV-2 virus is very similar to RAV-O virus, except that the 21 S mRNA is present as 8–12 $\times$ 10^3 copies per cell.

The viral gene src is closely related to the cellular gene sarc

Nucleotide sequences related to *src* (denoted *endogenous sarc*) are present in the DNA of chickens and may have been the progenitor of the viral gene. The gene *sarc* is present in one or two copies in a haploid genome. It is possible that RSV was formed by recombination between RAV-O or a similar virus with the cellular gene *sarc*. Experiments to determine whether indeed a nontransforming virus can recombine with the cellular *sarc* gene were done by H. Hanafusa and co-workers. Chickens were infected with the Schmidt-Ruppin strain of RSV (SR-RSV-A) that has a deletion of part of *src* and is incapable of transformation. Two to three months later, tumors appeared in different places in the chicken, but not at the site of inoculation. From the tumors, a virus was isolated that was found to resemble SR-RSV-A but was capable of transforming cells in culture. Injection of this virus into healthy chickens caused tumors within a few days. It is possible that the recovered sarcomogenic virus is a recombinant between the parental transformation-defective virus and endogenous *sarc* sequences of normal cells. These recovered avian sarcoma viruses produce the transforming protein $p60^{src}$ in infected cells in amounts comparable to those found in cells transformed by standard strains of avian sarcoma virus. The $p60^{src}$ protein present in normal cells at a low level is also associated with a protein kinase activity and is similar, both antigenically and chemically, to the $p60^{src}$ of avian sarcoma virus (Karess et al. 1979; Oppermann et al. 1979).

Evolution of avian leukosis viruses

Endogenous retroviruses invaded the germ line of chickens. Pheasant cells contain DNA sequences that have 20% homology with the RAV-O genome. Quail and turkey cells contain DNA that is homologous to 1–2% of the RAV-O genome. Other members of the same genus (*Giallus*) do not have related sequences and therefore were not likely to have been introduced after speciation.

THE ONCOGENE THEORY

The oncogene hypothesis (Huebner and Todaro 1969) originally postulated that viral *onc* genes are present in normal cells and may cause cancer if induced by carcinogens or other oncogenic agents. It was subsequently hypothesized that there are genetic elements in the cell with a potential of becoming viral *onc* genes, termed *protoviruses* (and later proto-*onc* genes) (Bishop 1981). After transduction by retroviruses, these proto-*onc* genes evolve into viral *onc* genes (Temin 1980). It has also been postulated that *onc* genes evolved from cellular prototypes in a multiphase process involving endogenous, defective retrovirus-like intermediates (Duesberg 1980). There is substantial evidence to indicate that the *onc* genes originated from the normal cellular genes by a recombination process between a parent nontransforming virus and host cellular DNA.

Over ten different transforming genes have been identified in the genomes of acute transforming retroviruses of avian and mammalian origin, and each of these viral onc (*v-onc*) genes has a normal cellular counterpart (*c-onc*). The normal cellular genes are generally expressed at low levels in normal, uninfected cells, but may be expressed at higher levels during certain stages of development (Hayward et al., 1981). It was shown by molecular hybridization that the cellular genes which gave rise to viral transforming genes are conserved throughout evolution. Evolutionary conservation of the cellular *onc* genes indicates that they probably play an essential role in cellular growth, differentiation, and/or development.

Hayward and associates (1981) have shown that avian leukosis virus induces neoplastic disease by activating the *c-myc* gene, the cellular counterpart of the transforming gene of MC29 virus (myelocytomatosis virus-29). Their data indicate that, as a rare event, the ALV provirus integrates adjacent to the *c-myc* gene and that transcription, initiating from a viral promoter, causes enhanced expression of *c-myc,* leading to neoplastic transformation.

It seems likely, therefore, that the role of both acute and slow RNA tumor viruses in neoplastic transformation is simply to provide a means for constitutively expressing a cellular gene that is normally expressed at low levels. These viruses accomplish this in slightly different ways. In the acute viruses, a cellular gene has been inserted into the viral genome (by recombination), where it comes under the control of the viral promoter. With slow transforming viruses such as ALV, the viral promoter is inserted into (or adjacent to) the cellular gene.

Thus, for the first time, it has been shown that a carcinogenic agent (namely ALV) acts by altering the expression of a normal cellular gene. Hayward and co-workers (1981) also raise the interesting possibility that activation of a *c-onc* gene may be the common initiation event in neoplastic transformation by both viral and nonviral agents.

In other studies, Eva and associates (1982) have shown that cellular genes analogous to retroviral *onc* genes are both present and frequently expressed in human neoplastic cells. Nonetheless, the genetic changes involved in human neoplastic transformation appear to be multiple, and our understanding of these mechanisms is still at an early stage.

MAMMALIAN RETROVIRUSES

Type C viruses have been recognized in mice, hamsters, rats, cats, cows, pigs, in several primates, and in humans.

Mouse leukemia viruses

Extracts of AK mouse embryos which were filtered to remove the cells and inoculated into newborn C3H mice caused leukemia in the recipients. This work by Gross (1951) showed that mouse strains like AK and C57 carry a latent virus that may be involved in spontaneous leukemia, and that can cause leukemia in other mouse strains, provided it is injected into newborn mice less than 48 hrs after birth. The Gross virus differs from the Friend, Moloney, and Rauscher MuLV designated as subgroup FMR. These are laboratory type C virus strains isolated from transplantable mouse tumors. Whether FMR viruses exist in nature is unclear.

The leukomogenic viruses of mice are divided into three groups: (1) Ecotropic viruses include N-tropic viruses that infect cells of NIH–Swiss strains of mice (N-type), but not cells of BALB/c mice, B-tropic viruses that infect cells of BALB/c mice (B-type) only, and NB-tropic viruses that infect both strains; (2) Xenotropic viruses (X-tropic) replicate only in cells from animals foreign to the host species and include the NZB virus from New Zealand Black mice. Although endogenous to the mouse, this virus could not productively infect mouse cells; and (3) Amphoteric viruses (A-tropic) share the host range of both E-tropic and X-tropic viruses, since they productively infect mouse cells as well as cells from heterologous species (Levy 1978).

Mouse genes determine sensitivity to leukemogenic viruses

In 1957, C. Friend reported the isolation from a 14-month-old Swiss mouse of a filterable agent that produced a leukemia-like disease in adult mice. This disease differs entirely from the thymus-dependent lymphoma typical of the Gross virus-induced disease of high leukemic virus strains. The Fv gene refers to a cellular locus that determines Friend virus susceptibility. Fv-1 is the gene that governs relative resistance to the original Friend virus strain (F-S), but with no influence on susceptibility to the BALB/c-adapted variant of Friend virus (F-B). Fv-2 is the gene that governs virtually absolute resistance to focus formation by both F-S and F-B viruses. The Fv-1 locus is not apparently linked to the Fv-2 or H_2 histocompatibility locus in the mouse genome (Lilly and Pincus 1973).

Radiation-induced leukemia (RadLV) in mice

Irradiation of adult mice strain C57BL leads to the appearance of leukemia. From these leukemic cells, radiation leukemia virus (RadLV) was isolated, which is able to induce leukemia in sensitive mice if they are irradiated at the time of virus inoculation. Virus must be injected directly into the thymus. The RadLV is a mixture of an ecotropic virus (B + N tropic) and a xenotropic virus.

Mammary tumor virus (MTV) in female mice

The mammary virus is transferred horizontally from the infected mother to the suckling mice via the milk. It has also been demonstrated that offspring of female mice with mammary tumors that were fed by a healthy mother also developed tumors, since the viral DNA is in the germ line. Thus newborn mice have the viral genes as a result of vertical transmission.

Mice of the GR strain which carry the mammary tumor virus genome in their cellular DNA have a high incidence of mammary tumors. An autosomal dominant cellular gene is responsible for the expression of the virus in the mouse and for the development of the mammary tumor.

Feline leukemia virus (FeLV)

Leukemia in cats has the form of lymphosarcoma and myeloproliferative disease. An exogenous type C retrovirus was isolated from the tumor cells. Injection of a homogenate made from leukemic cells from a spontaneous lymphoma (after filtration to remove cell debris) to a healthy cat leads to the development of leukemia. Thus the virus can be horizontally transmitted from one cat to another. Analysis of blood samples from healthy stray cats for antibodies to the cat leukemia virus showed that 50–60% have antibodies. The antibodies are against a cell-associated viral antigen designated FOCMA (feline oncornavirus-associated cell membrane antigen). Whereas only 3% of stray cats have viremia in the blood and excrete leukemia virus in the nasopharynx, 40% of cats living in human households have antiviral antibodies. Feline leukemia virus is congenitally (vertically) transmitted to offspring. The offspring of viremic mothers can also carry the viruses.

A significant percentage of cats (about 30%) with leukemia and lymphomas have no detectable virus or viral components. Cats may therefore be normal virus negative, normal virus positive, leukemic virus negative, or leukemic virus positive. However, it is suspected that exogenous FeLV does play a role in the virus-negative leukemias (Koshy et al. 1980).

FeLV is unrelated to the RD114 xenotropic cat virus that was identified during attempts to isolate a human retrovirus (Levy 1978). RD114 is an endogenous feline virus. It has been suggested that millions of years ago, FeLV-related genes were passed to cats by transspecies infection by a rat type C virus and that RD114 sequences were transferred to the cat by a primate virus.

Bovine leukosis in cattle and sheep

Leukocytosis in cows is expressed as a marked increase in the number of leukocytes in the peripheral blood (lymphocyte leukosis). The disease is caused by a retrovirus that is horizontally transmitted by infected saliva. Lymphocytes from an infected cow can be cultivated under in-vitro conditions and they can produce and release the BLV that can be cultivated in other cell lines.

The virus was isolated and shown to be a type C retrovirus. The virions contain 70 S RNA and the viral proteins p80, gp69, gp45, p24, p15, p12, and p10.

It is possible to screen for antibodies to the virus in sera from cattle in herds to determine the extent of virus infection using radioimmunoassay or enzyme linked immuno-sorbent assay (ELISA) (see chapter 27). Prior to the development of these tests, the diseased cows were diagnosed by hematological tests that determine the extent of leukocytosis.

The leukosis virus also infects sheep and causes a disease similar to that of cows. BLV does not infect humans. No anti-BLV antibodies were found in the blood of people (farmers and butchers) who are in contact with cattle or handle their meat (reviewed by Mussgay and Kaaden 1978).

Monkey retroviruses

Viral sequences in the cellular DNA of primates are related to those in baboons

Cocultivation of normal cells from various organs (like kidney, spleen, lung, testes, placenta) of different species of baboons (*Papio cynocephalus, P. anubis, P. papio, P. hemadryas*) with various sensitive cell lines led to the isolation of a virus from each organ studied. Characterization of virions showed that they have unique RNA and antigenicity. The viruses isolated from the various baboon species are related.

The conclusion of Todaro (1975) is that the baboon viruses contain endogenous viral sequences integrated into the genome that are related to those found in primates, including man.

Viruses isolated from monkeys of related genera have a high nucleic acid homology. The presence of an endogenous virus in anthropoid primates, and also in those species that evolved from them, indicate that the viruses evolved during a period of 30 to 40 million years as the species evolved (Benveniste and Todaro 1975).

Infectious type C retroviruses from monkeys

Infectious virus was isolated from a gibbon ape (kept in captivity) that developed leukemia and from a woolly monkey that developed fibrosarcoma. These viruses, named gibbon ape leukemia virus (GALV) and simian sarcoma virus (SSV), respectively, are transmitted horizontally from one monkey to another. The viruses are not endogenous but are infectious and can cause leukemia if infected into healthy monkeys. GALV was also isolated from brain tissue of a healthy gibbon ape, but this virus is related, yet not identical, to the GALV isolated from leukemic monkeys. The viruses isolated from these monkeys can be separated into four subgroups on the basis of the relatedness of the viral RNA (Todaro 1975).

Monkey mammary tumor virus (MoMTV): the Mason-Pfizer monkey virus

In a female rhesus monkey (*Macaca mulatta*) (an Old World monkey) a mammary tumor was diagnosed. Electron microscopy of the tumor cells showed the presence of virions in the cells and in the intercellular spaces. The tumor cells could be grown in vitro by cocultivation with monkey embryo cells. The mixed cultures of monkey tumor cells and embryo cells became chronically infected and continued to replicate the virus as permanent, established cell lines. The virus could also replicate in monkey and chimpanzee lung cells and could transform infected rhesus foreskin cells. Cocultivation of monkey cells infected with virus and human KC cells that were transformed in vitro by RSV causes the formation of syncytia within 24 hr that contain 40–50 nuclei. These syncytia were clearly distinguishable from those caused by simian foamy viruses.

The virions have a density in sucrose gradients of 1.16 g/ml, contain 60–70 S RNA as a genome, and reverse transcriptase. However, this virus is distinct from the type B and type C retroviruses and has been referred to as a type D virus particle. In some ways, such as in the developmental stages, MoMTV resembles the murine mammary tumor viruses.

Virus-transformed rhesus monkey foreskin cells inoculated into newborn rhesus monkeys caused tumor formation at the site of inoculation. These transformed cells also induced fibrosarcomas when inoculated subcutaneously into nude mice, but the absence of any tumor induction by cell-free virus remains a barrier to demonstrating that the virus itself causes cancer (reviewed by Chopra 1976).

The host range of this virus has been shown in tissue culture to include both primate and human cells. Viruses morphologically similar to the Mason-Pfizer monkey virus have been isolated from certain rhesus monkeys and human tissues and cell lines as well as from normal tissues of a langur monkey (another Old World monkey) and from squirrel monkeys (a New World monkey). Thus the Mason-Pfizer monkey virus is a member of a larger group of viruses of both Old and New World primates. There is interspecies cross reactivities between the major internal and external proteins of these viruses; they also share cross reactivity of their major external glycoproteins with those of type C baboon endogenous virus (Fine and Schochetman 1978).

HUMAN RETROVIRUSES

Type C retroviruses were isolated from human leukemic cells. Analysis of leukemic cells and serum taken from patients with chronic and acute leukemias and lymphomas revealed the presence of RNA protein complexes containing a reverse transcriptase. After induction of such leukemic cells, it was possible to isolate virions that had the properties of type C retroviruses. R.C. Gallo and his collaborators isolated a type C virus from a continuous cell line of human leukemic cells that had no cross reactivity with any known retrovirus. Antibodies to this virus (designated HTLV) were found in sera of some human leukemia patients (see chapter 28).

SARCOMA VIRUSES

These viruses, isolated from rodents (mice, rats), cats, and one monkey species (woolly monkey), are defective in their ability to replicate in infected cells, but can transform the infected cells. Such transformed cells can cause solid tumors when injected into animals. Moloney virus, Abelson virus, and BALB/c virus are sarcoma viruses isolated from mice, the Harvey and Kirsten strains were isolated from rats, while the Snyder-Theiler, Gardner, and McDonough strains are feline sarcoma viruses (FeSV). For their replication, the sarcoma viruses require coinfection with a type C helper virus. The original stocks of Harvey murine sarcoma virus (Ha-MSV) and Kirsten murine sarcoma virus (Ki-MSV) contained Moloney murine leukemia virus (Mo-MuLV) and Kirsten murine leukemia virus (Ki-MuLV), respectively, as helper viruses.

Sarcoma viruses developed as recombinant viruses whose genomes contain genetic information of a nontransforming C-type virus linked to sequences present in the chromosomal DNA of the host of origin. The genomes of Mo-MSV, Ha-MSV, Ki-MSV, FeSVs and Abelson murine leukemia virus (Ab-MuLV), contain sequences of helper-independent C-type virus located at the 5′ and 3′ termini of the viral RNA, separated by a stretch of inserted sequences of cellular origin. The C-type viral genetic information of the mammalian transforming viruses is required for proper replication of the viral genome and may provide a promoter region for transcription. In contrast, the inserted cellular sequences may be responsible for the sarcomogenic function specified by these viruses (Andersson 1980).

The transforming proteins of these viruses have been identified. The transforming protein of Ab-MuLV has protein kinase activity and is similar to the ASV *src* gene product, although they have different cellular targets for transformation (Witte et al. 1980).

MOLONEY MURINE LEUKEMIA AND SARCOMA VIRUSES

Moloney murine leukemia virus (Mo-MuLV) is a prototype mammalian type C virus originally isolated from a sarcoma passaged in BALB/c mice. This virus causes a generalized lymphocytic leukemia in newborn mice within three months of injection. The complete genetic structure of Mo-MuLV has been defined (Shinnick et al. 1981). The Moloney murine sarcoma virus (Mo-MSV) arose spontaneously during passage of Mo-MuLV in BALB/c mice and contains both Mo-MuLV and normal BALB/c mouse cell sequences. The leukemia virus sequences were replaced by about one kilobase of mouse cellular genetic information.

Cleavage of Mo-MSV DNA with the enzymes XhoI and BglII, which cut 5′ to the cellular insertion region (cell-derived sequences), resulted in a small reduction of the virus infectivity. The intact cellular insertion was needed for virus activity. Digestion of the viral DNA with the enzymes Hpa1 and HindIII, which cut within the cellular insertion in the viral DNA, also destroyed the biological activity of the viral DNA. The cellular sequences in the viral DNA thus appear to be essential for transformation (Anderssen et al. 1979; Blair et al. 1980). The acquired cell sequences found in sarcoma viruses (termed *src*) and the normal cell DNA sequences homologous to them (termed *sarc* or *onc* genes) appear to be identical (Oskarssen et al. 1980), but are not sufficient for cell transformation. Some leukemia-derived sequences located adjacent to the 5′ end of *src* are required for transformation, as well as a terminally redundant sequence of the integrated leukemia provirus (Blair et al. 1980).

Human cellular onc genes are related to simian sarcoma virus (SSV) transforming genes

Hybridization of cloned viral transforming genes to cellular DNA fragments cloned in bacteriophage λ led to the isolation of the cellular *onc* genes. These cellular *onc* genes are homologous with the viral transforming genes, but in the cell DNA they are interrupted by intron sequences.

The genomes of SSV, a defective transforming virus isolated from a fibrosarcoma of a woolly monkey and its helper virus, simian sarcoma-associated virus (SSAV), were cloned and characterized. The SSV genome was found to have a 1.2 kilobase insert not shared by SSAV, which represents the viral *onc* gene (*v-sis*). Molecular hybridization experiments have shown that *v-sis* is a unique viral *onc* gene originating from woolly monkey DNA (Wong-Staal et al. 1981).

In addition, it has been shown that the human *onc* gene (*c-sis*) is related to the *v-sis* transforming gene of SSV. The *c-sis* gene extends over a region of about 12 kilobases which includes the 1.2 kilobase region of the *v-sis*-related sequences interrupted by four intervening sequences (Dalla Favera et al. 1981).

VISNA VIRUS, A SHEEP RETROVIRUS

Visna (= wasting) in sheep was described in the 1930s as a slow virus disease with an incubation period of one year. The virus replicates in the CNS and infects sheep of both sexes. The disease starts with an effect on the hind legs, tremor of the lips, head movements and, in rare cases, blindness. The disease appeared in Iceland as an epidemic in sheep and was overcome only after complete removal of the sheep and repopulation with new stock.

The virus is a member of the subfamily Lentivirinae and is released from cells by a budding process. The RNA in the virions has a molecular weight of 10^7 and is made of two RNA subunits. The virions contain a reverse transcriptase and RNA molecules of 4 S and 5 S. The virus replicates in the sheep choroid plexus, and the biosynthetic process is similar to that of type C retroviruses. The cell DNA contains the viral DNA in an integrated state.

BIBLIOGRAPHY

Andersson, P. The oncogenic function of mammalian sarcoma viruses. *Adv. Cancer Res. 33*:109–171, 1980.

Andersson, P.; Goldfarb, M.P.; and Weinberg, R.A. A defined subgenomic fragment of in vitro synthesized Moloney sarcoma virus DNA can induce cell transformation upon transfection. *Cell 16*:63–75, 1979.

Bender, W.; Chien, Y.-H.; Chattopadhyay, S.; Vogt, P.K.; Gardner, M.B.; and Davidson, N. High-molecular-weight RNAs of AKR, NZB and wild mouse viruses and avian reticuloendotheliosis viruses all have similar dimer structures. *J. Virol. 25*:888–896, 1978.

Benveniste, R.E., and Todaro, G.J. Evolution of type C viral genes: 1. Nucleic acid from baboon type C virus as a measure of divergence among primate species. *Proc. Natl. Acad. Sci. USA 71*:4513–4518, 1975.

Bishop, J.M. Enemies within: the genesis of retrovirus oncogenes. *Cell 23*:5–6, 1981.

Bister, K., and Duesberg, P.H. Genetic structure and transforming genes of avian retroviruses. In: *Advances in Viral Oncology* (G. Klein, ed.), Raven Press, New York, 1982, forthcoming.

Blair, D.G.; McClements, W.L.; Oskarsson, M.K.; Fischinger, P.J.; and Vande Woude, G.F. Biological activity of cloned Moloney sarcoma virus DNA: terminally redundant sequences may enhance transformation efficiency. *Proc. Natl. Acad. Sci. USA* 77:3504–3508, 1980.

Bolognesi, D.P.; Montelaro, R.D.; Frank, H.; and Schäfer, W. Assembly of type C oncornaviruses: a model. *Science 199*:183–186, 1978.

Chopra, H.C. An oncorna-type virus from monkey breast tumor. *Prog. Med. Virol. 22*:104–122, 1976.

Coffin, J.M. Structure, replication and recombination of retrovirus genomes: some unifying hypotheses. *J. Gen. Virol. 42*:1–26, 1979.

Collett, M.S.; Dierks, P.; Parsons, V.; and Faras, A.J. RNase A hydrolyses of the 5′ terminus of the avian sarcoma virus genome during reverse transcription. *Nature 272*:181–184, 1978.

Dalla Favera, R.; Gelmann, E.P.; Gallo, R.C.; and Wong-Staal, F. A human *onc* gene homologous to the transforming gene (v-sis) of simian sarcoma virus. *Nature 292*:31–35, 1981.

Dhar, R.; McClements, W.L.; Enquist, L.W.; and Vande Woude, G.F. Nucleotide sequences of integrated Moloney sarcoma provirus long terminal repeats and their host and viral functions. *Proc. Natl. Acad. Sci. USA* 77:3937–3941, 1980.

Duesberg, P.H. Transforming genes of retroviruses. *Cold Spring Harbor Symp. Quant. Biol. 44*:13–29, 1980.

Duesberg, P.H., and Vogt, P.K. Differences between the ribonucleic acids of transforming and nontransforming avian tumor viruses. *Proc. Natl. Acad. Sci. USA 67*:1673–1680, 1970.

Eva, A.; Robbins, K.C.; Andersen, P.R.; Srinivasan, A.; Tronick, S.R.; Reddy, E.P.; Ellmore, N.W.; Galen, A.T.; Lautenberger, J.A.; Papas, T.S.; Westin, E.H.; Wong-Staal, F.; Gallo, R.C.; and Aaronson, S.A. Cellular genes analogous to retroviral *onc* genes are transcribed in human cells. *Nature 295*:116–119, 1982.

Fine, D., and Schochetman, G. Type D primate retroviruses: a review. *Cancer Res. 38*:3123–3139, 1978.

Gilboa, E.; Mitra, S.W.; Goff, S.; and Baltimore, D. A detailed model of reverse transcription and tests of crucial aspects. *Cell 18*:93–100, 1979.

Gross, L. "Spontaneous" leukemia developing in C_3H mice following inoculation in infancy with AK-leukemic extracts or AK-embryos. *Proc. Soc. Exp. Biol. Med. 76*:27–32, 1951.

Hayward, W.S.; Neel, B.G.; and Astrin, S.M. Activation of a cellular *onc* gene by promoter insertion in ALV-induced lymphoid leukosis. *Nature 290*:475–480, 1981.

Hill, M., and Hillova, J. Virus recovery in chicken cells treated with Rous sarcoma cell DNA. *Nature (New Biol.) 237*:35–39, 1972.

Huebner, R.J., and Todaro, G.J. Oncogenes of RNA tumor viruses as determinants of cancer. *Proc. Natl. Acad. Sci. USA 64*:1087–1094, 1969.

Karess, R.E.; Hayward, W.S.; and Hanafusa, H. Cellular information in the genome of recovered avian sarcoma virus directs the synthesis of transforming protein. *Proc. Natl. Acad. Sci. USA 76*:3154–3158, 1979.

Kawai, S.; Duesberg, P.H.; and Hanafusa, H. Transformation defective mutants of Rous sarcoma virus with *src* gene deletions of varying length. *J. Virol. 24*:910–914, 1977.

Keshet, E., and Shaul, Y. Terminal direct repeats in a retrovirus-like repeated mouse gene family. *Nature 289*:83–85, 1981.

Koshy, R.; Gallo, R.C.; and Wong-Staal, F. Characterization of the endogenous feline leukemia virus-related DNA sequences in cats and attempts to identify exogenous viral sequences in tissues of virus-negative leukemic animals. *Virology 103*:434–445, 1980.

Kung, H.-J.; Shank, P.R.; Bishop, J.M.; and Varmus, H.E. Identification and characterization of dimeric and trimeric circular forms of avian sarcoma virus-specific DNA. *Virology 103*:425–433, 1980.

Levy, J. Xenotropic type C viruses. *Curr. Top. Microbiol. Immunol. 79*:111–214, 1978.

Lilly, F., and Pincus, T. Genetic control of murine viral leukemogenesis. *Adv. Cancer Res. 17*:231–277, 1973.

Linial, M., and Neiman, P.E. Infection of chick cells by subgroup E viruses. *Virology 73*:508–520, 1976.

Montelaro, R.C.; Sullivan, S.J.; and Bolognesi, D.P. An analysis of type-C retrovirus polypeptides and their associations in the virion. *Virology 81*:19–31, 1978.

Mussgay, M., and Kaaden, O.-R. Progress in studies on the etiology and serologic diagnosis of enzootic bovine leukosis. *Curr. Top. Microb. Immunol. 79*:43–72, 1978.

Opperman, H.; Levinson, A.D.; Varmus, H.E.; Levintow, L.; and Bishop, J.M. Uninfected vertebrate cells contain a protein that is closely related to the product of the avian sarcoma virus transforming gene (*src*). *Proc. Natl. Acad. Sci. USA 76*:1804–1808, 1979.

Oskarsson, M.; McClements, W.L.; Blair, D.G.; Maizel, J.V.; and Vande Woude, G.F. Properties of a normal mouse cell DNA sequence (sarc) homologous to the src sequence of Moloney sarcoma virus. *Science 207*:1222–1224, 1980.

Reddy, E.P.; Smith, M.J.; Canaani, E.; Robbins, K.C.; Tronick, S.R.; Zain, S.; and Aaronson, S.A. Nucleotide sequence analysis of the transforming region and large terminal redundancies of Moloney murine sarcoma virus. *Proc. Natl. Acad. Sci. USA* 77:5234–5238, 1980.

Reynolds, R.K., and Stephenson, J.R. Intracistronic mapping of the murine type C viral *gag* gene by use of conditional lethal replication mutants. *Virology 81*:328–340, 1977.

Robinson, H.L. Inheritance and expression of chicken genes that are related to avian leukosis sarcoma virus genes. *Curr. Top. Microbiol. Immunol. 83*:1–36, 1978.

Shimotohno, K.; Mizutani, S.; and Temin, H.M. Sequence of retrovirus provirus resembles that of bacterial transposable elements. *Nature 285*:550–554, 1980.

Shinnick, T.; Lerner, R.A.; and Sutcliffe, J.G. Nucleotide sequence of Moloney murine leukemia virus. *Nature 293*:543–548, 1981.

Shoemaker, C.; Goff, S.; Gilboa, E.; Paskind, M.; Mitra, S.W.; and Baltimore, D. Structure of a cloned circular Moloney murine leukemia virus DNA molecule containing an inverted segment: Implication for retrovirus integration. *Proc. Natl. Acad. Sci. USA* 77:3932–3936, 1980.

Sutcliffe, J.G.; Shinnick, T.M.; Verma, I.M.; and Lerner, R.A. Nucleotide sequence of Moloney leukemia virus: 3′ end reveals details of replication, analogy to bacterial transposons, and an unexpected gene. *Proc. Natl. Acad. Sci. USA* 77:3302–3306, 1980.

Temin, H.M. The DNA provirus hypothesis: The establishment and implications of RNA-directed DNA synthesis. *Science 192*:1075–1080, 1976.

Temin, H.M. Origin of retroviruses from cellular movable genetic elements. *Cell 21*:599–600, 1980.

Temin, H., and Baltimore, D. RNA-directed DNA synthesis and RNA tumor viruses. *Adv. Virus Res. 17*:129–186, 1972.

Todaro, G. Evolution and modes of transmission of RNA tumor viruses. *Am. J. Pathol. 81*:590–605, 1975.

Van Beveren, C.; Goddard, J.G.; Berns, A.; and Verma, I.M. Structure of Moloney murine leukemia viral DNA: nucleotide sequence of the 5′ long terminal repeat and adjacent cellular sequences. *Proc. Natl. Acad. Sci. USA* 77:3307–3311, 1980.

Witte, O.N.; Goff, S.; Rosenberg, N.; and Baltimore, D. A transformation-defective mutant of murine leukemia virus lacks protein kinase activity. *Proc. Natl. Acad. Sci. USA* 77:4993–4997, 1980.

Wong-Staal, F.; Dalla Favera, R.; Gelmann, E.P.; Manzari, V.; Szala, S.; Josephs, S.F.; and Gallo, R.C. The *v-sis* transforming gene of simian sarcoma virus is a new *onc* gene of primate origin. *Nature 294*:273–278, 1981.

RECOMMENDED READING

Bader, M.P. Reproduction of RNA tumor viruses. *In: Comprehensive Virology* (H. Fraenkel Conrat and R.R. Wagner, eds.), Plenum Press, New York, 1975, Vol. 4, pp. 253–332.

Beemon, K.L. Oligonucleotide fingerprinting with RNA tumor virus RNA. *Curr. Top. Microbiol. Immunol. 79*:73–110, 1978.

Bishop, J.M. Retroviruses. *Annu. Rev. Biochem. 47*:35–88, 1978.

Fan, H. Expression of RNA tumor viruses at translation and transcription levels. *Curr. Top. Microbiol. Immunol. 79*:1–42, 1978.

Friis, R.T. Temperature-sensitive mutants of avian RNA tumor viruses: a review. *Curr. Top. Microbiol. Immunol. 79*:261–309, 1978.

Gardner, M.B. Type C viruses of wild mice: characterization and natural history of amphotropic, ecotropic and xenotropic MuLV. *Curr. Top. Microbiol. Immunol. 79*:215–259, 1978.

Haase, A.T. The slow infection caused by Visna virus. *Curr. Top. Microbiol. Immunol. 72*:101–156, 1975.

Hanafusa, H. Cell transformation by RNA tumor viruses. *In: Comprehensive Virology* (H. Fraenkel-Conrat and R.R. Wagner, eds.), Plenum Press, New York, 1977, Vol. 9.

Hayman, M.J. Transforming proteins of avian retroviruses. *J. Gen. Virol. 52*:1–14, 1981.

Hill, M., and Hillova, J. Genetic transformation of animal cells with viral DNA of RNA tumor viruses. *Adv. Cancer Res. 23*:237–297, 1976.

Hooks, J.J., and Gibbs, C.J., Jr. The foamy viruses. *Bact. Rev. 39*:169–185, 1975.

Moscovici, C. Leukemic transformation with avian myeloblastosis virus: Present status. *Curr. Top. Microbiol. Immunol. 71*:79–103, 1975.

Purchase, H.G., and Witter, R.L. The reticuloendotheliosis virus. *Curr. Top. Microbiol. Immunol. 71*:104–124, 1975.

Steeves, R., and Lilly, F. Interactions between host and viral genomes in mouse leukemia. *Annu. Rev. Genet. 11*:277–296, 1977.

Ter Meulen, R., and Hall, W.W. Slow virus infections of the nervous system: virological, immunological and pathogenic considerations. *J. Gen. Virol. 41*:1–25, 1981.

Vogt, P.K. Genetics of RNA tumor viruses. *In: Comprehensive Virology* (H. Fraenkel-Conrat and R.R. Wagner, eds.), Plenum Press, New York, 1977, Vol. 9, pp. 341–456.

Vogt, P.K., and Hu, S.S.F. The genetic structure of RNA tumor viruses. *Annu. Rev. Genet. 11*:203–238, 1977.

Wang, L.-H. The gene order of avian RNA tumor viruses derived from biochemical analyses of deletion mutants and viral recombinants. *Annu. Rev. Microbiol. 32*:561–592, 1978.

Weinberg, R.A. Structure of the intermediates leading to the integrated provirus. *Biochim. Biophys. Acta 473*:39–55, 1977.

E. UNCLASSIFIED VIRUSES

22. MARBURG AND EBOLA VIRUSES

The names given to these viruses were derived from the locations where an epidemic developed. Marburg virus was isolated in 1967 from monkeys that were shipped to the German city, Marburg, and from attendants of these monkeys who developed the disease or handled tissues obtained from monkeys. The Ebola virus is named after a river in the African State Zaire where an epidemic in humans was noted in 1976. This disease resembled the hemorrhagic fever found in patients with Marburg virus.

VIRUS MORPHOLOGY AND ANTIGENICITY

Electron microscopy revealed that these two viruses resemble each other morphologically. The virions are elongated or filamentous, sometimes coiled or branched, but in spite of the morphological similarity, Marburg and Ebola viruses differ antigenically. Antibodies to one virus do not neutralize the other. The classification of these viruses has not been fully determined.

The disease

Marburg

The disease in humans takes the form of hemorrhagic fever and was first recognized in Marburg (Germany) during an epidemic in 1967. A similar disease occurred simultaneously in Yugoslavia. In both countries, African green vervet monkeys, captured in the same district in Uganda, had been imported at about the same time. Out of 25 laboratory workers who developed the disease, 7 died, and 6 more contacts who developed the same disease recovered. Those who developed the disease in Marburg were laboratory personnel who had been in contact with tissues and blood of the monkeys that were found to harbor the virus.

In Uganda, in the district from which the monkeys had come, no virus-infected monkeys were found and no cases in humans were described. However, in 1975, two tourists in South Africa fell ill with the symptoms of Marburg. One of them, an Australian male, died of the disease, but his female companion recovered. A nurse who attended the sick woman contracted the disease and, after recovery, developed uveitis and anterior chamber infection with the virus. No further infections occurred.

Ebola

Two epidemics in Africa occurred close to the end of 1976: one in the north of Zaire and the other in the south of Sudan. During the epidemic in Zaire, 237 people fell ill with Ebola virus and 211 died. In Sudan, 300 people developed the disease and 151 died. In one hospital in Sudan, a patient with Ebola virus infected 76 members of the hospital personnel, of whom 41 died. Liver damage was noted in patients. There is also fever with a sharp onset, nausea, vomiting, and chest pains; a rash on the body appears between the fifth and seventh days of the disease.

The natural host of the virus

The virus reservoir in nature is the rat *Mastomys natalensis* [which is also the host of Lassa fever virus, a member of the Arenaviridae (chapter 16)].

Laboratory diagnosis

Because of the high pathogenicity of these viruses, research must be performed in special security laboratories from which the virus cannot escape; infection of personnel is prevented by specially designed outfits (resembling space suits) that preclude direct contact with the virus.

The virus can be isolated by injecting the patient's blood into guinea pigs, causing them to develop the fever, or in Vero cell cultures inoculated with blood. Due to its unique morphology, the virus can be diagnosed by electron microscopy.

International organization to combat the disease

WHO has organized a team of physicians who are constantly on call and can be flown to any country in the world that requires help after the discovery of Marburg or Ebola virus diseases. This team is equipped with plasma from Marburg and Ebola patients who recovered and developed neutralizing antibodies. Passive immunization of Marburg patients leads to a fast recovery. Human interferon can also be used for treatment of patients.

RECOMMENDED READING

Pattyn, S.R. (ed.) *Ebola Virus Haemorrhagic Fever.* Elsevier/North Holland and Biomedical Press, Amsterdam and New York, 1978.

23. SLOW VIRUS INFECTIONS OF THE CNS

DISEASES AFFECTING HUMANS

Kuru

Creutzfeldt-Jakob (C-J) disease (also called transmissible virus dementia [TVD])

After the initial discovery that human kuru could be transmitted to chimpanzees after an incubation period of 18–24 months, it was found that many species of primates developed both kuru and C-J disease after extraordinarily long, silent, asymptomatic incubation periods—lasting several years in some cases (Gibbs and Gajdusek 1973). These agents demonstrate unusual resistance to various chemical and physical agents that include formaldehyde, proteases, nucleases, heat (80°C) and are incompletely inactivated at 100°C. These properties separate them from all other microorganisms. In addition, the lack of infectious nucleic acids or antigenicity, and therefore the absence of antibody production or cytopathogenic effects in infected cells in vitro and no interferon production, further indicate the unusual nature of these agents (Gajdusek 1977). In sucrose gradients, these agents aggregate and are difficult to isolate. Kuru is regarded as a subviral agent. Cell-fusing activity has been demonstrated with brain suspensions from patients with C-J (Moreau-Dubois et al. 1979).

DISEASES AFFECTING ANIMALS

Scrapie virus in sheep and goats and transmissible mink encephalopathy (TME)

These agents were not isolated in pure form and their properties have not yet been fully established. It is thought that TME developed in mink as a result of their eating meat from sheep infected with scrapie. Treatment of hamster brain material containing scrapie agent with 1% sodium dodecyl sulfate inactivates the infectivity of the agent.

Visna virus of the Maedi/Visna group that causes brain infections in sheep belongs to the retrovirus group (chapter 21).

MANIFESTATIONS OF THE DISEASE

Kuru

The disease is manifested as a spongiform encephalopathy in humans and over 80% of cases were found in the Fore tribe in Papua, situated in the Eastern Highlands of New Guinea. This tribe came into contact with the first government patrol into North Fore in 1947. It was discovered that members of the tribe (mainly women and children) suffered from a neurological disorder. The wide and rapid dissemination of the disease throughout the population was attributed to the practice of cannibalism as a rite of mourning and respect for dead kinsmen, since it included feeding on the half-baked brain of the dead man and smearing the body with brain material. These practices resulted in conjunctival, nasal, and skin contamination with highly infectious brain tissue, mostly among women and children. Since total surveillance of kuru began in 1957, there has been a decline in the number of deaths from the disease, due to cessation of the practice of ritual cannibalism. (Gajdusek and Alpers 1972; Gajdusek 1977).

In-depth studies showed that the affected brain contained a virus-like agent that could be transmitted to chimpanzees by brain inoculation after incubation periods of 14–39 months. In one case, the incubation period in an inoculated monkey was 8.5 years. The kuru agent was also transmitted to monkeys by the intravenous route. The disease in monkeys is characterized by weakness and paralysis, followed by signs of ataxia (loss of balance), leading to death within 1 year. It is possible to infect New World monkeys (Capuchin, Woolly, Spider, and Squirrel) with the kuru agent by inoculating infected brain material. However, consumption of brain material by the monkeys did not lead to the development of the disease. The virus can be transmitted to mink and ferret.

Creutzfeldt-Jakob (C-J)

This disease is a subacute type of spongiform virus encephalopathy in humans, sporadically appearing in men and women ages 35 to 65. The disease begins with dementia, followed by paralysis, myoclonic seizures, and death within 6–24 months (Roos et al. 1973). Monkeys can be infected via the brain by inoculation of human brain material after autopsy, but not with material from internal organs. The agent was accidentally transmitted to two patients during surgery due to the use of silver electrodes previously implanted into the brain of a patient with C-J, even though the electrodes had been treated with formalin, 70% ethanol, and formaldehyde vapor (Gajdusek 1977). Thus formalin does not affect the agent, and it has to be destroyed on electrodes or surgical instruments by heating to 100°C or by treating with alkali or phenol.

Scrapie in sheep

The agent affects the CNS and causes behavioral changes in the affected animal. The animal shows signs of restlessness, lack of coordination, and dies within a few months. Skin itching causes the animal to scratch incessantly, hence the name *scrapie*.

The disease-causing agent is present in the CNS and can be transmitted by injection of infected brain material into a healthy sheep intracerebrally and also by the intraperitoneal route. Sheep can also be infected by eating the placentas of infected animals. Mice and hamsters are sensitive to scrapie agent and are used in laboratories for titration of the agent, but the disease takes 10–12 months to develop.

The scrapie agent can be inactivated by treatment with alkali (above pH 11), acid (below pH 2), hyperchlorite, 6M urea, 90% phenol, ether, boiling to 100°C for 30 min, or treatment with 1% sodium dodecyl sulfate. The agent was found to exist in an aggregated form. Studies by Prusiner and associates (1977, 1980) indicate that the scrapie agent is a hydrophobic protein moiety, probably with no nucleic acid. The monomeric form of the agent has a sedimentation coefficient of ≤ 40 S; it is smaller than any known animal virus or viroid.

Transmission mink encephalopathy (TME)

This disease was discovered in the 1950s in the United States. The incubation period for the disease is six months. The agent is not transmitted by contact, but minks feeding on meat from a scrapie-infected animal develop the disease. The disease appears in adult mink and not in the young. The agent can be transmitted to experimental animals like the golden hamster, racoon, goat, and Rhesus and squirrel monkeys. In monkeys, the incubation period is 22 months, whereas in the racoon it is 6 months.

The agents causing TME and scrapie are probably one and the same.

Transmission of the agents

Infection by these agents occurs mainly as a result of eating infected meat or brain material. C-J in humans may be due to the consumption of scrapie-infected mutton. Transmission of C-J to a patient via a corneal transplant has also been reported. After a neurosurgeon and two physicians became infected, the hazards of occupational infection due to exposure to infected human brain tissue during surgery or at postmortem examination were recognized (Gajdusek 1977; Brooks et al. 1979).

Pathology

C-J, scrapie, kuru, and TME all share identical pathology. One of the remarkable features of all the spongiform encephalopathies is the absence of any of the hallmarks of a virus infection: absence of cells in the spinal fluid, inflammation in the brain, or inclusion bodies in CNS cells.

BIBLIOGRAPHY

Brooks, B.R.; Jubelt, B.; Swarz, J.R.; and Johnson, R.T. Slow viral infections. *Annu. Rev. Neurosci.* *2*:309–340, 1979.

Gajdusek, D.C. Unconventional viruses and the origin and disappearance of kuru. *Science* *197*:943–960, 1977.

Gajdusek, D.C., and Alpers, M. Genetic studies in relation to kuru. I. Cultural, historical and demographic background. *Am. J. Hum. Genet.* *24,* suppl:S1–S38, 1972.

Gibbs, C.J., Jr., and Gajdusek, D.C. Experimental subacute spongiform virus encephalopathies in primates and other laboratory animals. *Science* *182*:67–68, 1973.

Moreau-Dubois, M.-C.; Brown, P.; and Gajdusek, D.C. Comparison of cell-fusing activity of brain suspensions from patients with Creutzfeldt-Jakob disease and other degenerative neurological diseases. *Proc. Natl. Acad. Sci. USA* *76*:5365–5367, 1979.

Prusiner, S.B.; Groth, D.F.; Bildstein, C.; Masiarz, F.R.; McKinley, M.P.; and Cochran, S.P. Electrophoretic properties of the scrapie agent in agarose gels. *Proc. Natl. Acad. Sci. USA* 77:2984–2988, 1980.

Prusiner, S.B.; Hadlow, W.J.; Eklund, C.M.; and Race, R.E. Sedimentation properties of the scrapie agent. *Proc. Natl. Acad. Sci. USA* *74*:4656–4660, 1977.

Roos, R.; Gajdusek, D.C.; and Gibbs, C.J., Jr. The clinical characteristics of transmissible Creutzfeldt-Jakob disease. *Brain* *96*:1–20, 1973.

RECOMMENDED READING

Carp, R.I.; Warner, H.B.; and Merz, G.S. Viral etiology of multiple sclerosis. *Prog. Med. Virol.* *24*:158–177, 1978.

Fuccillo, D.A.; Kurent, J.E.; and Sever, J.L. Slow virus diseases. *Annu. Rev. Microbiol.* *28*:231–264, 1974.

Gibbs, C.J., Jr., and Gajdusek, D.C. Atypical viruses as the cause of sporadic, epidemic, and familial chronic disease in man: slow viruses and human diseases. *Perspect. Virol.* *10*:161–194, 1978.

Prusiner, S.B., and Hadlow, W.J. (eds.) *Slow Transmissible Diseases of the Nervous System.* Academic Press, New York, 1979, Vols. 1 and 2.

Ter Meulen, V., and Katz, M. (eds.). *Slow Virus Infections of the Central Nervous System.* Springer-Verlag, Basel, 1977.

III. MEDICAL AND BIOLOGICAL CONSIDERATIONS

24. IMMUNIZATION AGAINST VIRUS DISEASES

For hundreds of years, man was powerless in the face of virus epidemics. It is understandable that one of the goals of science and medicine is to develop the means to protect against, prevent, or cure virus diseases. However, only during the eighteenth century was it noted that individuals who fell ill during an epidemic and recovered were not affected again by the same disease. In the year 1721, variolation against smallpox was initiated. Fluid from skin vesicles of patients suffering from smallpox was inoculated into the skin of a healthy person in an attempt to perform immunization. The natural infection that resulted was milder, but some people did develop a more severe case of smallpox, and some eventually died from the disease. Thus variolation was not safe, and a breakthrough occurred only when the English physician Edward Jenner (1798) noted that exposure of a person to cowpox leads to immunization against smallpox. His so-called vaccination procedure—namely, transfer of the cowpox virus to the broken skin of healthy individuals, led to safe immunization against smallpox. The same procedure has been used ever since to immunize people against smallpox and has led to the eventual eradication of the disease (chapter 5). The cowpoxvirus vaccine was the first live virus vaccine used for immunization of man.

VIRUSES ARE FOREIGN ANTIGENS THAT STIMULATE THE IMMUNE SYSTEM

The virions that invade the body of a human or an animal attach to receptors on the surface of cells, and after the initial replicative cycle in sensitive cells, the virion progeny is released into the blood stream. White blood cells migrate to the site of infection as part of the response of the immune system. Macrophages and lymphocytes of the T type (lymphocytes that mature during passage through the thymus) are the cells that encounter the virions. After attachment of the virions to the T cell membrane, they are presented to the B lymphocytes which mature in the bone marrow and are the antibody producing cells. A B lymphocyte that produces antibodies against a particular viral antigen, starts to replicate and produces a clone of cells, all of which produce the antiviral specific antibody. If the virions contain more than one viral antigen, a number of cell clones are produced, each producing a specific antibody against a specific viral antigen. The proliferation of the B cell clones is controlled by specific subclasses of T cells which are named *suppressor cells* and *killer cells*. These cells are capable of destroying unwanted lymphocytes.

MATURE T LYMPHOCYTES ARE REQUIRED FOR THE PRODUCTION OF ANTIVIRAL ANTIBODIES

Mature T cells are able to process the viral antigens and present them to the antibody synthesizing B cells. In the absence of mature T cells in the circulation, as in children with the genetic disorder ataxia telangiectasia (AT) who are born with a hypotrophic thymus, antibodies to viral antigens cannot be synthesized. AT-affected children cannot be immunized with viral vaccines.

The immune response

The combined response of mature T and B cells to a virus infection, or to live or killed virus vaccines, leads to the appearance in the bloodstream of antiviral antibodies—namely, gamma globulin of the IgM (macroglobulin) class. The synthesis of the IgM class of antibodies reaches a maximum within a few days and then declines. By an unknown mechanism, the antibody-producing cells switch over to the synthesis of gamma globulin of the IgG class and this continues for a period of a few months and then subsides. In addition, local antibodies of the IgA and IgE classes appear at the site of virus infection. These antibodies are important for the development of local immunity, especially in viral infections of the respiratory tract.

The secondary immune response depends on memory cells

Infection or immunization leads to a primary immune response that subsides after a few months, leaving behind memory cells. The secondary response to the same virus occurs quickly, and within a few days a high titer of specific virus neutralizing antibodies is reached. The ability to rapidly produce IgG antibodies when reinfection with the same virus occurs explains lifelong immunity.

Passive immunization with virus specific antibodies also protects against disease

In the face of a potentially fatal virus infection (chapter 22) or one that may cause paralysis (chapter 18), vaccination is not feasible since the primary immune response is too slow. One of the ways to combat the virus infection is by the passive administration of specific antibodies that are capable of neutralizing the infectious virus in the bloodstream. Such antiviral antibodies can be obtained by isolating the IgG fraction from the blood of people who recovered from the same virus infection. For example, Ebola and Marburg viruses are highly pathogenic, and for this reason WHO has a serum bank with sera from patients who have survived these infections. A team of physicians is on call to be dispatched to any country in which Ebola or Marburg may be diagnosed. Patients are treated by passive administration of antibodies. Passive immunization may also be useful in patients who develop poliomyelitis. Introduction of specific antibodies into the bloodstream at the viremic stage helps to neutralize the virus, decreases the severity of the disease, and enhances the chances for recovery.

Antibodies have a role in recovery and prevention of virus diseases

Specific antiviral antibodies interact with the antigens present on the surface of virions or infected cells, and this allows complement molecules to interact with the antigen-antibody complex, leading to the dissolution of the virions or the infected cells. Virions to which antibodies have attached can adsorb to cell membranes and are incorporated into the cells, but uncoating of the viral nucleic acid does not occur. Local antibodies in the immunized individuals limit the spread of the virus in the tissues and support the healing process in the body. In addition to antibodies, cell-mediated immunity is also involved in the prevention of virus spread.

The importance of such defense mechanisms in localizing the virus infection can be seen in individuals who lack these types of immune response, either due to genetic defects, such as agammaglobulinemia, or due to treatment with cytotoxic drugs, like those used in the treatment of cancer. In such patients, a local skin infection with herpes simplex virus can spread unchecked, causing damage to large skin areas.

MILESTONES IN THE DEVELOPMENT OF VIRAL VACCINES

Table 11 presents the time scale for the development of viral and bacterial vaccines. It is of interest that 87 years elapsed between the introduction of the smallpox vaccine by Edward Jenner and the introduction of rabies vaccine by Louis Pasteur. From 1885 to 1927, most of the attention was paid to recently discovered bacteria, and only subsequently (1927–1949) did progress in research on virus diseases of man and animals lead to the development of new viral vaccines. In 1937, the live vaccine against yellow fever (17D strain) was developed by Max Theiler, and an inactivated virus vaccine against influenza was produced. The breakthrough achieved by Enders, Weller and Robbins in 1949 with the cultivation of poliovirus in skin cells under in vitro conditions (chapter 1) led to large-scale cultivation of poliovirus in newly developed cell cultures, and this eventually led to the production of the Salk inactivated poliovirus vaccine (1954) and Sabin's live poliovirus vaccine (1957). Subse-

Table 11. Milestones in the development of virus vaccines

Year	Virus diseases	Bacterial diseases
1721	Variolation against smallpox	
1798	Jenner's vaccine against smallpox	
1885	Pasteur's vaccine against rabies	
1892		Hafkin's cholera vaccine
1913		Behring's diphtheria toxin/toxoid
1921		Calmette and Guerin's BCG vaccine against tuberculosis
1923		Diphtheria toxoid by Ramon and Glenny
1923		Madsen's vaccine against pertussis
1927		Ramon and Zaller's toxoid of tetanus
1937	Influenza inactivated vaccine; live virus vaccine (17D) against yellow fever by Theiler	
1949	Measles vaccine (inactivated)	
1954	Inactivated poliovirus vaccine (Salk)	
1957	Live attenuated poliovirus vaccine (Sabin)	
1960	Measles virus live vaccine (Enders)	
1962	Rubella virus live vaccine (Weller)	
1968		Meningococcus type C vaccine
1971		Meningococcus type A vaccine
1976	Hepatitis B (inactive) vaccine	
1978	Inactive rabies virus vaccine (grown in human diploid cells)	

(From "World Health", February–March 1977, p. 381. Reprinted by permission of World Health Organization, Geneva, Switzerland.)

quently, new vaccines were introduced to protect against measles and rubella, and in 1976 a vaccine against hepatitis B virus was produced.

In spite of these major accomplishments in the development of antiviral vaccines, certain virus diseases such as the common cold and infections of the CNS still prevail, and no effective drugs are available for treating them.

Vaccination against poliomyelitis: achievements and problems

As shown in table 11, five years elapsed from the time that Enders, Weller and Robbins demonstrated that poliovirus can replicate in skin cells under in vitro conditions to the time that inactivated poliovirus vaccine was produced on a large scale by J. Salk and his associates. This unique achievement should be evaluated according to what was known in the field of virology at that time. The three strains of poliovirus were grown on a large scale in cell cultures prepared from monkey kidneys and inactivated with formaldehyde to destroy virus infectivity without affecting viral antigenicity. Testing the safety of the vaccine was a crucial step, and this was done by inoculating the inactivated virus vaccine into the brains of monkeys. In spite of this, batches of vaccine containing some live virus were used for immunization, and as a result, children became infected with polio. Another problem was the presence of the monkey virus, SV40, in the vaccine preparation. Follow-up of children immunized with SV40-containing virus vaccine did not show that they suffered any untoward effects. The number of persons contracting polio markedly decreased within three years after the introduction of the Salk vaccine.

The poliovirus mutants developed by A. Sabin were isolated from infected cultures kept in an incubator in which the temperature accidentally rose to 41°C. The temperature-resistant poliovirus mutants were found to have lost their virulence after extensive passage in animals. Sabin's poliovirus mutants were administered to children during a poliomyelitis epidemic in Mexico and caused the immediate cessation of the epidemic. The vaccinated children developed antibodies, and the viruses excreted in the feces were found to be of the vaccine type only; no mutation occurred with the virus excreted by the immunized children. The results of the vaccination project were accepted by the USSR health authorities who decided to immunize five million Russian children with Sabin's live virus vaccine. The results of this immunization campaign showed that the vaccine was safe, easy to produce, and easy to administer. In addition, virus excreted from the immunized children was spread in the sewage, with the result that the wild type virus was replaced by the vaccine virus, and more people became immunized due to contact with the vaccine virus. Thus the attenuated virus strains were able to replace the wild type poliovirus in the gastrointestinal tract. Shortly after the Russian vaccination campaign, Sabin's live attenuated poliovirus vaccine was accepted in most countries of the world. Until recently, Sweden was the only country that continued to use the Salk inactivated virus vaccine to immunize children, but that too has now been replaced by the Sabin vaccine.

Immunization of humans with Sabin's vaccine is the method of choice

The widespread use of Sabin's vaccine and continuous monitoring of the properties of the poliovirus strains isolated from individuals all over the world has proved the safety and efficacy of this vaccine. The relatively low cost and the ease of administration of the vaccine have led to its use in developing countries as well. The use of Salk's killed poliovirus vaccines requires a number of injections, and it is also more expensive than Sabin's vaccine, which makes it less practical for use in developing countries.

Since live attenuated poliovirus vaccines are continuously being administered throughout the world, polioviruses excreted in feces must be constantly monitored by the health authorities. It became necessary to develop laboratory tests to distinguish between vaccine and wild type strains of poliovirus in order to know whether the vaccine strains may have changed in virulence as a result of infection of children and adults. Analysis of the capsid polypeptides of polioviruses, using electrophoresis in polyacrylamide gels, revealed differences between vaccine and wild type virus strains.

The use of Sabin's live polio vaccine is a very important accomplishment which has benefited almost every individual on earth, but health authorities responsible for national vaccination programs must be aware that, in some instances, a small number of individuals should be advised against vaccination. These include people with inherited genetic defects in the immune system and those who were not immunized during childhood for religious or other reasons.

Awareness of the dangers of immunization against poliomyelitis

The use of Sabin's vaccine started during the period when poliomyelitis epidemics were commonplace. In the United States, small-scale epidemics continued to occur even after the start of vaccination with the Salk inactivated vaccine. From 1955 to 1967, the incidence of poliomyelitis was reduced by 98.7% due to vaccination. Whereas 76,000 cases of poliomyelitis were reported in the United States, Canada, Australia, New Zealand, USSR, and 23 European countries in 1955, only 1,013 were reported in 1967.

The appearance of paralytic polio in 89 vaccinated children and in immunized adults in a vaccinated population of 370 million around the world led to the appointment of a committee by WHO in 1969 to investigate the nature of the disease in these individuals. This committee found that during a five-year period, 360 cases of paralytic poliomyelitis had been recorded in the participating countries. Of these cases, 155 were due to infection with wild type virus, and the remaining 205 cases were defined as possibly associated with the live poliomyelitis vaccines. This group of patients was subdivided into children who developed poliomyelitis after administration of live poliomyelitis vaccine and adults who came in contact with immunized children. The first subgroup of immunized children under five years of age showed a high association with type 3 poliomyelitis vaccine strain. In the second subgroup, the infectious virus was mostly type 2 poliovirus.

The calculated incidence of poliomyelitis in different countries in recipients of the live attenuated poliovirus vaccine per million doses administered ranged from 0.087 to 2.288. Contact cases and possible contact cases ranged from 0.135 per million doses up to 0.645 in most countries, except in two high incidence countries which reported 3.305 and 8.046 cases per million doses. These results clearly show the variance in the incidence of poliomyelitis in different countries. According to J.L. Melnick, the reason for the high incidence of poliomyelitis in recipients and contacts mentioned previously was the low quality of the live virus vaccine. The situation changed with the introduction of an improved vaccine, and the number of paralytic contact cases decreased.

The development of poliomyelitis in immunized and contact cases does not immediately imply that the vaccine virus strains are the cause of the disease. It is necessary to use specific criteria to define the disease and the virus that caused it. The following criteria have been recommended to identify the vaccine virus:

1. disease starts 4 to 40 days after oral vaccination
2. paralysis starts not before six days after vaccination
3. significant residual lower motor neuron paralysis
4. proper laboratory diagnosis of the infecting poliovirus strain
5. lack of damage to the upper motor neurons

Analysis of the children and adult contacts who developed paralytic poliomyelitis revealed a connection between a low immune status of the individual and the development of the disease. A child with hypogammaglobulinemia and an eight-month-old child born with an abnormal thymus (see earlier discussion) developed paralytic poliomyelitis after vaccination. Among the contacts, nonimmunized individuals developed the disease. Since the live attenuated poliovirus vaccine is now being administered to children in the absence of poliomyelitis in epidemic form, it is of great importance to protect those individuals who can be harmed by the live vaccine.

Families afflicted by poliomyelitis due to vaccination or contact have claimed indemnity from the vaccine-producing companies. U.S. courts have ruled that the vaccine producers are obligated to alert the population to the fact that there is a minimal danger connected with live attenuated poliovirus vaccination. American poliovirus vaccine producers enclose a leaflet with the vaccine providing basic information on the incidence of paralytic poliomyelitis cases per million administered doses. Unfortunately, not all poliovirus-vaccine producers use this practice. In the United States, informed consent of the parents is required prior to the immunization of a child.

It seems desirable for individuals known to suffer from immunological deficiencies or immune-compromised individuals (like cancer patients), as well as people not immunized in childhood, to be protected by vaccination with the inactivated Salk vaccine. Those who prefer the Salk vaccine should have the right to be immunized with it.

Immunization against influenza

Attempts started in 1937 to prepare influenza virus vaccines are still continuing since it is difficult to predict the antigenic changes in the influenza virus strains that will cause epidemics or pandemics in the future. The current influenza virus vaccines contain inactivated virus strains that are prevalent, and whenever a new strain of influenza virus appears, the vaccine of the following year will include the new viral antigen.

An infection with influenza virus is characterized by elevated fever, muscular pains, and congestion in the respiratory tract. The virus can be dangerous to the elderly and to those suffering from cardiopulmonary ailments. Most fatal influenza virus infections occur in old people and young children. Since the virus spreads fast by droplet infection, new mutants can easily spread through the world and cause new epidemics or a pandemic. In 1973–1974, which was not considered an epidemic year, there were about 61 million cases of influenza-like disease in the United States alone, although probably only a fraction of these was actually caused by influenza virus. Lack of accurate diagnosis results in influenza virus being implicated in many respiratory infections actually caused by other viruses or bacteria.

Influenza virus vaccines

The two antigens—hemagglutinin (HA) and neuraminidase (NA)—are responsible for the production of neutralizing antibodies in people recovering from influenza. The killed virus vaccines contain these two viral antigens from a number of influenza virus strains. The vaccine viruses are cultivated in the allantoic cavity of embryonated eggs, the virus multiplies in the allantoic membrane cells, and the harvested allantoic fluid contains the virus progeny which are then inactivated with formalin. Antigenic drift causes the appearance of new virus strains, and these must be included in the influenza virus vaccines (see chapter 15). New virus strains isolated from man that do not grow well in embryonated eggs are combined with another strain of influenza virus adapted to growth in chick embryos. From the resulting recombinant virus, a new strain is selected that contains the HA and NA antigens of the new virus. Thus a recombinant virus, well adapted to growth in embryonated eggs, is obtained, and the virus yield is high. This new recombination technique has made the production of a vaccine from a new virus strain a faster process than in the past.

Attempts have been made to prepare split virus vaccines containing only purified HA and NA antigens, but such vaccines are not yet available for general use.

The search for live attenuated vaccines is still in progress, using known virus strains after continuous propagation in cultured cells or embryonated eggs. Unfortunately, the genes controlling the virulence of influenza virus are not yet known, and selection of avirulent virus strains is hampered by the lack of criteria for virus attentuation.

Analysis of the viral HA and NA genes by cloning in bacterial plasmids could in the future lead to the large-scale production of polypeptides that elicit antibodies capable of neutralizing the HA or NA antigens of the influenza virions. Recently, the hemagglutinin gene of influenza virus was cloned in a bacterial plasmid; the nucleotide sequence was determined and compared with the amino acid sequence of the HA polypeptide (chapter 15). This procedure could lead to the cloning of the HA gene of all new influenza virus strains in bacteria, and, when the need arises, these bacteria could be used for producing viral antigens.

Problems encountered in the production of influenza virus vaccines

WHO influenza reference laboratories and central laboratories for the surveillance of influenza virus are responsible for detecting new mutant virus strains (see chapter 29). To prevent a pandemic of influenza, it is necessary to produce a vaccine that will include the new virus antigens. The problem is the time required for the production of such a vaccine. The following is an example of the time that was required for the production of the Hong Kong strain influenza vaccine in the 1968–1969 epidemic:

1. In July 1968, an influenza virus isolated from a patient was diagnosed as a new strain (H_3N_2).
2. Thirty days elapsed from the initial isolation until identification of the new virus strain was confirmed and the virus was passed on to the vaccine producers.
3. A recombinant producing a high titer in embryonated eggs took 50 days to produce following the isolation of the new virus strain.
4. One million doses of vaccine were prepared after 120 days had elapsed from the isolation of the virus. Each vaccine dose contained 400 CCA (chick cell agglutinating units).
5. After 140 days, five million doses had been prepared. At this stage, the virus had reached epidemic form.
6. By 165 days, there were 15 million doses.
7. By 180 days, there were 20 million doses.

By this time, there was enough vaccine to immunize 10% of the population in the United States, but the major wave of the Hong Kong influenza epidemic had subsided, and only a small amount of the vaccine had been used.

Swine influenza vaccine: immunization in the United States and its shortcomings

During the past 60 years, a new influenza virus mutant has appeared roughly at 10-year intervals, each time causing a new influenza pandemic. After the last pandemic caused by the H_3N_2 type of influenza virus in 1968, it was postulated that a new pandemic should be expected around 1978–1979. Previous attempts to prepare an influenza vaccine to immunize people in the United States against Asian influenza (caused by the H_2N_2 mutant in 1957) and Hong Kong influenza (caused by the H_3N_2 mutant in 1968) had failed, due to the slow production of the vaccine, and large quantities of vaccine were left unused. It was calculated in 1960 that about 20,000 people died of influenza during the years 1957–1958, and in the pandemic of 1968 about 60,000 out of an affected 51 million people succumbed to what was diagnosed as influenza. The cost of medical care during this epidemic was estimated to be 750 million dollars in the United

States, and the indirect economic damage was calculated to be 3.9 billion dollars.

These calculations led to the conclusion that the period between the 1968 epidemic and the following one (then projected for around 1978) should be used to organize a national program in the United States to prevent such an epidemic. The only problem was how to predict the nature of the mutation in the virus that would lead to the emergence of the new pandemic virus strain. Unfortunately, with all the available knowledge, such a prediction is still not possible.

When a new influenza virus strain related to swine influenza virus caused the death of a soldier at Fort Dix in the United States, it was concluded that this virus would develop into the new pandemic strain. The swine influenza virus strain isolated from persons with influenza was similar in its antigenic properties to the influenza virus strain that caused the 1918 pandemic at the end of the First World War. During February and March 1976, further information on the virus was collected, and it was found that this virus had appeared sporadically throughout the United States.

Evaluation of the possible alternatives led the U.S. health authorities to conclude that a national immunization program was necessary. On April 15, 1976, President Gerald Ford signed an allocation of 135 million dollars for the production of the new vaccine and for the vaccination program. This was necessary since the vaccine producers demanded a federal government guarantee to purchase all the vaccine produced and to cover damage claims from individuals adversely affected by the vaccine. Extraneous proteins in the vaccine can cause the Guillain-Barre syndrome, which is characterized by paralysis.

After the production of the vaccine and the implementation of the vaccination program, it turned out that the swine influenza remained sporadic and did not reach epidemic form. No epidemic occurred during the years 1978 and 1979, and there was a public outcry against unnecessary large-scale immunization programs. Several of the immunized individuals developed the Guillain-Barre syndrome. Thus the original question of how to predict the antigenic change that will occur due to influenza virus mutation or recombination is still unanswered.

Measles virus vaccine

Live attenuated measles virus vaccine was introduced in the United States in 1963, and during the next 13 years, 80 million children were immunized. The vaccination program caused the number of measles cases to decrease from half a million to 35,000 per year. The number of children contracting measles encephalitis also markedly decreased. Measles virus infections were noted only in unvaccinated children or in children vaccinated during the years 1962–1968 with a killed virus vaccine who did not develop immunity. The infectious diseases committee of the Academy of Pediatrics in 1977 proposed improvements in the use of the measles live, attenuated vaccine. In areas free of measles, children should not be immunized prior to the age of 15 months. However, in measles areas, children should be immunized at 6 months of age and then reimmunized at 15 months. The same applies to the combined measles-mumps live attenuated vaccine and attenuated rubella vaccine as to the monovalent measles vaccine. Pediatricians should decide on the best immunization method for each child in their care. There is no negative effect on the children after revaccination.

An investigation in England of children who were immunized at the age of 10 months to 2 years with a live attenuated measles vaccine revealed that the children were still protected against measles at the time of the investigation (12 years). Measles in immunized children (when it occurs) is less severe than in unimmunized children.

Mumps virus vaccine

A vaccine containing attenuated mumps virus was introduced as part of a trivalent vaccine against measles, mumps and rubella.

Rubella virus vaccine

Rubella virus causes a mild disease that usually ends without complications, except in the case of pregnant women. The Australian physician, Norman Gregg, was the first to note, in 1941, that mentally retarded or deaf children had been born to mothers who had had rubella virus during the first trimester of pregnancy. The rubella virus was isolated in cultured cells in 1961, and subsequent passage of the virus in green monkey kidney cells, dog kidney, and duck embryo cells led to the production of attenuated rubella virus strains (HPV-77 strain in the United States and Cendehill 51 in Belgium, by passage in rabbit kidney cells). These two strains were used for the vaccine that was marketed in 1969. The attenuated virus is included in the triple live vaccine (measles, mumps, rubella). The vaccine is administered to 15-month-old babies, and at 12 years of age to nonvaccinated girls.

Rabies virus vaccine

Studies by H. Koprowski and his associates in the United States have shown that rabies virus can be propagated in human diploid cells. The strain PM/WI 38 1503 from the Wistar Institute is used for the preparation of killed virus vaccine. The virus grown in diploid cells is inactivated by formalin treatment. Three or four injections intradermally lead to the development of an immune state in humans. Injection of 0.1 ml of the vaccine on the day someone is bitten, and on the third, fourth, and seventh days, leads to a high antibody titer. Today the vaccines contain killed purified rabies virus. The area of the bite by a rabid animal must be cleaned carefully with soap or detergent and water or alcohol (40–70%), iodine tincture or ammonium salts (0.1%). The medical treatment includes injection of antiserum into the bite and immunization as discussed earlier. Booster injections of the vaccine are given after 30 and 90 days.

Adenovirus vaccine

These vaccines were found to contain SV40, which resists the inactivation procedure used for adenovirus and therefore cannot be used.

Hepatitis B virus vaccine

The vaccine is still experimental and contains the hepatitis B surface antigen produced from plasma of carriers of hepatitis B virus. The vaccine was found to be effective in monkeys in protecting them against hepatitis B infection.

RECOMMENDED READING

Hilleman, M.R. Vaccines for the decade of the 1980s and beyond. *SA Med. J. 61*:691–698, 1982.
Kańtoch, M. Markers and vaccines. *Adv. Virus Res. 22*:259–325, 1978.
Krugman, S. Measles immunization, new recommendations. *JAMA 237*:366, 1977.
Melnick, J.L. Viral vaccines. *Prog. Med. Virol. 23*:158–195, 1977.
Nakano, J.H.; Hatch, M.Y.; Thieme, M.L.; and Nottay, B. Parameters for differentiating vaccine-derived and wild poliovirus strains. *Progr. Med. Virol. 24*:78–206, 1978.

25. VIRUSES AS SELECTIVE FORCES IN NATURE: EPIDEMICS

Since animal viruses are intracellular obligate parasites of vertebrates and invertebrates, in order to evolve in different hosts they had to maintain a balance between virulence in the host and the ability of the host to survive. When a virus is transmitted from one animal to another of the same species, the virulence remains unchanged, but when the same virus is transmitted from its natural host to a new species, the virulence may increase markedly due to the upset in ecological balance. Also when a virus is introduced into the same species in a new geographical location, an epidemic may ensue. All virus families display such changes in virulence. With the spread of colonization, new viruses infected people for the first time.

INTRODUCTION OF A NEW VIRUS INTO AN UNIMMUNIZED POPULATION

Destruction of the Aztec and Inca civilizations by smallpox

Invasion of Mexico by the Spaniards in the year 1519 led to the introduction of smallpox into a continent that had been completely free of this disease. The Spanish army came from a European country where smallpox was endemic, and some soldiers who had been infected before the voyage developed the illness when they reached Mexico. Most of the Spanish soldiers had acquired immunity due to natural exposure to the disease in Spain, but Mexican natives had had no experience of the disease, and the smallpox virus spread rapidly and caused many deaths. This was thought to be one of the reasons for the surrender of the Aztecs to the Spanish conquerors. A similar situation is thought to have developed during the conquest of Peru which led to the decline of the Inca civilization in the wake of the Spanish invasion (Fenner 1977).

Yellow fever in southern United States and Mexico

In 1648, a yellow fever epidemic was identified in Yucatan, Mexico. The virus, which was transmitted by the mosquito *Aedes aegypti,* was imported from Africa. For over 200 years, yellow fever caused epidemics in America; in 1905, New Orleans and other areas in the southern United States registered 5,000 yellow fever cases, with a 20% mortality rate.

Measles in Eskimos

Measles virus was introduced by white people coming into contact with Eskimos who had never encountered the virus before. A devastating epidemic with a high mortality rate ensued.

Measles in the court of Louis XIV of France

An epidemic of measles in 1712 resulted in the death of two heirs to the throne of France and thus changed the course of French history (Koprowski 1959).

Influenza pandemics

During the last 400 years, about 30 pandemics have been described. An outbreak in 1743 was as virulent as the 1918 pandemic. A pandemic in 1782 in Europe and Asia was also highly virulent.

Rabies

Rabid dogs were described by Democritus (500 B.C.) and Aristotle (322 B.C.). In Western Europe, rabies was recognized among wolves during the thirteenth century. Transmission of the disease by inoculation of saliva of a rabid dog to a healthy dog was achieved by Zinke in 1804. Galtier (1879) used domestic rabbits for the diagnosis of rabies. L. Pasteur and his associates in 1884 were the first to modify the pathogenicity of the virus by a series of intracerebral passages of the virus in a different species of host (the rabbit) (Johnson 1965).

Herpesvirus B of monkeys causes a fatal disease when transmitted to humans

This virus is latent in monkeys and does not cause any known disease except when the monkey that carries the virus in the saliva bites a human being. The B virus migrates to the brain of the bitten person, replicates, and causes fatal encephalitis. This is an example of the change in virulence when a virus encounters a new host.

Rift Valley fever virus

This virus, which belongs to the Bunyaviridae (chapter 17), infects sheep and is transferred by insects. The virus originated in Kenya and afterward appeared in Uganda, in southern Africa, and in the Sudan. It is now spreading in Egypt and causing encephalitis in people coming into contact with infected sheep and mosquitoes.

Marburg and Ebola viruses

These viruses, which appear in sporadic epidemics in Africa, reside in rats, and humans are accidentally infected. In a hospital in the Sudan, where a patient with hemorrhagic fever was hospitalized, medical staff members were infected by Ebola virus, and one person died of the disease (see chapter 22).

African swine fever virus (ASFV) in pigs

The virus, which infects wild boar in Central Africa (Coggins 1974), was introduced into the Iberian peninsula with devastating results among domestic pigs, indicating the high virulence of ASFV. The virus strains currently being isolated from the swine populations in Spain and Portugal appear to be less virulent, as compared to virus strains isolated in the past. However, the spread of ASFV to swine populations that have never been in contact with the virus (e.g., in Cuba and Malta) has led to the development of a highly virulent strain with rapid spread of the disease.

CHANGES IN THE VIRUS EXISTING IN THE POPULATION

Influenza viruses are constantly changing their antigenicity, and the slight changes are termed *antigenic drift* (see chapter 15). Changes in the amino acid sequence of the polypeptide chains of the hemagglutinin occur during antigenic drift (Webster et al. 1980). The virus that caused the 1918 influenza epidemic had the antigenic composition H_0N_1 (H – hemagglutinin and N – neuraminidase). The 1947 influenza epidemic was caused by a new virus mutant in which the mutation was in the hemagglutinin namely H_1N_1. In 1957, the mutation was in both antigens and the virus was designated H_2N_2.

A more drastic mutation called *antigenic shift* may occur by reintroduction of influenza viruses into a susceptible population or by recombination between a human influenza virus and a virus from a different host, like horse influenza or duck influenza. These viruses are excreted in the feces and can be found in water sources.

SLOW VIRUSES

The agents of Kuru and Creutzfeldt-Jakob (C-J)

These agents, which are regarded as slow viruses, affect humans and damage the CNS. The agent of Kuru was isolated by C. Gajdusek by injecting human brain samples from Kuru patients into monkeys' brains. An incubation period of several years duration passed before the typical disease symptoms appeared. Transmission of Kuru was traced to cannibalistic rituals in which brain material from a person who died of Kuru was consumed and spread on the skin by the women of the tribe (see chapter 23). The agent of C-J disease was transmit-

Table 12. Transfer of type C retroviruses from one animal species to another

Donor	Recipient	Genetic transfer in the recipient
Old World monkeys	Felis (ancestor of the domestic cat)	Yes
Mouse ancestor	Pig ancestor	Yes
Rat ancestor	Felis	Yes, but also horizontal transmission to the cat (Felis catus)
Rodent	Primates	No

(After Todaro et al. 1976.)

ted to a normal person by electrodes that were used in the brain of a patient with the disease. The electrodes had been sterilized with formaldehyde, which does not kill the agent (see chapter 23).

It is thought that mink encephalopathy developed in the mink after they fed on meat from a sheep that died of scrapie, another slow virus disease.

VIRUSES CAPABLE OF INTRODUCING THEIR GENES INTO THE CHROMOSOMAL DNA OF THE HOST

Retroviruses (chapter 21)

These viruses are remarkably well adapted to their hosts. They reside in the chromosomal DNA of the host cell. The viral genetic information is transmitted vertically with the germ cells to the next generation. The sarcoma viruses in birds, mice, cats, and monkeys are transferred vertically. All these species contain at least one virogene. The virus can also be transferred from host to host by horizontal transmission (table 12).

BIBLIOGRAPHY

Coggins, L. African swine fever virus. Pathogenesis. *Progr. Med. Virol. 18*:48–63, 1974.

Fenner, F. The eradication of smallpox. *Progr. Med. Virol. 23*:1–21, 1977.

Johnson, H.N. Rabies virus. In: *Viral and Rickettsial Infections of Man*. 4th ed. (F.L. Horsfall and I. Tamm, eds.), J.B. Lippincott Co., Philadelphia, Montreal. pp. 814–840, 1965.

Koprowski, H. Viruses 1959. *Trans. NY Acad. Sci. 22*:176–190, 1959.

Todaro, G.J.; Benveniste, R.E.; and Sherr, C.J. Interspecies transfer of RNA tumor virus genes: implications for the search for "human" type C viruses. *In: Animal Virology* (D. Baltimore, A.S. Huang, C.F. Fox, eds.), Academic Press, New York, San Francisco, London, 1976, pp. 369–384.

Webster, R.G.; Laver, W.G.; Air, G.M.; Ward, C.; Gerhard, W.; and van Wyke, K.L. The mechanism of antigenic drift in influenza viruses: analysis of Hong Kong (H_3N_2) variants with monoclonal antibodies to the hemagglutinin molecule. *Ann. NY Acad. Sci. 354*:142–161, 1980.

26. ANTIVIRAL DRUGS AND CHEMOTHERAPY OF VIRAL DISEASES OF MAN

THE SEARCH FOR ANTIVIRAL AGENTS

The discovery of antibacterial agents, such as penicillin and streptomycin, led to the expectation that among the fermentation products of molds, substances with antiviral properties would be found. The search for naturally occurring antiviral agents revealed that many compounds are able to inhibit virus replication in cultured cells in vitro. Unfortunately, almost all of these antiviral substances were found to be unsuitable for treatment of virus diseases in animals and man. Several chemically synthesized compounds were found to be effective in inhibiting virus replication in mammalian cells in vitro and a few of these are currently in use for treatment of virus infections in man.

Many of the available antiviral drugs have been used in furthering our understanding of the molecular processes involved in virus replication. However, the ability of most antiviral compounds to interfere with the molecular processes of the host cell has been a major obstacle in the development of drugs suitable for treating virus diseases. The ideal antiviral drug would interfere only with a virus-coded process and would not affect uninfected cells.

Antiviral drugs can be divided into six categories: (1) antivirals that interfere with cellular processes required by the virus for its replication; (2) antivirals that selectively bind to virus-coded enzymes and thus inhibit their function; (3) antivirals that bind to the virus nucleic acid and inhibit its expression; (4) antivirals that prevent the processing of viral precursor polypeptides; (5) antivirals that interfere in the assembly of virus particles; and (6) antivirals that modify the proteins on the surface of the viral envelope and therefore prevent the virus from infecting new cells.

Two major approaches were used in the research and development of antiviral drugs. First, synthetic and natural antivirals are selected according to their ability to interact with virus-coded enzymes in infected cells, but these drugs can also affect normal cells. The second approach consists of producing antivirals in the form of proantiviral drugs, or prodrugs, that can only be activated by a virus-coded enzyme in the infected cells. The prodrug form of the antiviral agent does not affect normal cells in the patient.

The first approach meant a systematic testing of all available natural and synthetic compounds for their antiviral activity against different viruses grown in cell cultures. The structure of a number of antiviral compounds is presented in figure 76. A list of natural substances with antiviral properties is presented in table 13 and the modes of action of natural and synthetic antivirals against herpesviruses are presented in table 14. Such compounds were tested not only for the inhibition of virus replication but also for their ability to inhibit a virus-specified enzyme under in vitro conditions. The value of an antiviral drug depends on its therapeutic index—namely, a high antiviral activity and low toxicity. The therapeutic index of any compound can be tested in virus-infected animals. If certain compounds show promising antiviral activity, it is possible to synthesize a series of derivatives for further screening of antiviral activity.

In the second approach, a number of modified nucleotides have been chemically synthesized and tested in the search for the ultimate proantiviral drug that would be activated only in virus-infected cells, preferably by a specific virus-coded enzyme. Thus only in the infected cell would the prodrug interfere with virus-coded molecular processes and inhibit virus replication without affecting the metabolism of uninfected cells and tissues. The antiherpes virus drug acycloguanosine (acyclovir; figure 76) has the properties of a prodrug (table 15).

VIRUS-CODED ENZYMES AS TARGETS FOR THE DEVELOPMENT OF ANTIVIRAL DRUGS

Among the viral enzymes, the viral DNA polymerase has received much of the attention of those trying to develop synthetic and natural antivirals. A number of sites on the DNA polymerase of herpes simplex virus could serve as targets for antiviral drugs (figure 77):

1. The gamma-phosphate binding site can be interfered with by phosphonoacetic acid (PAA) or phosphonoformate (PFA).
2. The binding site for the sugar moiety of the nucleoside triphosphate can be interfered with by acycloguanosine (which has a modified sugar moiety).
3. The nucleotide binding site can be interfered with by the binding of a modified nucleotide.

4. Inhibitors that bind to, or intercalate into, the DNA template (e.g., distamycin, ethidium bromide) can interfere with the DNA template binding site on the enzyme.
5. The metalloenzyme (e.g., Zn^{++}) binding site can be interfered with by adding compounds that remove the metal ion from its position on the enzyme molecule.

Various substances (table 15) have the ability to interfere with the enzyme binding sites and new substances can be designed. Unfortunately, when inhibitors of the DNA polymerase were studied, it became evident that the virus has the ability to mutate in the presence of an inhibitor and to give rise to drug-resistant mutants. In the future, it will be necessary to use two antiviral drugs for the treatment and cure of virus infections in man.

In table 15, a number of enzymes coded for by different viruses are listed, including the DNA-dependent RNA polymerase responsible for the synthesis of the viral messenger RNA and the DNA polymerases that serve as possible targets for antiviral substances.

MODE OF ACTION OF NATURAL ANTIVIRALS (TABLE 13)

Natural substances with antiviral activity are fermentation products of molds: Distamycin and netropsin inhibit the transcription of viral messenger RNA and the synthesis of the viral DNA, thus inhibiting virus replication. Novobiocin binds to a virus-coded enzyme and is capable of inhibiting viral DNA synthesis. Rifampicin prevents the formation of the viral envelope in poxvirus-infected cells.

Distamycin, the structure of which is shown in figure 76, acts by binding to the viral DNA, thus becoming an obstacle to the attachment of the viral DNA and RNA polymerases to the DNA template. Distamycin binds to A-T rich sequences on the viral DNA, inhibiting virus replication, but removal of the drug from the infected cells results in a reversal of inhibition.

Figure 76 illustrates the structure of rifampicin, which is a semisynthetic drug. Addition of rifampicin to poxvirus-infected cultures at any time during the virus growth cycle inhibits virus replication. It was found that rifampicin inhibits cleavage of a polypeptide inserted into the viral envelope, and as a result, the viral envelope cannot become functional. Thus the synthesis of infectious virions is completely inhibited.

Unfortunately, these antiviral substances are not selective for virus processes and have toxic effects on uninfected cells. Therefore, although they are very useful for studies on the molecular processes of poxvirus replication, they cannot be used as effective antiviral drugs for the treatment of infected persons. The significance of smallpox has by now markedly declined because of the worldwide vaccination campaign of WHO, especially in the endemic areas of India and Africa. However, herpesviruses are a very important group of

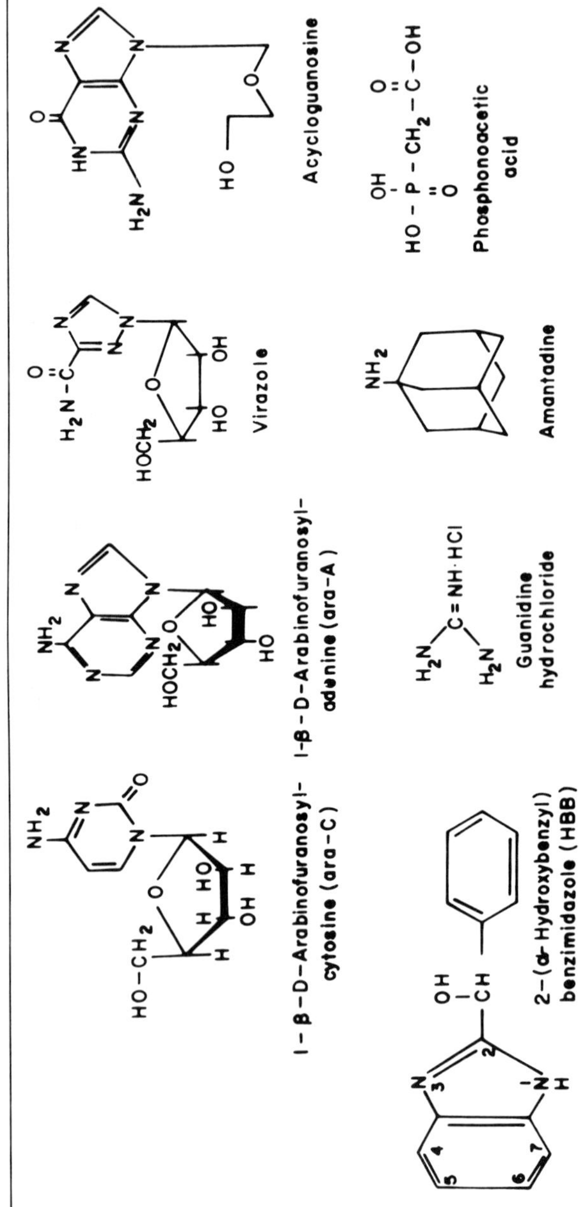
1-β-D-Arabinofuranosyl-cytosine (ara-C)
1-β-D-Arabinofuranosyl-adenine (ara-A)
Virazole
Acycloguanosine
2-(α-Hydroxybenzyl) benzimidazole (HBB)
Guanidine hydrochloride
Amantadine
Phosphonoacetic acid

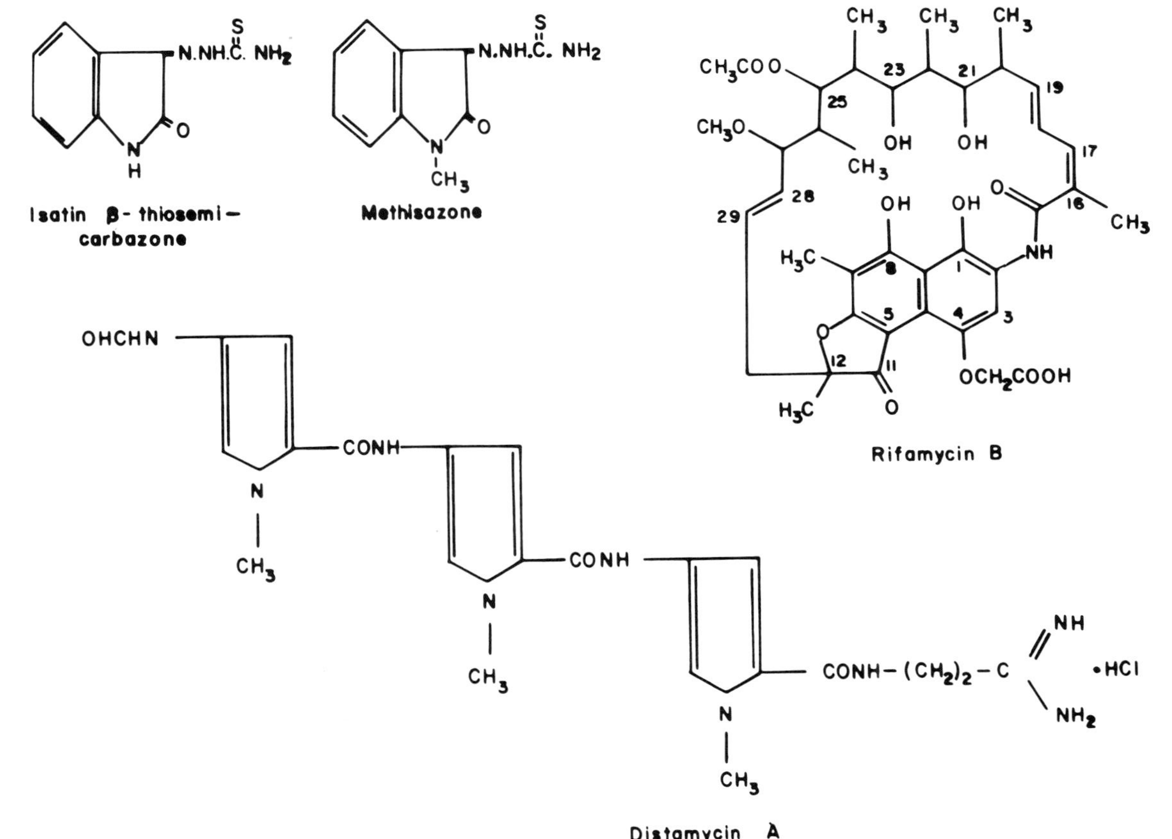

Figure 76. Antiviral drugs

Table 13. Natural substances inhibiting animal viruses

Antiviral agent	Virus group affected (-viridae)	Specific viruses	Mode of action (inhibition of)
Actinomycin D	Rhabdo	rabies	Virus replication dependent on cellular DNA
Adriamycin (see daunomycin)			
α-Amanitin	Herpes	HSV	Cellular RNA polymerase II
Amphotericin B	Rhabdo	VSV	Infectivity due to interaction with sterols in virus membrane causing morphological changes
	Toga	Sindbis	
	Herpes	HSV	
	Pox	vaccinia	
	Paramyxo	NDV	
Aphidicolin	Herpes	HSV	Cellular DNA synthesis and virus replication
	Pox	vaccinia	
Aranotin (see gliotoxin)			
Bacterial extracts	Herpes	HSV	Virus replication
	Adeno		
Bleomycin	Pox	vaccinia	Virus replication
cAMP	Adeno		Virus replication
	Herpes	HSV type 2	
		EBV	Virus induction
Caffeine	Retro		Virus replication
Camptothecin	Pox	vaccinia	DNA and RNA synthesis
	Herpes	HSV	
	Adeno		
	Papova	SV40	
Chromomycin (see olivomycin)			
Concanavalin A	Herpes	HSV	Virus detachment from host cell membrane
	Rhabdo	VSV	
	Orthomyxo	influenza	

Congocidine (see netropsin)			
Cordycepin	Adeno		Poly(A) attachment to viral mRNA
	Herpes	HSV	mRNA synthesis
	Pox	vaccinia	mRNA synthesis
	Paramyxo	NDV	Protein synthesis?
		Sendai	(accumulation of 50 S RNA)
	Picorna	polio	RNA synthesis
		rhinoviruses	RNA synthesis
	Toga	WEE	RNA synthesis
	Retro	MSV	Transformation induced by IUdR
Cyanidanol		hepatitis B	
Cytochalasin B	Herpes	HSV	Synthesis of virus envelope glycoproteins
Daunomycin, adriamycin	Retro	RLV, AMV, MuLV	Reverse transcriptase
2-Deoxy-D-glucose (and/or glucosamine)	Herpes	HSV	Glycolipid synthesis
	Paramyxo	NDV	
	Orthomyxo	Influenza	
		Fowl plague	
	Toga	Sindbis	
		Semliki	
		Jap. B. enceph.	
	Retro	ASV	
Distamycin A	Adeno		Virus replication by binding to DNA
	Pox	vaccinia	
		Shope fibroma	
	Herpes	EBV	
		HSV	Viral DNA polymerase by binding to DNA template
	Retro	ASV, MSV	Reverse transcriptase
		Friend leukemia	
Erythromycin A	Retro	RSV	Reverse transcriptase
Fagaronine	Retro		Reverse transcriptase
Filipin	Rhabdo	VSV	Infectivity by interaction with virus envelope

Table 13 (continued)

Antiviral agent	Virus group affected (viridae)	Specific viruses	Mode of action (inhibition of)
Gliotoxin, aranotin, and related compounds	Picorna	polio	RNA synthesis
		Coxsackie A21	
		rhinoviruses	
	Myxo	influenza	RNA polymerase
		parainfluenza	
Human complement	Arena	LCM	Virus replication
	Paramyxo	NDV	
	Rhabdo	VSV	
Human and cow's milk	Paramyxo	NDV	Virus replication
	Orthomyxo	Influenza B	
	Picorna	Rhinoviruses	
Litmomycin	Retro	AMV	Reverse transcriptase
		MuLV	
Long chain alcohols	Herpes	HSV	(affects lipid membranes)
Marine algae	Herpes	HSV	Virus replication
	Pox	vaccinia	
	Rhabdo	VSV	
Mithramycin	Myxo	influenza	Host cell RNA polymerase II
	Herpes	pseudorabies	
Netropsin, Congocidin	Pox	vaccinia	(like Distamycin A)
		Shope fibroma	
Neuramin	Paramyxo	NDV	Neuraminidase
Olivomycin and chromomycin	Retro	RMLV	(Similar to actinomycin D)
Plant flavonoids	Herpes	HSV	Virus replication
		H. suis	
	Paramyxo	Parainfluenza-3	
	Toga	Sindbis	
	Retro	RSV	Reverse transcriptase

Pokeweed antiviral protein	Picorna	Polio	Elongation of proteins
	Orthomyxo	Influenza	
Rifampicin	Pox	vaccinia	Virus maturation; induction of viral nucleotide phosphorylase, RNA polymerase and two virus coded DNases.
	Retro	MuLV	Reverse transcriptase
		MSV	
		ASV	
Sinefungin	Paramyxo	NDV	Virus mRNA (Guanine -7-) methyltransferase (mRNA methylation prevented)
	Pox	vaccinia	
Streptovaricins	Pox	vaccinia	Early mRNA synthesis
	Retro	MuLV	Reverse transcriptase
Sulfated polysaccharides	Picorna	EMC	Virus adsorption to cell membrane
		Echo 6	
	Orthomyxo	Influenza	
	Toga	Dengue	
	Rhabdo	Rabies	
	Herpes	HSV	
Toyocamycin	Adeno		mRNA metabolism due to prevention of polyadenylation and methylation of internal adenosine residues
Tunicamycin	Rhabdo	VSV	Glycosylation and virus replication
	Retro	ASV	
	Toga	Sindbis	Glycosylation and virus replication
		Semliki	
	Orthomyxo	Influenza	Glycosylation and virus replication
	Paramyxo	NDV	Glucosamine incorporation
Zinc	Herpes	HSV	Viral DNA polymerase
	Picorna	Rhinoviruses	Cleavage of virus precursor polypeptide
		polio	
		EMC	
		FMDV	

Becker, 1980. Reprinted by permission from *Pharmacol. Therap.* 10, Table 6, p. 149. Copyright 1980 by Pergamon Press Inc., New York.

Table 14. Action of antiviral agents in the replication of herpesviruses in permissive cells[a]

Function	Inhibitor	Mode of action
A. *Virion attachment and penetration*		
1. Electrostatic binding	Heparin	Prevention of virus attachment
	p-Fluorophenylalanine	
2. Membrane interaction: Membrane fusion and virus interaction with receptor	Nonionic surfactants	Damage to lipid-containing virus membrane
	Unsaturated monoglycerides and alcohols	
B. *Early transcription*		
1. Viral DNA-protein complexes	Distamycin A	Binding to sequences in minor groove of viral DNA; cellular DNA also affected
	Congocidin	
	Netropsin	
	Camptothecin	Fragmentation of viral DNA
2. Interaction of cellular RNA polymerase II with DNA	α-Amanitin	Binding to cellular RNA polymerase II and inhibition of transcription products
	Amantadine	
	Cycloheximide	
	Actinomycin D	
3. Transcription of viral DNA; polymerization of polyribonucleotide chains (precursors); polyadenylation of 3′ end	Cordycepin	Incorporated into RNA; inhibits transcription
	Cycloheximide	Only α-transcripts synthesized
4. Processing of viral HnRNA by ribonuclease (site-specific)		
5. Methylation of the viral mRNA cap at 5′ end	5′S-isobutyl-adenosine (SIBA)	Inhibits methylation of mRNA; especially that of 5′ cap
6. Transport of viral mRNA from nucleus to cytoplasm		
C. *Translation of viral mRNA*		
1. Interaction of mRNA with cellular ribosomal subunits	Interferon	Inhibition of viral mRNA translation
2. Addition of initiation factors to complex	dsRNA	Binds the initiation factor
	Interferon and dsRNA	

3. Synthesis of polypeptides	Modified amino acids	Abnormal proteins synthesized
4. Processing of precursor polypeptides		
5. Synthesis of viral thymidine kinase (TK) activity	Acycloguanosine	Phosphorylated by a virus-coded TK; antiviral activity dependent on this phosphorylation; more active against viral DNA polymerase than against similar host cell enzyme
	Deoxythymidine analogs	Alternative substrates of viral TK
6. Viral DNA polymerase activity	Phosphonoacetate (PAA) Phosphonoformate (PFA)	Binds to enzyme and inhibits function; interacts at a pyrophosphate binding site on DNA polymerase
	Zinc	Binds to DNA polymerase
7. Semiconservative synthesis of viral DNA—replication intermediates—elongation of viral DNA	Virazole (Ribavirin)	Inhibitor of inosine monophosphate dehydrogenase
	Ara-A (Vidarabine)	Incorporated into DNA; rapidly deaminated to hypoxanthine nucleoside in cell; ara-ATP inhibits viral DNA polymerase
	Arabinosyl-N^6-hydroxyadenine (ara-HA)	Resistant to deamination; more active than ara-A in vivo
	Ara-C	Acts on viral and cellular DNA; deaminated to uridine arabinoside; toxic to cell
	5-Methyl-arabinosylcytosine (5-methyl ara-C)	Precursor to ara-T in cells possessing deoxycytidine deaminase only
	Ara-T	Incorporated into viral DNA
	Arabinofuranosylthymine (ara-Thd)	Inhibition of viral DNA polymerase by ara-TTP (formed in cell from ara-Thd)
	Bromodeoxycytidine	Phosphorylated by normal cells but molecule is deaminated; no deamination in infected cells
	5′-substituted pyrimidines (IUdR, BUdR, FUdR)	Incorporated into viral DNA; competitive inhibition of thymidine; FUdR blocks transcription of portion of viral genome

Table 14. (continued)

Function	Inhibitor	Mode of action
	Trifluorothymidine (F_3T)	
	5′-halo-5′-amino deoxyribonucleosides (AIU)	Inhibits TK; not phosphorylated; not incorporated into DNA in vivo; not effective for HSV-2 or CMV; affects zoster
	Azauridine (AzUrd)	AzUrd converted to AzUMP which inhibits orotidylic acid decarboxylase
	Proflavine Neutral red Acridine orange	Dye intercalates with base pairs in DNA in presence of light; fragmentation of DNA due to excision of guanine: can cause cell transformation
D. *Translation of late transcripts*		
Interaction of mRNA with ribosomes	Isoprinosine (Inosiplex)	Stimulation of cell-mediated immunity; inhibition of RNA synthesis by interaction with ribosomes
	Levamisole	

[a]References for mode of action of each antiviral drug are given in the original table.

Becker and Hadar, 1980. Reprinted by permission from *Prog. Med. Virol.* 26, Table 1, pp. 3–7.

human pathogens all over the world, and effective antiviral agents against these virus infections are paramount.

Antivirals against herpesviruses

Table 16 shows a number of antiviral substances developed during the last 15 years in attempts to find drugs that would not only cure herpesvirus infections in the eye, but also herpesvirus infections in the skin, the brain, and other organs. Most of these antivirals were designed to inhibit the viral DNA polymerase, with the idea that this would prevent virus replication and result in the cure of the virus disease.

Iododeoxyuridine and trifluorothymidine

Halogenated pyrimidines, such as iododeoxyuridine (IUdR) (Prusoff 1959, 1967; Kaufman 1965) and trifluorothymidine (F_3T) (Kaufman and Heidelberger 1964) represent effective antiherpes drugs. Incorporation of IUdR into the viral genome causes breaks in the DNA template and prevents DNA replication by the viral DNA polymerase (table 14). This drug is also incorporated into cellular DNA, making it too toxic to be used systemically, and it is both mutagenic and teratogenic (Prusoff et al. 1965).

Zinc

Zinc ions are also able to bind to the viral DNA polymerase, which is a zinc metallo-enzyme, and thus prevent its activity.

Ara-A (adenine arabinoside, 9-β-D-arabinofuranosyladenine)

This compound is converted to a triphosphate form by cellular kinases and is reported to inhibit the viral DNA polymerase preferentially. However, the different values reported for the differential effect of Ara-A triphosphate on the viral and cellular DNA polymerases would appear to be dependent on virus strain, enzyme purity, and other experimental conditions. Ara-A is incorporated into both cellular and viral DNA. It is also rapidly deaminated in vivo by adenosine deaminase to arabinosylhypoxanthine that has a much lower antiviral activity (reviewed by North and Cohen 1979).

Phosphonoacetate (PAA) and phosphonoformate (PFA)

These drugs bind to the same site on the enzyme as the terminal γ-phosphate of the nucleoside triphosphate. As a result, the DNA polymerase cannot cleave the pyrophosphate from the nucleoside triphosphate, and the activity of the enzyme is inhibited.

PAA and PFA (table 14) were found to interfere with the phosphate binding site of HSV DNA polymerase and are very efficient antivirals for this virus, both in vitro and in vivo. The difficulty with these substances is the rapid development of resistant virus mutants from the population of wild type virus. The mutation is in the DNA polymerase, and the resistant virus is no longer

Table 15. Virus-coded enzymes as possible targets for antiviral drugs

Virus	Viral enzyme	Antiviral agent	Mode of action
DNA viruses			
Poxviruses	DNA-dependent RNA-polymerase	Distamycin, Netropsin	Bind to DNA template and inhibit viral enzyme
	DNA polymerase	Distamycin, Netropsin	Bind to DNA template and inhibit viral DNA polymerase
	Gyrase (topoisomerase)	Novobiocin	Binds to enzyme and inhibits DNA synthesis
		Rifamycin	Inhibits cleavage of precursor peptide, abnormal envelope formed
Iridovirus	DNA polymerase	Phosphonoacetate	Binds to DNA polymerase and inhibits DNA synthesis
Herpes simplex	DNA polymerase	Distamycin, Netropsin	Bind to viral DNA, prevent DNA synthesis
		F_3T	Modification of template
		Zn^{++}	Binds to enzyme
		Phosphonoacetate, Phosphonoformate	Bind to enzyme site for γ-phosphate of nucleoside triphosphate

		Acycloguanosine (acyclovir)	*Prodrug;* phosphorylated by viral thymidine kinase to form triphosphate and inhibits viral DNA polymerase. TK^- virus is resistant to drug
	Gyrase (as above)		
RNA viruses			
Influenza	RNA polymerase	Actinomycin D Amantadine	Inhibition of RNA polymerase initiation
	Neuraminidase	FANA	Inhibition of enzyme activity
Poliovirus	RNA polymerase	HBB Guanidine	Inhibition of replication complex
	(Cellular) peptidases	Zn^{++}	Inhibition of processing of precursor peptides
Retroviruses	RNA-dependent DNA polymerase	Distamycin A Netropsin	Bind to viral DNA
		Rifamycins Streptovaricin Alkaloids	Bind to viral enzyme
		Oligothymidylate derivatives Vinyl analogs Partially thiolated poly C	Template-primer analogs

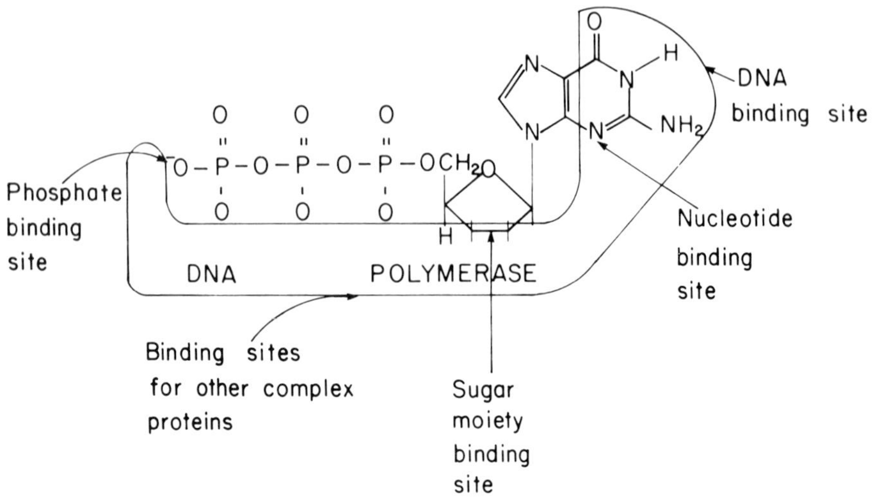

Figure 77. Scheme for possible inhibition of the DNA polymerase by antiviral compounds.

sensitive to the drug. It is possible to isolate HSV mutants that produce a PAA-resistant DNA polymerase, and in cells infected with PAA resistant mutants, the enzyme is resistant to PAA over a large range of concentrations. Phosphonoacetate is active in animal models, but its deposition in bone has delayed consideration of clinical trials (Boezi 1979). Phosphonoformate is currently being tested in humans by topical application for herpes labialis.

Acycloguanosine (acyclovir)

One of the newest antiviral substances, acycloguanosine (figure 76) has a distorted sugar moiety and closely resembles deoxyguanosine. Acycloguanosine can be phosphorylated by the HSV-coded thymidine kinase to form acycloguanosine triphosphate (the active form of the drug) only in virus-infected cells and is then able to inhibit the viral DNA polymerase (tables 14 and 15). This is the first drug that can be called an antiviral prodrug, since even though it may enter all the cells of the body, activation of the drug can occur only in cells infected with HSV. The difficulty with this drug is the emergence of resistant mutants—mainly, thymidine kinase minus (TK^-) mutants.

In summary, the antiviral substances that inhibit the herpesvirus DNA polymerase interfere with three important sites on the DNA polymerase: the metallo-enzyme site, the phosphate binding site, and the sugar-base binding site. Anti-herpesvirus drugs can be obtained by modification of the purine and pyrimidine bases or by the use of analogs that can be phosphorylated only in cells that contain the viral TK, or phosphorylated analogs that are capable of binding to a key viral enzyme.

Antivirals to RNA viruses

Table 15 describes some antiviral substances effective against RNA viruses, such as HBB and guanidine, that inhibit the poliovirus replication complex, and zinc, that inhibits the peptidases that process the viral precursor peptides. Concerning influenza virus, we know that actinomycin D and amantadine inhibit the function of the virus RNA-dependent RNA polymerase. Recently, it was demonstrated that these anticellular substances prevent the synthesis of cellular messenger RNA molecules, and these are required for the production of the dinucleotides necessary for the initiation of influenza virus RNA polymerase activity. The substance 2-deoxy-2,3-dehydro-N-trifluoroacetylneuraminic acid (FANA) was found to inhibit the viral enzyme neuraminidase, and thus prevent the virus from infecting cells.

Antivirals were also tested on the retrovirus RNA-dependent DNA polymerase (tables 13 and 15). The antiviral substances distamycin A and rifampicin are able to inhibit the RSV RNA-dependent DNA polymerase, preventing the synthesis of the viral double-stranded DNA molecules. Unfortunately, distamycin and its related substances are also toxic to uninfected cells, and therefore it has not been possible to use them in chemotherapy. Derivatives of rifampicin were also quite effective in inhibiting these viruses. Template-primer analogs (table 15) and other inhibitors of the retrovirus DNA polymerase are reviewed by Chandra and colleagues (1977).

ANTIVIRAL DRUGS IN MEDICAL USE

Antivirals used in treating patients

Idoxuridine (5-iodo-2'-deoxyuridine; IUdR)

This compound was synthesized by W. Prusoff in 1959 and was used by H. Kaufman in the early 1960s for the treatment of herpetic keratitis. This was essentially the first effective human antiviral drug. Since the drug was used for external treatment and was effective in inhibiting the replication of HSV-1 in the eye, its clinical use was permitted by the U.S. Federal Drug Agency. In the early years after the introduction of idoxuridine, the compound was highly effective, but it soon became apparent that drug resistant mutants of HSV-1 were appearing, with the result that the efficiency of this compound markedly decreased. Today other antivirals are used in cases of herpetic keratitis that do not respond to treatment with idoxuridine.

The success achieved in the treatment of herpetic keratitis with idoxuridine led to its use for the treatment of herpetic encephalitis in humans by intravenous injection. It was soon found that idoxuridine was not effective and caused brain damage in encephalitis patients who had no herpesvirus etiology.

Ara-A (Vidarabine) and Ara-C (cytosine arabinoside; Cytarabine)

The drug ara-A is currently being used for treating encephalitis patients after a diagnosis of HSV-1 has been established by virus-specific antigens found in a

Table 16. Efficacy of antiviral agents against herpesviruses[a]

Compound	In animals	In man
Acycloguanosine (Acyclovir)	Mice: highly effective against HSV-1 encephalitis; nontoxic; not effective against established latent infections; can prevent latency Rabbits: as effective in herpetic keratitis as IDU and F_3T; safe and effective systemically against encephalitis	Effective in curing generalized herpes infections in children Effective in 12 patients[b] with dendritic corneal ulcers; no adverse effects
Amantadine hydrochloride		Herpes zoster in 100 patients[b] no effect on rate of healing or appearance of new lesions
Ara-A (Vidarabine)	Mice: antiviral effect linked to host immune response; partial protection against latency (also ara-AMP); not effective against HSV-2 intravaginal infection (also ara-AMP); reduces severity of orofacial HSV-1 infections (also ara-AMP)	Effective in keratitis, iritis and corneal stromal disease; same potency and toxicity as IDU; used systemically for encephalitis or keratouveitis, does not inhibit bone marrow or host immune system to great degree; results with confirmed herpes encephalitis Toxic side effects: nausea, vomiting; ara-A breaks chromosomes in vivo; acts on host enzyme systems. Herpesvirus genitalis: refractory to treatment
Ara-AMP (ara-A 5′-monophosphate)	Rabbits: more soluble and more effective than ara-A	
Ara-C (Cytarabine)	Effective in rats	Some improvement in herpetic gingivostomatitis; varying results in clinical trials; ineffective for herpes zoster
AIU (5-iodo-5′-amino-2′5′-dedeoxyuridine)	Rabbits: varying effects on herpesvirus keratitis depending on virus strain	
F_3T (trifluorothymidine; 5-trifluoromethyl-2′-deoxyuridine)	Mice: effective in HSV-1 encephalitis Rabbits: potent therapeutic activity in cornea, more potent than IDU	As good as IDU and Ara-A for local treatment of herpetic eye infections, various degrees of toxicity

IDU (IUdR; Idoxuridine)		Effective topically in herpetic eye infections; some toxicity causing corneal epithelial damage and retarding stromal wound healing; not always effective in HSV-1 encephalitis with high percentage of treatment failures and marked toxicity of drug; drug-resistant strains
Interferon and interferon inducers	Mice: suppression of viral replication in genital tract	Favorable results with herpetic keratitis; herpes zoster in cancer patients
Isoprinosine (Inosiplex)	Hamsters: large doses of drug required for significant protection against HSV-1 encephalitis	Zoster in childhood cancer—no discernible therapeutic effects; some positive results in therapeutic treatment of HSV
Levamisole		Some improvement in herpesvirus genitalis recurrences; not considered appropriate drug for HSV treatment
Neutral red and light	Oncogenic capability of HSV-1 and HSV-2	170 episodes of recurrent HSV infections in 96 patients—no beneficial effects
Phosphonoacetate (PAA) Phosphonoformate (PFA)	Mice: effective in intravaginal HSV-2 and orofacial HSV-1 infections; cannot eradicate latent infection; irritating effect on skin Hamsters: intravaginal treatment with PAA or ara-A more effective than either ara-C or IDU Rats: PAA reduces HSV-1 in brain Rabbits: PAA as effective as IDU in herpetic keratitis; nontoxic Cebus monkeys: skin irritation	
Ribavirin (Virazole)	Mice and rabbits: effective for HSV-1 and HSV-2	

[a]See original table for references
[b]Double-blind placebo-controlled study.
Becker and Hadar, 1980. Reprinted by permission from *Prog. Med. Virol.* 26, Table 2, pp. 11–15. Copyright 1980 by S. Karger AG, Basel, Switzerland.

brain biopsy. Treatment of patients with herpetic encephalitis early in the development of the brain infection with ara-A has indicated that some patients may be cured without brain damage. Unfortunately, most encephalitis patients are treated late in the infection, with the result that they die or recover with severe brain damage.

Ara-C was effective in the treatment of herpetic keratitis in humans, but the toxicity of the drug is quite high and the drug is also immunosuppressive.

Acycloguanosine (Acyclovir)

This compound, which is a guanosine derivative, is activated by a specific viral enzyme and is very effective in inhibiting both HSV-1 and HSV-2 in infected cells. Only the HSV-specified thymidine kinase can phosphorylate acycloguanosine from the monophosphate form that is rapidly converted to the deoxynucleoside di- and triphosphate. These molecules interfere with the action of the virus-induced DNA polymerase. The conversion of acycloguanosine monophosphate to the diphosphate is catalyzed by the cellular enzyme, guanylate kinase (Miller and Miller 1980). Several cellular enzymes catalyze the subsequent conversion to the triphosphate form. Acyclovir inhibits the biosynthesis of HSV and EBV DNA (Furman et al. 1979; Datta et al. 1980). The drug was found to be effective in curing herpetic encephalitis in mice without toxic side effects and also proved useful in the treatment of herpetic keratitis in rabbits. Acycloguanosine did not achieve the desired results against latent herpesvirus infections in mice when the virus is already established in the ganglia (Field and Wildy 1981). Clinical trials are in progress to test the efficacy of acyclovir in ophthalmic, topical, and intravenous formulations. The efficacy of the drug against herpes keratitis in man has been excellent (Coster et al. 1980; Collum 1980). The drug is being tested topically and intravenously in genital HSV infections, in immuno-compromised patients, and in herpes labialis, with good results. Parental acyclovir therapy has also been effective in trials.

It should be noted that acyclovir-resistant mutants of HSV have made their appearance (Field et al. 1981), but these TK^- mutants have a low pathogenicity in adult mice.

Acycloguanosine is probably the first drug to be utilized that is activated by a viral enzyme and is therefore effective only in virus-infected cells. The more recently discovered bromovinyldeoxyuridine (De Clercq et al. 1979) is another potent antiherpes virus drug that is activated by the viral thymidine kinase.

Prophylactic drugs

Amantadine hydrochloride

This compound was found to be useful for preventing influenza A infections in humans when given prophylactically. It is given to adults in doses of 200 mg per day. It has been demonstrated that amantadine binds to the channel protein adjacent to the acetylcholine receptor on the cell surface and not to the receptor itself (Tsai et al. 1978). Studies on the mode of action of amantadine on influenza A virus in cultured cells in vitro revealed that, in the presence of the drug, the viral RNA-dependent RNA polymerase responsible for the synthesis of the viral mRNA is unable to function. It is not yet known if amantadine affects the viral enzyme or the host cell metabolism.

Because of the inefficiency of vaccine production in the face of a new influenza epidemic, it has been suggested that the use of amantadine-HCl should be permitted. This drug, which is used in the treatment of Parkinsons's disease, does, however, produce neurological side effects; therefore, there are counter-demands for proof that the drug would be effective against a new influenza A virus strain causing the epidemic. A number of other antiviral agents against influenza virus are given in table 17.

Marboran

This drug, which is N-methyl isatin-β-thiosemicarbazone, was developed by D.J. Bauer and was experimentally used in the treatment of smallpox patients and uninfected attendants in smallpox endemic areas in India (Bauer 1973). It was found that in the marboran-treated, uninfected individuals, the extent of smallpox was significantly lower than that in untreated individuals. This compound inhibits the synthesis of late viral proteins.

INTERFERON: A CELLULAR PROTEIN WITH ANTIVIRAL ACTIVITY

Isaacs and Lindenman (1957) noted that virus-infected cells secrete molecules that protect uninfected cells against infection with the same or another virus. This glycoprotein, which was named *interferon*, is produced by cells of almost all vertebrates when exposed to certain viruses but is species-specific: human interferon protects human cells, while mouse interferon protects mouse cells only. Induction of the virus-resistant state of cells by interferon is associated with the de novo synthesis of several proteins—in particular a protein kinase, an oligoisoadenylate synthetase, and a phosphodiesterase. The genes responsible for the synthesis of interferon are present in chromosomes 2 and 5. Interferon requires a specific receptor on the cell surface, in order to bind to the cell membrane. This receptor is coded by a gene in chromosome 21 in human cells (see chapter 4).

Two major species of human interferon were found: leukocyte interferon and fibroblast interferon, produced by stimulated leukocytes and fibroblasts, respectively. These two interferons differ not only antigenically, but also in their target cell specificity: both induce a virus-resistant state in human cells, but leukocyte interferon is also very active on bovine, porcine and feline cells, whereas fibroblast interferon is not. The two human interferons were purified and analyzed by electrophoresis in sodium dodecyl sulfate/polyacrylamide gels. Fibroblast interferon yielded one peak of activity, with an equivalent molecular weight of about 20,000, whereas leukocyte interferon consists of

Table 17. Efficacy of antiviral agents against influenza[a]

Compound	In animals	In man
Amantadine hydrochloride	Mice and squirrel monkeys: prophylactic and therapeutic effects Ferrets: enhanced effect by aerosol administration	Safe in children with cystic fibrosis; associated with vision loss; therapeutic effect in man less readily demonstrable and less striking than prophylactic effect; 70–100% prevention of clinical influenza A but not effective against influenza B; prophylaxis tends to decrease frequency and severity of illness; conflicting reports of field trials; difficulties of diagnosis
Rimantadine hydrochloride	Mice: mortality reduced from 80 to 10%	Better tolerated at higher doses than amantadine·HCl; not necessarily more effective than amantadine·HCl
D-Glucosamine	Mice: intraperitoneal treatment increased survival	Prevented influenza A virus shedding, seroconversion and symptoms; problems for use in man
Interferon and interferon inducers (9-MS)	Mice: protective action of 9-MS against influenza A due to interferon production	
Isoprinosine (Inosiplex)	Mice: suppression of mortality	Prophylactic plus therapeutic drug treatment only moderately effective; antibody production enhanced
Ribavirin (Virazole)	Mice: significant protection, particularly by aerosol route Ferrets: concentrations of drug effective against influenza also had immunosuppressive effects	Marginal effectiveness against influenza A and B

[a]See original table for references.
Becker and Hadar, 1980. Reprinted by permission from *Prog. Med. Virol.* 26, Table 4, pp. 26–27.

two components with apparent molecular weights of 21,000–22,000 and 15,000–18,000 (Vilcek et al. 1977; Zoon et al. 1979).

Fibroblasts synthesize a single species of interferon, but lymphoblasts synthesize both fibroblast and leukocyte interferon (Cavalieri et al. 1977). Leukocyte, lymphoblastoid and fibroblast interferon represent the type I interferons, whereas mitogen-induced immune interferon is a type II interferon. Immune interferon which has been partially characterized (Taniguchi et al. 1981) is also referred to as γ interferon, and leukocyte and fibroblast interferons, as α and β interferons, respectively.

Leukocyte and fibroblast interferons are encoded by separate mRNAs from two structured genes (Cavalieri et al. 1977), but these two interferons are structurally related. The coding sequences of the complementary DNAs of cloned human leukocyte interferon I and human fibroblast interferon show homologies of 45% at the nucleotide and 29% at the amino acid level (figure 78). Taniguchi and associates (1980) concluded on the basis of these findings that a common ancestral gene developed into the two interferon genes.

Interferon acts on cells by binding to receptors on the cell membrane

After binding to the specific receptor in the cell membrane, interferon induces a series of events that lead to the establishment of the antiviral state of the cell.

The mode of action of interferon was studied by using cell homogenates of interferon treated cells. Interferon affects the translation of mRNA into proteins by the following mechanisms:

1. A series of 2′5′-linked oligoadenylic acid triphosphate inhibitors of protein synthesis are synthesized from ATP by the enzyme oligoisoadenylate synthetase E that is activated in interferon-treated cell extracts by double-stranded RNA. The oligonucleotide pppA2′p5′A2′p5′A is the predominant product. The oligonucleotide enhances mRNA degradation in cell extracts. This was demonstrated by the isolation of RNase F, whose nucleolytic activity is completely dependent on the continued presence of pppA2′p5′A2′p5′A (Kerr and Brown 1978). The RNase F was isolated from both interferon-treated and untreated cells.
2. Interferon induces the protein kinase PK-i that is also activated by double-stranded RNA and ATP. This enzyme phosphorylates the initiation factor 2 in eukaryotic cells and decreases Met-$tRNA_f^{met}$ binding to the 40 S ribosomal subunit.
3. Interferon inhibits translation of mRNA by affecting polypeptide chain elongation, and this process, which does not require double-stranded RNA can be reversed by the addition of tRNA to the extract. An interferon-induced phosphodiesterase was shown to inhibit mRNA translation by degrading the C-C-A terminus of tRNA, and thus reducing the amino acid acceptance of tRNA in cell-free extracts (Kerr and Brown 1978; Schmidt et al. 1979; Slattery et al. 1979; Hovaressian and Wood 1980).

```
                                                                                    S1                                S10
G23CT CTA GGT TCA GAG TCA CCC ATC TCA GCA AGC CCA GAA GTA TCT GCA ATA TCT ACG ATG GCC TCG CCC TTT GCT TTA CTG ATG GTC CTG GTG GTG CTC AGC
                                                                              MET ALA SER PRO PHE ALA LEU LEU MET VAL LEU VAL VAL LEU SER

                                                                              MET THR ASN LYS CYS LEU LEU GLN ILE ALA LEU LEU LEU CYS PHE
                                                                      GTC AAC ATG ACC AAC AAG TGT CTC CTC CAA ATT GCT CTC CTG TTG TGC TTC

            S20        1                          10                                  20
TGC AAG TCA AGC TGC TCT CTG GGC TGT GAT CTC CCT GAG ACC CAC AGC CTG GAT AAC AGG AGG ACC TTG ATG CTC CTG GCA CAA ATG AGC AGA ATC TCT CCT TC
CYS LYS SER SER CYS SER LEU GLY CYS ASP LEU PRO GLU THR HIS SER LEU ASP ASN ARG ARG THR LEU MET LEU LEU ALA GLN MET SER ARG ILE SER PRO SE

SER THR THR ALA LEU SER MET SER TYR ASN LEU LEU GLY PHE LEU GLN ARG SER SER ASN PHE GLN CYS GLN LYS LEU LEU TRP GLN LEU ASN GLY ARG LEU GL
TCC ACT ACA GCT CTT TCC ATG AGC TAC AAC TTG CTT GGA TTC CTA CAA AGA AGC AGC AAT TTT CAG TGT CAG AAG CTC CTG TGG CAA TTG AAT GGG AGG CTT GA
                            1                          10                                  20

    30                          40                                  50                                  60
TCC TGT CTG ATG GAC AGA CAT GAC TTT GGA TTT CCC CAG GAG GAG TTT GAT GGC AAC CAG TTC CAG AAG GCT CCA GCC ATC TCT GTC CTC CAT GAG CTG ATC CAG
SER CYS LEU MET ASP ARG HIS ASP PHE GLY PHE PRO GLN GLU GLU PHE ASP GLY ASN GLN PHE GLN LYS ALA PRO ALA ILE SER VAL LEU HIS GLU LEU ILE GLN

TYR CYS LEU LYS ASP ARG MET ASN PHE ASP ILE PRO GLU GLU ILE LYS GLN LEU GLN GLN PHE GLN LYS GLU ASP ALA ALA LEU THR ILE TYR GLU MET LEU GLN
TAT TGC CTC AAG GAC AGG ATG AAC TTT GAC ATC CCT GAG GAG ATT AAG CAG CTG CAG CAG TTC CAG AAG GAG GAC GCC GCA TTG ACC ATC TAT GAG ATG CTC CAG
30                              40                                  50                                  60

                    70                                  80                                  90
CAG ATC TTC AAC CTC TTT ACC ACA AAA GAT TCA TCT GCT GCT TGG GAT GAG GAC CTC CTA GAC AAA TTC TGC ACC GAA CTC TAC CAG CAG CTG AAT GAC TTG
GLN ILE PHE ASN LEU PHE THR THR LYS ASP SER SER ALA ALA TRP ASP GLU ASP LEU LEU ASP LYS PHE CYS THR GLU LEU TYR GLN GLN LEU ASN ASP LEU

ASN ILE PHE ALA ILE PHE ARG GLN ASP SER SER SER THR GLY TRP ASN GLU THR ILE VAL GLU ASN LEU LEU ALA ASN VAL TYR HIS GLN ILE ASN HIS LEU
AAC ATC TTT GCT ATT TTC AGA CAA GAT TCA TCT AGC ACT GGC TGG AAT GAG ACT ATT GTT GAG AAC CTC CTG GCT AAT GTC TAT CAT CAG ATA AAC CAT CTG
                    70                                  80                                  90
```

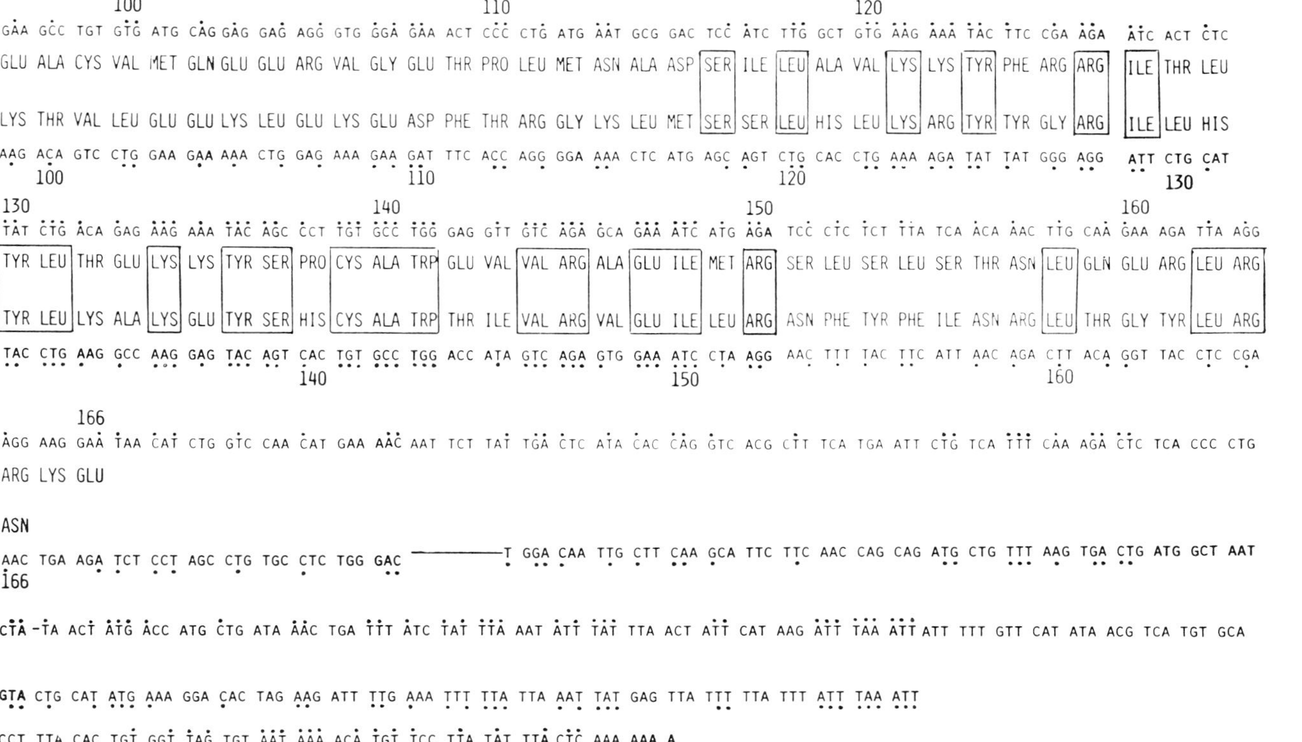

Figure 78. Comparison of the nucleotide sequences of human leukocyte interferon I (Le-IF I) and human fibroblast interferon cDNA and of the derived amino acid sequences. They were aligned to give maximal homology without introducing gaps in the coding sequence. Identical amino acids are framed; identical nucleotides are marked by a dot. S1-S23 indicate the amino acids of the putative signal sequence and 1-166 the amino acids of the interferon polypeptides.

(Taniguchi et al. 1980. Reprinted by permission from *Nature* 285, Fig. 1, p. 584. Copyright © 1980 Macmillan Journals Limited.)

Induction of interferon production in animals and man

Interferon is found in tissues and serum of patients with virus diseases. The body reacts to the virus invasion by the production of interferon as the initial defense mechanism before the production of antibodies. Nevertheless, the natural interferon is not sufficient to prevent or cure the disease, and attempts were made to use interferon inducers before or after infection in the hope that it would lead to the production of high interferon titers and to an early cure of the virus disease. Since double-stranded RNA molecules were found to induce interferon in cells, a synthetic polymer, polyriboinosinic acid:polyribocytidylic acid (poly(I:C)), which has a double-stranded conformation was devised. It is an effective interferon inducer in rodents, but not in cattle, subhuman primates, or humans. In humans it causes side effects, such as a rise in temperature and adverse effects on white blood cells. In addition, nucleases are present in the serum which degrade the synthetic polymer. H.B. Levy and his colleagues have modified the poly(I:C) polymer by attaching a poly-1-lysine molecule to the synthetic RNA and then adsorbing the complex to carrier carboxymethylcellulose. Such a complex was found to be resistant to the nucleases in the serum and effective in producing interferon in monkeys and chimpanzees. Rhesus monkeys did not exhibit serious side effects (Sammons et al. 1977).

Treatment of virus infections with exogenous interferon

Since interferon is species-specific, human interferon must be used for treating patients. K. Cantell used leukocytes removed from blood collected at the Finnish Blood Bank to produce interferon. The interferon is partially purified and has been used for experimental therapy in herpes virus, influenza virus, and hepatitis B virus chronic infections (reviewed by Dunnick and Galasso 1979). T. Merigan has demonstrated that relatively high interferon titers are needed to inhibit the appearance of hepatitis B Dane particles in the serum of the patient. In some instances, the Dane particles reappeared in the patient's serum after termination of the course of interferon treatment. Attempts are being made to produce lymphoblast interferon on a large scale for human therapy (Christofinis et al. 1981). Successful results in inhibiting the development of osteosarcoma (tumor of the bone) in humans by treatment with interferon was reported. Interferon is being used in the experimental treatment of cancer (Dunnick and Galasso 1979; Marx 1979; Krim 1980).

Cloning of human interferon genes in bacterial plasmids: a new approach to production of interferon

Isolation and purification of human leukocyte interferon on a commercial basis requires that there should be a constant supply of human leukocytes for induction with an inactivated virus. In addition, there is fear that, if no tests are done for the presence of viruses in human leukocytes, the crude interferon preparations used for treating cancer patients or patients with chronic virus diseases like infectious hepatitis may be harmful. Also, the interferon preparations used in recent years contained far more foreign proteins than interferon and had undesirable side effects on the patients. With all these reservations, it became obvious that cloning interferon DNA in bacterial plasmids may open the way for large-scale production of human interferon. Methods involving the use of high-performance liquid partition chromatography have been developed for purification of human interferon (Rubinstein et al. 1979). Monoclonal antibodies prepared in mice against human interferon are currently being used in affinity chromatography to obtain homogeneous preparations of interferon.

C. Weissmann's group isolated the 12 S fraction of poly(A)-RNA from interferon-producing human leukocytes and synthesized the complementary DNA (cDNA) of the interferon mRNA. The interferon cDNA was cloned in the plasmid pBR322 of *E. coli*, and bacterial clones were selected that carried the hybrid plasmid. One of the resulting clones had a 910-base pair insert that hybridized to interferon mRNA and was responsible for the production of a polypeptide with biological interferon activity. One gram of cells in some of the clones produced up to 10,000 units of interferon (Nagata et al. 1980). A similar procedure was used independently for the cloning of the human fibroblast interferon gene by the group of W. Fiers. The total nucleotide and amino acid sequence of human fibroblast interferon was deduced (Derynck et al. 1980).

BIBLIOGRAPHY

Bauer, D.J. Antiviral chemotherapy: the first decade. *Brit. Med. J. 3*:275–279, 1973.

Becker, Y. Antiviral agents from natural sources. *Pharmacol. Ther. 10*:119–159, 1980.

Becker, Y., and Hadar, J. Antivirals 1980—an update. *Prog. Med. Virol. 26*:1–44, 1980.

Boezi, J.A. The antiherpesvirus action of phosphonoacetate. *Pharmacol. Ther. 4*:231–243, 1979.

Cavalieri, R.L.; Havell, E.A.; Vilcek, J.; and Pestka, J. Synthesis of human interferon by *Xenopus laevis* in oocytes: two structural genes for interferons in human cells. *Proc. Natl. Acad. Sci. USA* 74:3287–3291, 1977.

Chandra, P.; Steel, L.K.; Ebener, U.; Woltersdorf, M.; Laube, H.; Kornhuber, B.; Mildner, B.; and Gotz, A. Chemical inhibitors of oncornaviral DNA polymerases: biological implications, and their mode of action. *Pharmacol. Ther. 1*:231–287, 1977.

Christofinis, G.J.; Steel, C.M.; and Finter, N.B. Interferon production by human lymphoblastoid cell lines of different origins. *J. Gen. Virol. 52*:169–171, 1981.

Collum, C.M.T.; Benedict-Smith, A.; and Hillary, I.B. Randomized double-blind trial of acyclovir and idoxuridine in dendritic corneal ulceration. *Br. J. Ophthalmol. 64*:766–769, 1980.

Coster, D.J. A comparison of acyclovir and idoxuridine as treatment for ulcerative herpetic keratitis. *Br. J. Ophthalmol. 64*:763–765, 1980.

Datta, A.K.; Colby, B.M.; Shaw, J.E.; and Pagano, J.S. Acyclovir inhibition of Epstein-Barr virus replication. *Proc. Natl. Acad. Sci. USA* 77:5163–5166, 1980.

De Clercq, E.; Descamps, J.; De Somer, P.; Barr, P.J.; Jones, A.S.; and Walker, R.T. (E)-5-(2-bromovinyl)-2′-deoxyuridine: a potent and selective antiherpes agent. *Proc. Natl. Acad. Sci. USA 76*:2947–2951, 1979.

Derynck, R.; Content, J.; De Clercq, E.; Volckaert, G.; Tavernier, J.; Devos, R.; and Fiers, W. Isolation and structure of a human fibroblast interferon gene. *Nature 285*:536–542, 1980.

Dunnick, J.K., and Galasso, G.J. Clinical trials with exogenous interferon: summary of a meeting. *J. Infect. Dis. 139*:109–123, 1979.

Field, H.; McMillan, A.; and Darby, G. The sensitivity of acyclovir-resistant mutants of herpes simplex virus to other antiviral drugs. *J. Infect. Dis. 143*:281–285, 1981.

Field, H.J., and Wildy, P. Recurrent herpes simplex: the outlook for systemic antiviral agents. *Br. Med. J. 282*:1817–1908, 1981.

Furman, P.A.; St. Clair, M.H.; Fyfe, J.A.; Rideout, J.L.; Keller, P.M.; and Elion, G.B. Inhibition of HSV-induced DNA polymerase activity and viral DNA replication by 9-(2-hydroxyethoxymethyl)-guanine and its triphosphate. *J. Virol. 32*:72–77, 1979.

Hovanessian, A.G., and Wood, J.N. Anticellular and antiviral effect of pppA(2′p5′A)$_n$. *Virology 101*:81–90, 1980.

Isaacs, A., and Lindenmann, J. Virus interference. I. The interferon. *Proc. R. Soc. Lond. [Biol.] 147*:258–267, 1957.

Kaufman, H.E. Problems in virus chemotherapy. *Progr. Med. Virol.* 7:116–159, 1965.

Kaufman, H.E., and Heidelberger, C. Therapeutic antiviral action of 5-trifluoromethyl-2′-deoxyuridine. *Science 145*:585–586, 1964.

Kerr, I.M., and Brown, R.E. pppA2′p5′A2′p5′A: an inhibitor of protein synthesis, synthesized with an enzyme fraction from interferon-treated cells. *Proc. Natl. Acad. Sci. USA 75*:256–260, 1978.

Krim, M. Towards tumor therapy with interferons, part II. Interferons: in vivo effects. *Blood 55*:875–884, 1980.

Marx, J.L. Interferon (1): on the threshold of clinical application. *Science 204*:1183–1186, 1979.

Miller, W.H., and Miller, R.L. Phosphorylation of acyclovir (acycloguanosine) monophosphate by GMP kinase. *J. Biol. Chem. 255*:7204–7207, 1980.

Nagata, S.; Taira, H.; Hall, A.; Johnsrud, L.; Streuli, M.; Ecsödi, J.; Bell, W.; Cantell, K.; and Weissmann, C. Synthesis in E. coli of a polypeptide with human leukocyte interferon activity. *Nature 284*:316–320, 1980.

North, T.W., and Cohen, S.S. Aranucleosides and aranucleotides in viral chemotherapy. *Pharmacol. Ther. 4*:81–108, 1979.

Prusoff, W.H. Synthesis and biological activities of iododeoxyuridine an analog of thymidine. *Biochim. Biophys. Acta 32*:295–296, 1959.

Prusoff, W.H. Recent advances in chemotherapy of viral diseases. *Pharmacol. Rev. 19*:209–250, 1967.

Prusoff, W.H.; Bakkle, Y.S.; and Sekely, L. Cellular and antiviral effects of halogenated deoxyribonucleosides. *Ann. NY Acad. Sci. 130*:135–150, 1965.

Rubinstein, M.; Rubinstein, S.; Familletti, P.C.; Miller, R.S.; Waldman, H.A.; and Pestka, S. Human leukocyte interferon: production, purification to homogeneity and initial characterization. *Proc. Natl. Acad. Sci. USA 76*:640–644, 1979.

Sammons, M.L.; Stephen, E.L.; Levy, H.B.; Baron, S.; and Hilmas, D.E. Interferon induction in cynomolgus and rhesus monkeys after repeated doses of a modified polyriboinosinic-polyribocytidylic acid complex. *Antimicrob. Agents Chemother. 11*:80–83, 1977.

Schmidt, A.; Chernajovsky, Y.; Shulman, L.; Federman, P.; Beriss, H.; and Revel, M. An interferon induced phosphodiesterase degrading (2′-5′)oligoisoadenylate and the C-C-A terminus of tRNA. *Proc. Natl. Acad. Sci. USA 76*:4788–4792, 1979.

Slattery, E.; Ghosh, N.; Samanta, H.; and Lengyel, P. Interferon, double-stranded RNA and RNA degradation: activation of an endonuclease by (2′-5′)A$_n$. *Proc. Natl. Acad. Sci. USA 76*:4778–4782, 1979.

Taniguchi, T.; Mantei, N.; Schwarzstein, M.; Nagata, S.; Muramatsu, M.; and Weissmann, C. Human leukocyte and fibroblast interferons are structurally related. *Nature 285*:547–549, 1980.

Taniguchi, T.; Pang, R.H.L.; Yip, Y.K.; Henriksen, D.; and Vilcek, J. Partial characterization of γ (immune) interferon mRNA extracted from human lymphocytes. *Proc. Natl. Acad. Sci. USA 78*:3469–3472, 1981.

Tsai, M.-C.; Mansour, N.A.; Eldefrawi, A.T.; Eldefrawi, M.E.; and Albuquerque, E.X. Mechanism of action of amantadine on neuromuscular transmission. *Mol. Pharmacol. 14*:787–813, 1978.

Vilcek, J.; Havell, E.A.; and Yamazaki, S. Antigenic, physicochemical and biologic characterization of human interferons. *Ann. NY Acad. Sci. 284*:703–710, 1977.

Zoon, K.C.; Smith, M.E.; Bridgen, P.J.; zur Nedden, D.; and Anfinsen, C.B. Purification and partial characterization of human lymphoblastoid interferon. *Proc. Natl. Acad. Sci. USA 76*:5601–5605, 1979.

RECOMMENDED READING

Alford, C.A., and Whitley, R.J. Treatment of infections due to herpesvirus in humans. A critical review of the state of the art. *J. Infect. Dis.* [Suppl.] 133:A101–A108, 1976.

Becker, Y. *Antiviral Drugs, Mode of Action and Chemotherapy of Viral Infections of Man*. Monographs in Virology No. 11, 1976, S. Karger, Basel.

Colby, B.M.; Furman, P.A.; Shaw, J.E.; Elion, G.B.; and Pagano, J.S. Phosphorylation of acyclovir [9-(-hydroxymethoxymethyl)guanine] in Epstein-Barr virus-infected lymphoblastoid cell lines. *J. Virol. 38*:606–611, 1981.

Couch, R.B., and Jackson, G.G. Antiviral agents in influenza—summary of influenza workshop VIII. *J. Infect. Dis. 134*:516–527, 1976.

Elion, G.B. The chemotherapeutic exploitation of virus-specified enzymes. *Adv. Enzyme Regul. 18*:53–66, 1980.

Gordon, J., and Minks, M.A. The interferon renaissance: molecular aspects of induction and action. *Microbiol. Rev. 45*:244–266, 1981.

Grumert, R.R. Search for antiviral agents. *Annu. Rev. Microbiol. 33*:335–353, 1979.

Heidelberger, C., and King, D.H. Trifluorothymidine. *Pharmacol. Ther. 6*:427–442, 1979.

Saral, R.; Burns, W.H.; Laskin, O.L.; Santos, G.W.; and Lietman, P.S. Acyclovir prophylaxis of herpes-simplex-virus infections. A randomized, double-blind, controlled trial in bone-marrow-transplant recipients. *N. Engl. J. Med. 305*:63–67, 1981.

Scholtissek, C. Inhibition of the multiplication of enveloped viruses by glucose derivatives. *Curr. Top. Microbiol. Immunol.* 7:101–119, 1975.

Sidwell, R.W.; Robins, R.K.; and Hillyard, I.W. Ribavirin: an antiviral agent. *Pharmacol. Ther. 6*:123–146, 1979.

Stewart, W.E., II. *The Interferon System*. Springer-Verlag, Berlin, 1979.

Stewart, W.E., II, and Lin, L.S. Antiviral activities of interferons. *Pharmacol. Ther. 6*:443–512, 1979.

Stringfellow, D.A. *Interferon and Interferon Inducers. Clinical Applications*. Modern Pharmacology-Toxicology, Vol. 17, 1980. Marcel Dekke, Inc., New York and Basel.

Tamm, I., and Sehgal, P.B. Halobenzimidazole riboside and RNA synthesis of cells and viruses. *Adv. Virus Res. 22*:187–258, 1978.

27. LABORATORY DIAGNOSIS OF DISEASE-CAUSING VIRUSES

Rapid diagnosis of virus diseases is important for the treatment of patients (e.g., ara-A treatment of patients with herpetic encephalitis; passive immunization for Ebola patients) and from the epidemiological point of view for protecting medical personnel and the general public (e.g., smallpox patients who must be quarantined). Rapid isolation and identification of new influenza virus strains is essential for detecting a new virus mutant that may spread and cause an epidemic or pandemic. The rapid diagnosis of a new influenza A mutant would allow the early production of a killed virus vaccine for the immunization of children and older people, as well as those suffering from chronic diseases (see chapter 24). Diagnosis of rubella in pregnant women is necessary before making a decision regarding continuation of the pregnancy.

Before the establishment of cell-culture techniques, viruses were isolated by injecting the clinical material into test animals. However, the introduction of tissue culture techniques in the early 1950s greatly facilitated the process of virus isolation from clinical specimens. For example, blood or feces from a patient suspected of a virus infection is injected into experimental animals or cell cultures of different types. The presence of a virus in the sample may be detected by the appearance of cytopathogenic effects (CPE) in the cultured cells or the use of immunological techniques to identify viral antigens. The virus can be identified by a neutralization test in which medium from cells used to culture the virus is incubated with known antisera, and the mixture is inoculated into new cultures or animals. If the virus is neutralized, it will not replicate or kill the cells or the animals. However, these techniques require a number of days for identification of the virus and therefore will not influence the treatment of the patient. New technologies that have been developed for rapid virus diagnosis include electron microscopy, radioimmunoassay (RIA), and the enzyme-linked immunosorbent assay (ELISA). These techniques are fast and effective, as well as being accurate and objective. It is also possible to use RIA and ELISA methods for the detection of antibodies in sera.

RAPID DIAGNOSIS OF VIRUSES WITH THE AID OF THE ELECTRON MICROSCOPE

The classification of viruses into different families is based on the morphology of the virions (chapter 2). Thus the electron microscope can be used to identify the virions present in tissues, body fluids, or excretions of infected individuals, although the exact diagnosis of the virus strain within the virus family requires the use of additional techniques like neutralization with specific antibodies, RIA or ELISA. Nonetheless, the initial diagnosis of a virus in a clinical sample can be obtained within hours, using electron microscopy.

Diseases with fever and vesicular skin rashes may be caused by smallpox virus (Poxviridae) or herpes zoster virus (Herpesviridae). To rapidly identify the virus, a sample of the fluid is removed with a syringe from a skin vesicle and mixed with a solution of phosphotungstic acid to obtain negative staining of the virus. This staining technique allows the virus to be viewed in the electron microscope as a clear particle on a black background. The smallpox and varicella virions differ markedly from each other, and this allows an initial rapid diagnosis to be made.

To be able to treat a patient suspected of having herpetic encephalitis, it is necessary to confirm the presence of herpes simplex virus antigens or virions in a brain biopsy (chapter 26). Since it is essential to treat the patient as early as possible so as to increase the chances of recovery with as few sequelae as possible, it is imperative that the virus be recognized with the utmost speed. Brain biopsy material is viewed in the electron microscope to identify virus particles, and the immunofluorescent technique is used to identify viral antigens by staining with specific fluorescent antibodies to HSV-1. The stained material is viewed in a special ultraviolet microscope and the presence of fluorescence in the brain cells confirms the herpesvirus infection in the brain.

Electron microscopy of viruses concentrated by ultracentrifugation from stool samples of patients suffering from diarrhea led to the discovery of the Rotaviruses (chapter 12) that are responsible for infantile gastroenteritis. Detection of adenoviruses in stools by electron microscopy indicates that they may also cause certain types of diarrhea in humans. Virions of hepatitis A can also be identified by this technique.

Mixing specific antibodies with the specimen obtained from patients leads to the appearance of virion aggregates that can be spun down and examined by electron microscopy to aid in diagnosis. This technique is referred to as immune electron microscopy (IEM).

ISOLATION AND CHARACTERIZATION OF VIRUSES FROM CLINICAL MATERIALS

The sample of clinical material must be transferred as fast as possible to the diagnostic laboratory in closed containers to prevent the virus from spreading. Shipment of clinical samples is subject to special rules issued by WHO. In the diagnostic laboratory, the samples are used for isolation of the virus in cultured cells, embryonated eggs, or laboratory animals. Most viruses replicate in cultured cells in vitro and cause cytopathogenic effects (CPE). The medium of the affected cells is then transferred into fresh cultures to ensure that a virus, and not a toxic substance, caused the CPE.

Embryonated eggs and primate kidney cell cultures are used for the isolation of influenza A viruses. Throat washings from influenza patients are inoculated into the amniotic sac of the 10-day old chick embryos and also into primary cultures of monkey kidney cells. The amniotic and allantoic fluids are harvested after three days at 33–35°C and tested for hemagglutination with guinea pig or human erythrocytes. CPE is minimal in the monkey kidney cultures which should show hemadsorption of erythrocytes at 4°C after three–seven days at 33–35°C. The hemagglutination inhibition (HI) test (see the following) is used for identification of the virus.

Newborn mice are used for the isolation and characterization of arboviruses, rabies virus, and some Coxsackie A viruses, in addition to isolation in cultured cells. Some Coxsackie A viruses infect newborn mice but not cell cultures (see chapter 18).

IMMUNOLOGICAL RESPONSE IN PATIENTS

In almost all the virus diseases (except arenaviruses which cause a depression in the immune response), the antibody-producing B cells respond to viral antigens (processed by T lymphocytes) by the synthesis of virus-specific antibodies. The IgM class of antibodies is produced initially, followed by the production of IgG immunoglobulins with the same antiviral specificity. The antibodies appear in measurable quantities in the blood of the patients, usually within 10–14 days. To determine the immunological response to a virus infection, paired blood samples are taken from the patient at the onset of fever and 10–14 days later. The paired sera are tested for a rise in the amount of specific antibodies, using preparations of suspected viruses as antigens. An increase in antibody titer in the second blood sample provides proof as to the identity of the virus. Humoral antibodies IgM, IgG, and local antibodies IgE and IgA can be used for diagnosing the etiological viral agents. The presence of IgM or IgG antibodies makes it possible to differentiate between a virus infection and antiviral immunity. The presence of IgM antibodies indicates an active or recent virus infection, whereas the presence of IgG antibodies implies that the individual is immune. IgG antibodies arise as a secondary immune response to a virus infection. For example, the presence of IgG antibodies to rubella virus in a pregnant woman shows that she is immune; no antibodies to rubella means a lack of immunity, and the presence of IgM antibodies indicates an active rubella virus infection. The macroglobulin antibody molecule (IgM) is sensitive to treatment with 2-mercaptoethanol, whereas IgG molecules are resistant. Treatment of the serum with 2-mercaptoethanol prior to performing an immunological test and comparison with the untreated serum sample shows whether IgM or IgG antiviral antibodies are present in the patient's serum.

TESTS AVAILABLE FOR THE DETECTION OF VIRAL ANTIBODIES

Neutralization test

This test is based on the neutralization of virus infectivity by specific antibodies that attach to antigens on the surface of the virus. The sera are heated at 56° for 30 min to destroy nonspecific inhibitors, and serial dilutions of the sera are mixed with a constant dose of virus. After incubation for about 60 min, the mixtures are inoculated into cultured cells, embryonated eggs, or laboratory animals. The titer is defined as the reciprocal of the highest dilution of serum to inhibit virus multiplication.

A color test based on the ability of antibodies to inhibit virus replication is used as an assay. If infection occurs, the cells die and the culture medium remains at neutral pH or becomes alkaline, and the neutral red indicator present in the medium remains red. If the serum samples (in serial dilution) contain neutralizing antibodies that interact with the virus and prevent its replication in the cells, the color of the indicator in the medium will change to yellow. Multiple samples can be tested in cultures in plastic plates, and the titer of neutralizing antibodies can be determined with relative ease. This is called the metabolic inhibition test.

In the plaque reduction test, cell monolayers are inoculated with virus-antiserum mixtures and overlaid with suitable medium containing agar. After an incubation period to allow countable plaques to develop, the endpoint is determined as the highest dilution of serum to reduce the number of plaques by at least 50%.

Hemagglutination inhibition (HI) test

Certain viruses (e.g., influenza virus) are able to agglutinate erythrocytes. If a serum sample contains antibodies that block the specific viral antigens responsible for hemagglutination, this phenomenon will be inhibited. Serial dilutions of serum are mixed with the virus, erythrocytes are added, and the HI titer, defined as the reciprocal of the highest dilution of serum causing inhibition, can be determined.

Complement fixation (CF) test

This test consists of two parts: (1) the virus, specific antibodies, and complement, and (2) erythrocytes and specific antibodies to the erythrocytes. The reaction is based on the fact that each antibody-antigen complex can bind serum complement. Binding of complement to the erythrocyte-antibody complex causes lysis of the cells, and the hemoglobin gives a clear red color to the solution. In the first part of the test, a known viral antigen (two to four units) is incubated overnight at 4°C with dilutions of the patient's acute and convalescent serum (heated to 56°C for 30 min to inactivate complement) (or, alternatively, a known antibody preparation is incubated with the unknown viral antigen) in the presence of two units of complement derived from guinea pig serum. If viral antigen-antibody complexes are formed, the complement binds to the complexes, and no free complement remains to lyse the erythrocytes that are subsequently added. Sheep erythrocytes that are sensitized by the addition of rabbit antiserum against them (hemolysin) are used. If the viral antigen and antibodies do not combine, the complement remains unbound and causes lysis of the erythrocyte complexes; the solution will attain a red color from the released hemoglobin. In the absence of free complement, the erythrocytes are not lysed by the hemolysin and settle at the bottom of the test tube.

A number of controls are essential in this test to check whether the serum or antigen preparation is not anticomplementary—namely, they fix complement even in the absence of antigen or antibody, respectively.

Immunofluorescence for detection of viral antigens in infected cells

Identification of the virus isolated in cell cultures or in tissues of infected persons can be done using fluorescein-labeled antibodies. Sera are prepared in rabbits against a number of viruses, and the specific immunoglobin in each serum is tagged with fluorescein or rhodamine dye to yield fluorescent antibodies that can be seen with ultraviolet or blue light. Infected cells grown on a cover slip or a section of a tissue are fixed with acetone or methanol, and the fluorescein-tagged antibodies are allowed to interact with the virus antigens, followed by washing to remove unbound antibody molecules. This is called the direct immunofluorescence technique. Fluorescence in the infected cells can be seen in a special light microscope with a xenon or mercury vapor lamp. Special filters are used that make the specimen appear black, except for those areas where the fluorescein-labeled antibodies interact with the antigens of the specific virus and emit a greenish-yellow light. This technique requires that the diagnostic virus laboratory have fluorescent antibody preparations against all known human viruses.

The indirect immunofluorescent technique utilizes antibodies prepared in rabbits against human γ-globulin that are tagged with fluorescein. The antiviral antibody, which is untagged, is allowed to react with the viral antigen, and only if they combine will the fluorescent IgG antibodies attach to the preparation. Fluorescence therefore indicates that viral antigen is present in the cell.

The same immunofluorescence techniques can be used with known viral antigens to detect the presence of antibodies in sera of patients.

Radioimmunoassay (RIA) and enzyme-linked immunosorbent assay (ELISA)

In addition to tagging antibodies with fluorescein, it became possible to label antibodies with radioactive iodine or with an enzyme.

In the RIA test, the antigen is immobilized by binding to a solid surface to which the antibody attaches. ^{125}I-labeled goat or rabbit antibodies to human γ-globulin are then added to bind to the antigen-antibody complex. For the solid phase in RIA, paper discs or polystyrene balls may be used. Purified antigen can be adsorbed onto the polystyrene balls by submerging the balls in antigen diluted in buffer overnight. The balls are dried and can be stored for further use. Serum dilutions are pipetted into disposable plastic tubes and one virus-coated ball is added to each tube. After incubation, the serum dilutions are aspirated and the balls washed twice with tap water. At each step, the excess of unbound antibodies is removed by washing. ^{125}I-labeled anti-human γ-globulin is then added to each tube. After further incubation, the balls are washed twice, placed in clean tubes, and the radioactivity measured in a gamma counter. The amount of labeled anti-human γ-globulin in each tube is determined by the amount of antibody that has combined with the antigen.

In the ELISA test for quantitative determination of antigen or antibodies, an enzyme is used instead of the radioactive isotope. This test is also referred to as the enzyme immunoassay (EIA). The enzyme, horseradish peroxidase, is coupled to the antiglobulin used in the test by means of periodate. An alkaline phosphatase conjugate prepared from goat antiserum to human γ-globulin is also used. This assay can be done in special polystyrene microtiter plates to which the antigen is adsorbed. Dilutions of test and control sera are added, and after a suitable incubation period, the plates are washed and the conjugated antiglobulin added. The plates are again incubated and, after washing, a suitable substrate (5-aminosalicylic acid or p-nitrophenyl phosphate) is added to give the color reaction used to determine the endpoint of the titration, either by spectrophotometry or with the naked eye.

Detection of viral nucleic acids

The presence of viral nucleic acid in human and animal tissues can be used to determine if a particular virus (e.g., HSV-2 in cervical carcinoma and EBV in Burkitt's lymphoma) is present in the cancer cells. Frozen sections of biopsy tissue cells grown on coverslips or cell suspensions affixed to coverslips are used for hybridization in situ. The nucleic acids in the cells are hybridized with radioactive probes prepared from the viral nucleic acid. After the hybridization process is completed, excess nucleic acid is removed by washing, and the extent of hybridization is determined by radioautography. Hybridization between HSV-2 DNA and mRNA in cells of cervical carcinoma biopsies was reported (Maitland et al. 1981). However, this technique is not absolutely specific, and without adequate controls the results may be subject to misinterpretation.

DNA fingerprinting with the aid of restriction endonuclease enzymes is now being used to differentiate between the various serotypes of HSV-1 and HSV-2 (Buchman et al. 1980). In this technique, cell cultures are infected with each virus and the viral DNA extracted after 24 hours of incubation. The DNA is purified, cut with restriction enzymes, and subjected to electrophoresis in agarose gels. After staining with ethidium bromide, the gels are photographed and the electrophoretic profiles of the stained DNA bands from each virus are compared.

BIBLIOGRAPHY

Buchman, T.G.; Simpson, T.; Nosal, C.; and Roizman, B. The structure of herpes simplex virus DNA and its application to molecular epidemiology. *Ann. NY Acad. Sci. 354*:279–290, 1980.

Maitland, N.J.; Kinross, J.H.; Busuttil, A.; Ludgate, S.M.; Smart, G.E.; and Jones, K.W. The detection of DNA tumour virus-specific RNA sequences in abnormal human cervical biopsies by in situ hybridization. *J. Gen. Virol. 55*:123–137, 1981.

RECOMMENDED READING

Arstila, P.; Vuerimaa, T.; Kalimo, K.; Halonen, P.; Viljanen, M.; Granfors, K.; and Toivanen, P. A solid-phase radioimmunoassay for IgG and IgM antibodies against measles virus. *J. Gen. Virol. 34*:167–176, 1977.

Berthiaume, L.; Alain, R.; McLaughlin, B.; Payment, P.; and Trepanier, P. Rapid detection of human viruses in faeces by a simple and routine immune electron microscopy technique. *J. Gen. Virol. 55*:223–227, 1981.

Cameron, C.H., and Dane, D.S. Viruses. In radioimmunoassay and saturation analysis. *Brit. Med. Bull. 30*:90–92, 1974.

Evans, A.S. (ed.). *Viral Infections of Humans. Epidemiology and Control.* Plenum Medical Book Co., New York and London, 1976.

Fenner, F.J.; and White, D.O. *Medical Virology*. 2nd ed. Academic Press, New York, San Francisco, London, 1976.

Forghani, B., and Schmidt, N.J. Antigen requirements, sensitivity and specificity of enzyme immunoassays for measles and rubella viral antibodies. *J. Clin. Microbiol. 9*:657–664, 1979.

Hermann, J.E.; Hendry, R.M.; and Collins, M.F. Factors involved in enzyme-linked immunoassay of viruses and evaluation of the method for identification of enteroviruses. *J. Clin. Microbiol. 10*:210–217, 1979.

Hsiung, G.D.; Fong, C.K.Y.; and August, M.J. The use of electron microscopy for diagnosis of virus infections: an overview. *Prog. Med. Virol. 25*:133–159, 1979.

Lennette, E.H., and Schmidt, N.J. (eds.) *Diagnostic Procedures for Viral, Rickettsial and Chlamydial Infections.* 5th ed. American Public Health Association, Inc., Washington, D.C., 1979.

McCracken, A.W., and Newman, J.T. The current status of the laboratory diagnosis of viral diseases of man. CRC *Crit. Rev. Clin. Lab. Sci. 5*:331–363, 1975.

Stephens, R. Comparative studies on EBV antigens by immunofluorescent and immunoperoxidase techniques. *Int. J. Cancer 19*:305–312, 1977.

Volken, R.H. Enzyme-linked immunosorbent assay (ELISA): a practical tool for rapid diagnosis of viruses and other infectious agents. *Yale J. Biol. Med. 53*:85–92, 1980.

28. VIRUSES AND HUMAN CANCER

Cancer is a general term used when a group of cells escapes from the tissue control mechanisms and begins to proliferate at an abnormal rate in the body. Leukemias and lymphomas are lymphoproliferative disorders of the bone marrow and the blood; osteosarcoma involves the bones; and hepatoma involves the liver cells. Essentially, cancer cells can arise in all tissues as a response to intrinsic or extrinsic factors that are largely unknown. It is well documented that chemical carcinogens and radiation damage can lead to the transformation of a normal cell into a cancer cell capable of unlimited development and tumor formation. Cigarette smoking was shown to be associated with cancer of the lung. Environmentalists demand the removal of chemical carcinogens from the environment to reduce the incidence of cancer in humans, but this is an unrealistic approach. As regards viruses, some retroviruses are endogenous in humans, as has been shown with breast cancer, but the factors that control sensitivity or resistance to cancer are largely unknown. This is because genetic markers for determining these parameters are not yet available. Thus certain individuals develop cancer, while most of the population remains unaffected. In persons with Down's syndrome (chromosome 21 trisomy; mongoloids) and with chromosome-breakage syndromes like ataxia-telangiectasia (A-T), the incidence of cancer is significantly higher than in the general population. The nature of the defective gene which is linked with the increased incidence of cancer is not yet known.

Several viruses were found to cause tumors naturally in animals, while others cause tumors experimentally under laboratory conditions (chapters 7, 8, 10, and 21). Experimental evidence was first provided by Ellerman and Bang in 1908 and by Rous in 1911 that some viruses of chickens are capable of cell transformation (chapter 1). Stemming from the discovery by H. Temin and D. Baltimore in 1970 of the retrovirus RNA-dependent DNA polymerase, studies during the last ten years have shown that a gene in the retrovirus genome (which is essentially similar to cellular DNA) is responsible for the transformation of the infected cell by insertion of this DNA sequence into the DNA of the cell (chapter 21).

In spite of the marked advances in studies on the role of viruses in animal cancer, the exact role of viruses in human cancers is still unknown. Herpesvirus DNA can be demonstrated in human tumor material, or retroviruses can be rescued from human breast cancer or leukemic cells, but the etiological role of such viruses in human cancer remains a puzzle.

HERPESVIRUSES AND CANCER

Epstein-Barr virus (EBV), Burkitt's lymphoma (BL) and nasopharyngeal carcinoma (NPC)

EBV was discovered by A. Epstein in cultured lymphoma cells obtained from tumors of lymphoma patients. It was noted in epidemiological studies that this type of tumor is restricted to central equatorial Africa and New Guinea. The herpesvirus rescued from BL cells was used to test for specific antibodies in other populations, and it was found that EBV is a ubiquitous virus but is not associated with other human lymphomas or leukemias. To correlate between EBV infection of African children in the high incidence areas of BL and the incidence of tumors, a group of 42,000 children was studied during a five-year period. The extent of EBV infection, as well as malaria infection, and the incidence of BL were studied. It was found that all the 42,000 children in the survey had been exposed to malaria and EBV early in life and had developed EBV antibodies. However, only 32 children from this group developed BL, which shows that only in rare instances does infection of a child with EBV lead to the development of cancer.

The carcinogenic potential of EBV was shown by injecting the viral DNA into owl monkeys which did, indeed, develop lymphoma. There is also a correlation between infection of Southern Chinese with EBV in the nasopharynx and the incidence of NPC in this population. A high incidence of NPC was also found in North Africa.

Infection of B lymphocytes from normal individuals with EBV can be achieved under in vitro conditions. Cells thus infected become immortalized and are capable of unlimited growth in vitro. The ability of EBV to affect the cells in such a way also indicates its carcinogenic potential.

Herpes simplex virus type 2 (HSV-2) and cervical cancer

Experiments by F. Rapp and D. Duff (chapter 7) demonstrated that partial inactivation of HSV-2 DNA with ultraviolet light causes damage to several genes required for virus replication and leads to the expression of those viral functions that cause hamster cells to become transformed. Injection of such in vitro transformed cells into hamsters leads to the development of tumors. In cells isolated from such tumors and cultivated in vitro, it was possible to demonstrate that a fragment of the herpes-virus DNA is present in the chromosomal DNA.

The involvement of HSV-2 with cervical carcinoma is under investigation. A correlation between the incidence of cervical carcinoma and a high HSV-2 antibody titer in the serum of such patients has been noted (Aurelian and Strand 1976; Nahmias et al. 1973; Rawls et al. 1973). Current studies by J.K. McDougall and his associates, using in-situ nucleic acid hybridization, suggest that cervical carcinoma cells contain HSV-2 nucleic acid, but these findings require substantiation (McDougall et al. 1979; Galloway et al. 1979).

Hepatocarcinoma and hepatitis B virus

In patients with chronic hepatitis caused by hepatitis B virus, a high incidence of hepatomas was noted. A cell line obtained by Alexander and associates (1976) from a human hepatoma was found to contain integrated fragments of hepatitis B virus DNA in the tumor cell DNA (Chakrabarty et al. 1980; Edman et al. 1980). The integrated viral DNA has been cloned together with the flanking cellular DNA sequences. In about 5% of hepatocarcinoma patients, hepatitis B virus DNA sequences were not detectable in the tumor cells. Thus it is not clear whether a chronic infection with the virus leads to cirrhosis and the spontaneous appearance of tumor cells that are then infected with the virus or whether the virus is the cause of cell transformation into a neoplastic cell (P. Hofschneider, NATO International Advanced Study Institute on Biochemical and Biological Markers of Neoplastic Transformation, Corfu Island, Greece, 1981).

RETROVIRUSES AND HUMAN CANCER

The first indication that human adenocarcinomas of the breast, leukemias, sarcomas, and lymphomas contain retrovirus information was provided by S. Spiegelman and his associates who showed that in such cells complexes exist that contain RNA sequences, related to a mouse retrovirus, and a viral reverse transcriptase (Baxt et al. 1973). Subsequently, type-C virions were isolated from cells of leukemic patients and also from a number of cell lines. R.C. Gallo and his colleagues isolated a retrovirus from peripheral blood leukocytes of a patient with acute myelogenous leukemia that were cultured in vitro. The isolated type-C virus particles had nucleotide sequences related to the genome of SSV (Reitz et al. 1976). Human T-cells isolated from a patient with a cutaneous T-cell leukemia and a patient with a cutaneous T-cell lymphoma were continuously propagated in vitro using T-cell growth factor. From these cancer cells, a retrovirus was isolated that did not cross react with antigens of any known primate or other animal retrovirus. This virus was designated human T leukemia virus (HTLV). This might be a virus causing human leukemia, since antibodies to HTLV internal structural antigen p24 were found in sera of some leukemia patients (Poiesz et al. 1981; Kalyanaraman et al. 1981).

Retrovirus particles were observed in human milk, but their role in the etiology of human breast cancer is not known. It was shown by S. Spiegelman and his associates that the tumor cells in a biopsy of human breast cancer contain new antigens that cross react with antigens of a mouse mammary tumor retrovirus (Mesa-Tejada et al. 1978). The normal cells in the biopsy do not contain such antigens.

BIBLIOGRAPHY

Alexander, J.J.; Bey, E.M.; Geddes, E.W.; and Lecatsas, G. Establishment of a continuously growing cell line from primary carcinoma of the liver. *S.Afr. Med. J. 50*:2124–2128, 1976.

Aurelian, L., and Strand, B. Herpesvirus type 2 related antigens and their relevance to humoral and cell mediated immunity in patients with cervical cancer. *Cancer Res. 36*:810–820, 1976.

Baxt, W.; Yates, J.W.; Wallace, H.J., Jr.; Holland, J.F.; and Spiegelman, S. Leukemia-specific DNA sequences in leukocytes of the leukemic member of identical twins. *Proc. Natl. Acad. Sci. USA 70*:2629–2632, 1973.

Chakrabarty, P.R.; Ruiz-Opazo, N.; Shouval, D.; and Shafritz, D.A. Identification of integrated hepatitis B virus DNA and suppression of viral RNA in an HBsAg-producing human hepatocellular carcinoma cell line. *Nature 286*:531–533, 1980.

Edman, J.C.; Gray, P.; Valenzuela, P.; Rall, L.B.; and Rutter, W.J. Integration of hepatitis virus sequences and their expression in a human hepatoma cell. *Nature 286*:535–538, 1980.

Galloway, D.A.; Fenoglio, C.; Shevchuk, M.; and McDougall, J.K. Detection of herpes simplex RNA in human sensory ganglia. *Virology 95*:265–268, 1979.

Kalyanaraman, V.S.; Sarngadharan, M.G.; Bunn, P.A.; Minna, J.D.; and Gallo, R.C. Antibodies in human sera reactive against an internal structural protein of human T-cell lymphoma virus. *Nature 294*:271–273, 1981.

McDougall, J.K.; Galloway, D.A.; and Fenoglio, C.M. *In-situ* cytological hybridization to detect herpes simplex virus RNA in human tissues. *In: Antiviral Mechanisms in the Control of Neoplasia* (P. Chandra, ed.), Plenum Press, New York, 1979, pp. 233–240.

Mesa-Tejada, R.; Keydar, I.; Ramanapayanan, M.; Ohne, T.; Fenoglio, C.; and Spiegelman, S. Detection in human breast carcinomas of an antigen immunologically related to a group-specific antigen of mouse mammary tumor virus. *Proc. Natl. Acad. Sci. USA 75*:1529–1533, 1978.

Nahmias, A.J.; Naif, Z.M.; and Josey, W. E. Herpesvirus hominis type 2 infection associated with cervical cancer and prenatal disease. *Perspect. Virol. 8*, 73–88, 1971.

Poiesz, B.J.; Ruscetti, F.W.; Reitz, M.S.; Kalyanaraman, V.S.; and Gallo, R.C. Isolation of a new type C retrovirus (HTLV) in primary uncultured cells of a patient with Sézary T-cell leukaemia. *Nature 294*:268–271, 1981.

Rawls, W.E.; Adam, E.; and Melnick, J.L. An analysis of seroepidemiological studies of herpes-virus type 2 and carcinoma of the cervix. *Cancer Res. 33*:1477–1482, 1973.

Reitz, M.S.; Miller, N.R.; Wang-Staal, F.; Gallagher, R.E.; Gallo, R.C.; and Gillespie, D.H. Primate type-C virus nucleic acid sequences (woolly monkey and baboon types) in tissues from a patient with acute myelogenous leukemia and in viruses isolated from cultured cells of the same patient. *Proc. Natl. Acad. Sci. USA 73*:2113–2117, 1976.

RECOMMENDED READING

Epstein, M.A., and Achong, B.G. *The Epstein-Barr Virus*. Springer-Verlag, Berlin, Heidelberg, New York, 1979.

Gallo, R.G. (ed.) *Recent Advances in Cancer Research, Cell Biology, Molecular Biology and Tumor Virology*. CRC Press, Inc., Florida, USA, 1980.

Klein, G. *Viral Oncology*. Raven Press, New York, 1982.

Nahmias, A.J.; Dowdle, W.R.; and Schinazi, R.F. (eds.) *The Human Herpesviruses*. Elsevier, New York, 1981.

Rapp, F. *Oncogenic Herpesviruses*. CRC Press, Inc., Florida, USA, 1980.

Zur Hausen, H. The role of viruses in human tumors. *Adv. Cancer Res. 33*:77–107, 1980.

29. SOCIAL, ECONOMIC, AND JURIDICAL ASPECTS OF VIRUS DISEASES

Viruses that cause diseases in man, animals and plants are of great international concern. Because of their tendency to spread rapidly, viruses that harm livestock or plants could threaten food resources, and major efforts must be made to prevent large-scale damage. Laboratory facilities are required for speedy detection and identification of disease-causing viruses as well as systems to prevent spread of new viruses across state borders. Surveillance systems have been developed for early detection of viruses, and vaccine factories must be able to produce vaccines in time. The economic effect can be calculated in terms of working days lost due to viral infections, the cost of hospitalization and medical care, the cost of vaccine production and its administration to the public, the cost of livestock losses, the cost of purchasing meat from other sources to replace the loss, the cost of plant products damaged by virus diseases, and so on.

Juridical problems may arise from claims by previously healthy individuals for damages allegedly incurred due to vaccination or medical treatment.

SOCIAL ASPECTS

Herd immunity to viral diseases

The use of killed or live attenuated virus vaccines to protect people against epidemics requires a continuous immunization program to include all those born after the start of the epidemic. In countries with large populations, it is virtually impossible to immunize everyone, and only a situation of herd immunity can be attained in which most individuals are immune to a particular virus. Some people do not become immune in spite of vaccination, due to immunological deficiency, and others refuse to be immunized for religious or other reasons.

Vaccination of entire populations in areas endemic for smallpox by WHO led to the gradual eradication of the disease. Immunization of large populations with live attenuated polio, measles, mumps, or rubella vaccines markedly decreased the incidence of these diseases. However, the immune state of the population must be under constant surveillance to ensure that the vaccines are still effective in providing protection.

Monitoring of viruses in water reservoirs

One of the easiest ways for a virus epidemic to develop is by contamination of drinking water due to sewage leaks into the water system. Enteric viruses that inhabit the gut and are released with the feces are stable and retain their viability in sewage. Thus it is necessary to monitor sewage as well as recycled water for the presence of infectious viruses. The viruses must be concentrated from large volumes of water prior to identification (chapter 27). Use of sewage to irrigate fields also requires the monitoring of airborne infectious virus.

International virus diagnosis systems

In order to prevent influenza pandemics, WHO has organized two international diagnostic laboratories: one in Europe and one in America. These laboratories are stocked with all the known influenza virus antigens and can determine if an influenza virus from any country is an old or a new virus strain. National diagnostic laboratories that are connected with the WHO centers isolate influenza virus strains for identification. If a new virus strain is recognized, it is sent to the WHO center for final identification. The appearance of a new influenza virus strain is announced publicly, and the vaccine-producing companies are provided with the virus to facilitate early production of a vaccine containing the new viral antigen (chapter 24).

The Food and Agriculture Organization (FAO) of the United Nations monitors the spread of viruses that cause African swine fever, foot and mouth disease, and Rift Valley fever, all of which affect domestic cattle. FAO offices in Rome distribute the information but depend on the national diagnostic laboratories for identification of the new virus isolates.

Problems of virus vaccine production

Vaccines to immunize human or animal populations are produced either by national health authorities or by private companies, depending on the health authorities in each country. However, countries with small populations must purchase their vaccines from large, dependable firms that distribute viral vaccines around the world. This is true mainly for human vaccines that must be of the highest quality and standards of safety to comply with the strict rules of the

U.S. Food and Drug Administration (FDA). The development of a new vaccine requires quality testing and field trials that need substantial financial investments. Safe and effective viral vaccines are also required for poultry and domestic animals. The vaccines for poultry must be prepared in specific pathogen-free (SPF) eggs to prevent the spread of endogenous chicken viruses or their nucleic acids in the vaccines.

Active research in the field of virology has led to constant improvements in vaccine production. Adaptation of rabies virus to growth in human diploid fibroblast cells by H. Koprowski and his colleagues (see chapter 13) made it possible to develop a safe killed rabies vaccine, replacing the vaccine made from infected rabbit brain which produced neurological side effects. The live poliovirus vaccine strains, which are very effective in immunization, need to be constantly checked for changes that might lead to increased virulence. Since 1969, it has been noted that a small number of children vaccinated with the Sabin live, attenuated vaccine developed paralytic poliomyelitis (the estimate is one out of four million children). WHO appointed a committee to evaluate the live poliovirus vaccine, and in the United States the vaccine producers must include a warning in brochures accompanying the poliovaccine that there is a remote possibility of danger to immunized children (see chapter 24).

Responsibility for individuals affected by a viral disease

Virus diseases in most individuals pass without sequelae, but some cases of poliomyelitis result in paralysis, and patients with herpetic encephalitis frequently suffer brain damage. In rare instances, individuals vaccinated against smallpox or rabies develop acute disseminated encephalomyelitis with permanent neurologic disease. Victims of neurological diseases may be disabled and then should become the responsibility of a welfare system that provides hospitalization and possible rehabilitation.

ECONOMIC ASPECTS

An influenza virus epidemic can cause the loss of millions of working days. In the winter of 1973–1974—a so-called ordinary year without an influenza epidemic—it was calculated in the United States that 51 million workers were absent from work for an average of four sick days.

Epidemics among animals used as a source of food have an immediate effect on the economy—not only because the diseased animals must be destroyed but also because of reduced trade between the affected country and other countries due to the fear that the virus might be spread by contaminated meat products. When African swine fever entered Cuba, the entire swine population on the island was sacrificed, and a new population of pigs was introduced. A similar situation occurred on the island of Malta.

The economic losses from human and animal virus diseases are huge in comparison with the investment needed for research on viruses. Such research could help prevent these losses and thus save the money paid to the farmers in compensation.

JURIDICAL ASPECTS

The demand by national health authorities that the population be immunized against virus diseases requires strict measures to ensure that the population will indeed be vaccinated. In many countries, vaccination of children with viral vaccines is a compulsory part of the national health program. However, it is known that some individuals do have depressed immune systems, and sometimes they are detected only after being harmed by the vaccine. Individuals suffering ill effects due to vaccination, especially in the United States, have claimed damages in the courts from vaccine-producing companies. The ruling in a series of cases was that the vaccine producers were liable for damages for not warning the public that use of their vaccine might entail a certain, although minimal, danger to the immunized individual.

During the period of vaccination against swine influenza which was conducted as a national vaccination campaign in the United States (chapter 24), the vaccine-producing firms demanded—and obtained—from the national health authority an agreement that all claims against them for damages incurred through vaccination would be indemnified by the health authorities.

Early identification of individuals sensitive to vaccination would help to reduce the numbers of those suffering ill effects due to the vaccine.

IV. SUMMARY

30. SUMMARY

CHAPTER 1 STEPS IN THE DEVELOPMENT OF VIRUS RESEARCH

Viruses as disease-causing agents.
Viruses existed even before the dawn of human history.
Attempts to immunize humans against smallpox and measles as early as the eighteenth century.
The concept *virus*, meaning poison, was initiated by Edward Jenner in 1798.
The concept of virus during the period of Pasteur (end of nineteenth century) was that of a filterable agent not visible in the light microscope.
Discovery in 1887 that the mosaic disease of tobacco plants is caused by a virus.
Discovery in 1898 that foot-and-mouth disease in cattle is caused by a virus.
Investigations on filtration of homogenates of tumors revealed that viruses cause cancer.
Viruses infect bacteria: the discovery of the bacteriophages.
Viruses cause diseases in test animals.
Yellow fever virus (YFV) was attenuated by passage in chick embryos: development of the vaccine.
Viruses replicate in the infected host in selected tissues.
Animal viruses can replicate in isolated tissues in culture.
Plaque assay for animal viruses allowed quantitative analysis of viruses.
The electron microscope made possible the visualization of virus structure.
Viruses cause mouse leukemia.

The molecular basis of virology is that the viral nucleic acids are the carriers of the viral genes.
Tobacco mosaic virus (TMV) genes are present in viral RNA.
Experimental evidence provided the proof that viral DNA genomes contain the viral genes.
The viral DNA is the template for mRNA for the synthesis of viral proteins.
Virus defined as a particle containing either DNA or RNA that carries all the viral genetic information.
Subviruses that include viroids are infectious agents.
Genetic engineering technology makes it possible to isolate and study individual genes.

CHAPTER 2 VIRUS CLASSIFICATION

The classification of viruses is an ongoing process.

CHAPTER 3 MOLECULAR CONSIDERATIONS OF VIRUS REPLICATION AND VIRUS-CELL INTERACTIONS.

Viruses are obligate parasites of cells.
Viral genes code for different groups of functional proteins.
The viral capsomeres assemble into capsids.
Virion formation requires capsid-nucleic acid interactions.
Cells can be transfected with naked viral nucleic acids.
Virions must be released from cells to initiate new infections.
Viruses are transferred from one host to another.
An infection starts with the attachment of virions to receptors in the cell membrane.
RNA viral genomes can serve as plus or minus nucleic acids.
The one-step growth cycle in vitro requires spontaneous infection of all cells.
Enzymes are contained in virions of some virus families.
The mechanisms utilized for virus replication in infected cells depend on the viral nucleic acid.
Cells respond differently to different viruses.

CHAPTER 4 GENES IN HUMAN CELLS DETERMINING VIRUS SUSCEPTIBILITY

Different genes in the human chromosomes determine the sensitivity of human cells to virus infection.

CHAPTER 5 THE POXVIRUS FAMILY

Relatedness between poxviruses can be determined by DNA-DNA reassociation techniques.
The DNA genome of the orthopoxviruses contains more than 150 genes.
Poxvirions contain structural proteins and enzymes.
The mechanism of cell infection with a poxvirus is a multistage process.

The virions are phagocytized by infected cells.
Uncoating of the virions leads to release of the viral cores.
Virus infection inhibits nuclear processes.
Expression of early viral genes leads to the synthesis of early viral mRNA molecules.
The biosynthesis of viral DNA takes place in discrete cytoplasmic sites.
Expression of late viral genes in infected cells takes place after viral DNA synthesis.
Morphogenesis of the poxvirions: the final step in virus infection of a cell.
Specific antipoxvirus agents inhibit or prevent virus infection.
Marboran is the prophylactic drug against smallpox.
Rifampicin and distamycin inhibit poxviruses.
Interferon inhibits virus replication.
Smallpox, a disease in humans, starts with a respiratory infection.
Differential diagnosis of the disease is important.
Virulent smallpox viruses in research laboratories constitute a danger around the world.
Humans are infected by monkeypox virus.
Animal poxviruses are yabavirus of monkeys and myxomatosis in rabbits.

CHAPTER 6 THE IRIDOVIRUS FAMILY

Iridoviruses include African swine fever virus, frog viruses, and lymphocystis virus in fish.

CHAPTER 7 THE HERPESVIRUS FAMILY

Viruses belonging to the herpesvirus family occur in numerous hosts.
The icosahedral virus capsid is made of 162 capsomeres.
The viral DNA is double-stranded.
DNAs of different members of the herpesvirus group differ in density.
The viral DNA contains repetitive sequences.
The virions contain subpopulations of viral DNA.
Viral DNA is infectious for permissive cells by transfection.
Partial homology exists between DNAs of different herpesviruses.
Virion proteins are antigenic.
Cloning of HSV-DNA restriction fragments in bacterial plasmids allows analysis of viral genes.
Herpesvirus replication in cells is controlled by the cell and the virus.
The viral DNA is uncoated by cellular enzymes.
Early transcription of viral DNA is carried out by the host cell RNA polymerase II.
Viral mRNA is transcribed from a particular group of genes in the viral DNA.
Virus infection causes the disaggregation of the nucleoli.
Synthesis of viral proteins is a process regulated by virus-coded proteins.
DNA binding proteins are involved in the synthesis of viral DNA.
A cellular gene function determines the initiation of viral DNA synthesis.

Synthesis of HSV DNA is semiconservative and initiates at three possible sites.
Replication of a viral DNA molecule takes 20 min.
Mutants of HSV defective in DNA synthesis are used to characterize the viral enzymes.
Recombinants between HSV-1 and HSV-2 were developed.
Defective HSV is due to an error in viral DNA biosynthesis.
HSV causes latent infections in humans and animals.
Inactivated HSV transforms cells in vitro.
Herpesviruses affect humans and animals and spread in the blood and along nerve axons.
HSV-2 is connected with cervical carcinoma.
HSV-coded thymidine kinase gene biochemically transforms TK^- cells.
Chemotherapy of herpesvirus infections with phosphonoacetic acid, IUdR, Ara-A, and acyclovir.
Herpesviruses like varicella-zoster and cytomegalovirus cause diseases in humans.
Epstein-Barr virus (EBV) is a ubiquitous virus associated with Burkitt's lymphoma (BL) and causes infectious mononucleosis in humans.
EBV receptors are found on human B lymphocytes.
EBV might have a role in human cancer.
Herpesvirus papio is a monkey virus related to EBV.
Herpesvirus saimiri and H. ateles cause malignant lymphomas in monkeys.
Herpes B virus is a latent monkey virus but causes fatal infections in man.
Pseudorabies is a herpesvirus that causes Aujesky's disease in pigs.
Herpesviruses infect cattle, horses, and dogs.
Herpesvirus in frogs causes Lucké renal carcinoma.
A herpesvirus from turkeys immunizes chickens against a herpesvirus that causes Marek's diseases (lymphoma and neural damage).

CHAPTER 8 THE ADENOVIRUS FAMILY

Human and mammalian adenoviruses belong to the genus mastadenovirus.
Adenoviruses of birds belong to the genus aviadenovirus.
The adenovirions are made up of 252 capsomeres.
The virions attach to the cell membrane and are incorporated into the cytoplasm by pinocytosis.
Adenovirus DNA is 20–25 $\times$ 10^6 daltons and is infectious.
There are at least five separate transcription units for early mRNA in the viral genome.
Intermediate and late mRNA synthesis are coupled to the onset of DNA replication.
Viral DNA synthesis marks the late phase of virus replication.
Virions assemble in the nuclei of infected cells.
Removal of arginine from the medium of infected cells results in abortion of virus replication.

Adenoviruses are divided into four subgroups on the basis of their oncogenicity for newborn hamsters.
Cells can be transformed by adenovirus DNA.
No evidence is available for specific integration sites for adenovirus DNA in the cell chromosomes.
Adenovirus-SV40 hybrid viruses appear when the two viruses infect the same cell.
Adenovirus type 7 can cause epidemic outbreaks of respiratory disease in different parts of the world.

CHAPTER 9 THE PAPOVAVIRUSES

Genus papillomavirus infects vertebrates and causes benign papillomas
Genus polyomavirus includes polyoma virus of mice, SV40 of monkeys, BK and JC viruses of man, and other related viruses.
The host cell determines the type of virus infection.
Viral DNA is superhelical and is a minichromosome.
The nucleotide sequence of SV40 was reported.
Under certain conditions of infection, SV40 DNA might contain cellular DNA.
Structural viral proteins and DNA-bound histones constitute the proteins of the virions.
The viral DNA contains two strands: one coding for early and the other for late functions.
Products of the early (E) genes are the T and t antigens.
Transcription of SV40 late (L) genes starts immediately after the initiation of DNA replication.
The transcription of polyoma virus DNA resembles that of SV40.
The initiation sequence for DNA replication is mapped in 0.67 map units and contains 27 base pairs that form a palindrome.
The synthesis of viral DNA is semiconservative and bidirectional.
Cells are transformed by SV40 as a result of viral DNA integration into chromosomal DNA.
SV40 is released from transformed cells by fusion with permissive cells.
SV40 mutants were isolated.
Polyoma viruses were isolated from brains of progressive multifocal leukoencephalopathy (PML) patients.
Papilloma virus causes human warts.

CHAPTER 10 HEPATITIS B VIRUS

The discovery of Australia antigen led to the discovery of hepatitis B virus in humans and related viruses in vertebrates.
The surface antigen of hepatitis B virus (HBsAg) is present in the blood of hepatitis patients.

The virion-like Dane particles in the patient's blood contain a circular DNA genome.
Primary hepatic carcinoma cells in humans contain integrated viral DNA in the cell chromosomes.

CHAPTER 11 THE PARVOVIRUS FAMILY

The genus parvovirus contains animal parvoviruses that are able to replicate without a helper virus.
The adeno-associated viruses require human or simian adenoviruses for their replication.
The genus densovirus contains insect viruses.
The genome of parvoviruses is a single-stranded DNA molecule.

CHAPTER 12 THE REOVIRUSES

The virions contain a fragmented double-stranded RNA genome.
Ten species of mRNA molecules are made in the infected cell by a virion-bound RNA polymerase.
Each mRNA species is translated to a viral protein.
The viral mRNA is the template for the double-stranded genome in the virions.
Rotaviruses cause gastroenteritis in humans.

CHAPTER 13 THE RHABDOVIRUSES

The genus vesiculovirus (vesicular stomatitis virus), genus lyssavirus (rabies virus), and genus sigmavirus (Drosophila σ virus) constitute the rhabdovirus family.
The rabies virions have a bullet-shaped structure.
When humans are bitten by rabid animals, virus is introduced into the wound and migrates to the nervous system.
A rabies vaccine prepared from virus-infected human fibroblasts inactivated by formaldehyde is used to immunize humans.
Vesicular stomatitis virus molecular structure and mode of replication was investigated.

CHAPTER 14 THE PARAMYXOVIRUSES

Genus paramyxovirus contains Newcastle disease virus (NDV) of birds, mumps virus of humans, and parainfluenza viruses.
Genus morbillivirus includes measles virus of humans.
Pneumovirus subgroup contains the respiratory syncytial virus of humans.
Two types of spikes are present on the envelope of paramyxoviruses: neuraminidase acid and hemagglutinin.
The virions attach to the neuraminic acid-containing receptor.
The virion genomic RNA has a sedimentation coefficient of 50S in sucrose gradients, but cytoplasmic viral RNA contains subgenomic information.

Arginine must be present in the medium of cells infected with NDV.
Incomplete virions are formed in the course of virus synthesis.
Paramyxoviruses cause persistent infections.
Several types of paramyxovirus mutants are known.
A number of human diseases are caused by paramyxoviruses.
Subacute sclerosing panencephalitis (SSPE) is associated with the presence of measles virus in the human brain.

CHAPTER 15 THE ORTHOMYXOVIRUSES

Genus influenzavirus contains influenza A, B, and C viruses.
The viral genome is RNA negative and is made up of eight genes—each on a separate, single-stranded RNA molecule.
The virions contain an RNA-dependent RNA polymerase responsible for mRNA synthesis.
The influenza virus hemagglutinin and neuraminidase genes were cloned in bacterial plasmids and their sequence determined.
The initiation of RNA synthesis is dependent on the host cell.
Incomplete defective virions (von Magnus effect) are formed.
Recombinants of influenza virus were constructed.
Influenza in man is a respiratory infection.
Inactivated influenza virus vaccines are available for use in man.
A new approach to the development of a vaccine: cloning of the viral hemagglutinin gene in bacterial plasmids.
Amantadine is a possible prophylactic drug (see chapter 24).

CHAPTER 16 THE ARENAVIRUSES

The virions contain cellular ribosomes.

CHAPTER 17 THE BUNYAVIRUSES

Single-stranded viruses with a circular fragmented genome.
The viral genome is divided into three minus circular molecules (L, M, S); ribonucleoprotein complexes can be isolated.
RNA-dependent RNA polymerase is present in the virions.
Diseases like RVF are transmitted by a mosquito-borne virus.

CHAPTER 18 THE PICORNAVIRUSES

The genus enterovirus includes human pathogens like poliovirus and Coxsackie virus.
The viral genome is RNA plus and serves as mRNA.
A gene in chromosome 19 of the human cell codes for the poliovirus receptor on the cell surface.
Synthesis of viral RNA is done by a virus-coded RNA polymerase.
Viral mRNA is monocistronic and is translated into a long peptide that is processed into structural peptides.

Phenotypic mixing occurs in cells infected with two picornaviruses.
Poliovirus still exists in nature; laboratory diagnosis of attenuated and virulent poliovirus strains is necessary.
Rhinoviruses cause the common cold in humans.
Foot-and-mouth disease virus in cattle and its prevention by vaccination.

CHAPTER 19 THE TOGAVIRUSES

Alphavirus, flavivirus, rubivirus, and pestivirus are the various genera of this family of RNA plus viruses.
Subgenomic mRNA species are found in infected cells.
Togaviruses include viruses like yellow fever, St. Louis encephalitis, and rubella.

CHAPTER 20 THE CORONAVIRUSES

These viruses affect pigs, rats, and mice.

CHAPTER 21 THE RETROVIRUSES

Oncovirinae, spumavirinae, and lentivirinae are the subfamilies.
The viral genome is made up of two hydrogen-bonded RNA^+ molecules.
Each RNA molecule contains four genes (gag-pol-env-onc).
The nucleotide sequence at the 5′ end includes a cap.
The nucleotides at the 3′ end include a unique sequence, a repeat sequence, and a poly(A) sequence.
The primer for the synthesis of viral DNA is $tRNA^{trp}$.
The RNA-dependent DNA polymerase is responsible for the synthesis of the double-stranded viral DNA.
The viral DNA resembles the bacterial transposons.
Viral RNA transcripts from the integrated viral DNA are produced by the cellular DNA-dependent RNA polymerase.
Several species of viral mRNA exist in the infected cell.
The integrated viral DNA can be infectious.
Retrovirus DNA can be incorporated into the germ line.
The H_2 histocompatibility locus might affect leukemogenesis.
Retroviruses have distinct evolutionary pathways.
Porcine retrovirus was acquired from an ancient rodent.
Monkey retrovirus originated from an ancient Asian rodent.
In avian sarcoma viruses (ASVs), the *src* gene is the transforming gene.
Endogenous viruses replicate only in chicken cells.
The viral gene *src* is closely related to the cellular gene *sarc*.
Endogenous retroviruses invaded the germ line of chickens.
The oncogene theory postulated that viral *onc* genes present in normal cells may cause cancer.
Mammalian type C retroviruses have been recognized in mice, hamsters, rats, cats, pigs, in several primates, and in humans.
Mouse genes determine sensitivity to leukemogenic viruses.

Radiation-induced leukemia virus (RadLV) is mouse-associated.
Mammary tumor virus (MTV) in female mice is transferred horizontally.
Feline leukemia virus (FeLV) was isolated from a lymphosarcoma in cats.
Bovine leukosis virus (BLV) was isolated from leukemic cattle.
Viral sequences in the cellular DNA of primates are related to sequences in baboons.
Infectious type C retroviruses were isolated from a gibbon ape and a woolly monkey.
Mason-Pfizer monkey virus was isolated from a breast tumor in a rhesus monkey.
Human retroviruses were isolated from human leukemic cells.
Sarcoma viruses are defective in their ability to replicate in infected cells and require a helper virus.
Human cellular *onc* genes are related to simian sarcoma transforming genes.

CHAPTER 22 UNCLASSIFIED VIRUSES: MARBURG AND EBOLA VIRUSES

The disease in humans takes the form of hemorrhagic fever.

CHAPTER 23 SLOW VIRUS INFECTION OF THE CNS

Kuru and Creutzfeldt-Jakob (C-J) diseases in humans are caused by a virus-like or subviral agent not yet characterized.
Transmissible mink encephalopathy is caused by eating meat from sheep infected with the scrapie agent.

CHAPTER 24 IMMUNIZATION AGAINST VIRUS DISEASES

Viruses are foreign antigens that stimulate the immune system.
Mature T-lymphocytes are required for the production of antiviral antibodies.
The secondary immune response depends on memory cells.
Passive immunization with virus-specific antibodies also protects against disease.
Antibodies have a role in recovery and prevention of virus diseases.
The development of viral vaccines is an ongoing process.
Vaccination against poliomyelitis: achievements and problems.
Immunization of humans with Sabin's vaccine is the method of choice.
Awareness of the dangers of immunization against poliomyelitis.
Immunization against influenza and problems encountered in the production of influenza virus vaccines.
Swine influenza vaccine: immunization in the United States and its shortcomings.
Other virus vaccines in current use.

CHAPTER 25 VIRUSES AS SELECTIVE FORCES IN NATURE: EPIDEMICS

Destruction of civilizations and populations as a result of the introduction of a new pathogenic virus into an unimmunized population.

Viruses are capable of introducing their genes into the chromosomal DNA of the host cell.

CHAPTER 26 ANTIVIRAL DRUGS AND CHEMOTHERAPY OF VIRAL DISEASES OF MAN

The search for antiviral drugs.
Virus-coded enzymes are targets for the development of antiviral drugs.
Mode of action of natural antivirals.
Antivirals against herpesviruses.
Antivirals against RNA viruses.
Antiviral drugs in medical use.
Prophylactic drugs.
Interferon: a cellular protein with antiviral activity.
Interferon acts on cells by binding to cell membrane receptors.
Cloning of human interferon genes in bacterial plasmids.

CHAPTER 27 LABORATORY DIAGNOSIS OF DISEASE-CAUSING VIRUSES

Rapid diagnosis of viruses can be done with the aid of the electron microscope.
Isolation and characterization of viruses from clinical materials.
The immunological response in the patient.
Tests available for the detection of viral antibodies.
Immunofluorescence is used for the detection of viral antigens in infected cells.
Radioimmunoassay (RIA) and enzyme-linked immunosorbent assay (ELISA) are used for the detection of viral antigens and antibodies.
Viral nucleic acids can be detected in infected cells.

CHAPTER 28 VIRUSES AND HUMAN CANCER

Epstein-Barr virus and Burkitt's lymphoma.
Nasopharyngeal carcinoma.
Herpes simplex virus type 2 and cervical cancer.
Hepatocarcinoma and hepatitis B virus.
Retroviruses and human cancer.

CHAPTER 29 SOCIAL, ECONOMIC, AND JURIDICAL ASPECTS OF VIRUS DISEASES

Social aspects: herd immunity to viral diseases.
Environmental aspects: monitoring of viruses in water reservoirs.
International virus diagnosis systems.
Problems in virus vaccine production.
Responsibility for individuals affected by a virus disease or vaccination.
Economic aspects.
Juridical aspects of vaccination of humans.

INDEX